Vegetation and climate interactions in semi-arid regions

Advances in vegetation science 12

Vegetation and climate interactions
in semi-arid regions

Edited by
A. HENDERSON-SELLERS AND A. J. PITMAN

Reprinted from Vegetatio, volume 91

Kluwer Academic Publishers
DORDRECHT/BOSTON/LONDON

Library of Congress Cataloging-in-Publication Data

Vegetation and climate interactions in semi-arid regions / edited by
A. Henderson-Sellers, A.J. Pitman.
     p.   cm. -- (Advances in vegetation science ; v. 12)
   ISBN 0-7923-1061-6 (alk. paper)
   1. Arid regions plants--Climatic factors. 2. Vegetation and
climate. 3. Land use--Planning. 4. Arid regions--Management.
5. Arid regions plants--Australia--Climatic factors. 6. Vegetation
and climate--Australia. 7. Land use--Australia--Planning. 8. Arid
regions--Australia--Management.   I. Henderson-Sellers, A.
II. Pitman, A. J. III. Series.
QK938.D4V44 1991
581.5'2652--dc20   90-22597

Published by Kluwer Academic Publishers,
P.O. Box 17, 3300 AA Dordrecht, The Netherlands.

Kluwer Academic Publishers incorporates
the publishing programmes of
D. Reidel, Martinus Nijhoff, Dr W. Junk and MTP
Press.

Sold and distributed in the U.S.A. and Canada
by Kluwer Academic Publishers,
101 Philip Drive, Norwell, MA 02061, U.S.A.

In all other countries, sold and distributed
by Kluwer Academic Publishers Group,
P.O. Box 322, 3300 AH Dordrecht, The Netherlands.

printed on acid free paper

Printed in Belgium

Contents

Dedication

For Eve Laura

Vegetatio **91**: VII, 1991.
A. Henderson-Sellers and A. J. Pitman (eds).
Vegetation and climate interactions in semi-arid regions.

Preface

This book represents a selection of the papers presented at a conference held at Macquarie University, Sydney, Australia in January 1990 entitled *Degradation of vegetation in semi-arid regions: Climate impact and implications.* As the conference title suggests, the aim of the meeting was to bring together those working in land degradation with researchers in climate and climatic change. As such, the themes were intentionally varied and wide ranging within the framework of section topics: measurement; modelling and management. Cross disciplinary awareness and linkage are not easy to achieve but they are vital for the understanding of processes upon which we all depend. This book records a first step in inter-disciplinary exchange and interaction.

The first three chapters of this book provide the basic framework for the rest of the book. Verstraete and Schwartz discuss desertification at the global scale and in the context of global change; Roberts reviews the importance of planning in semi-arid areas while Nicholls discusses one of the important driving forces for drought and land degradation in much of the southern hemisphere, El Niño.

The rest of the book is organised in three further sections which describe the basic themes covered at the conference: measurement, modelling and management.

The second section is broadly about measurement. Measuring the rate of land degradation and identifying first if, and then why, it is increasing is of considerable concern to all semi-arid land managers. This section reviews the significance of field measurements, areal measurements from satellites and the importance of assessing the sinks and sources of greenhouse gases. A final chapter reviews and comments on models of soil erodability and suggests an improved modelling strategy.

The third section covers models and the methods of modelling semi-arid regions at several spatial scales. The global numerical simulation and prediction of drought and the possibilities of incorporating an interactive biosphere model into Atmospheric General Circulation Models (AGCMs) is discussed while a method of improving the hydrological simulations from climate models is described. Modelling surface processes at regional or catchment scales is also described in two chapters.

The final section discusses the crucial role of land management in semi-arid regions with a particular emphasis on Australia. Vegetation and soil changes, management, and the development of public perception of drought are all described. The evolution of government and public responses to drought are also discussed in the context of future planning and policy endeavours.

The editors would like to thank Macquarie University for hosting this conference which was supported by a grant from the Australian Department of Industry, Technology and Commerce. This book contains a selection of the conference papers presented, all of which have been reviewed and revised before incorporation. We are most grateful for the efforts of the referees and the authors during this period. Our grateful thanks also go to Ms. P. Lack for her considerable and enthusiastic efforts in organising this conference and to Mrs. E. Jones who helped edit the submitted manuscripts.

A. Henderson-Sellers & A. J. Pitman, Sydney, Australia, August 1990

Overview

Introduction

The chapters in this section place the problems of vegetation and climate interactions in semi-arid regions into the context which recur throughout the book. First, Verstraete and Schwartz review desertification as a process of global change evaluating both the human and climatic factors. The theme of human impact and land management is discussed further by Roberts whose review focuses on semi-arid land-use planning. In the third and final chapter in this section we return to the meteorological theme. Nicholls reviews the effects of El Niño/Southern Oscillation on Australian vegetation stressing, in particular, the interaction between plants and their climatic environment.

Vegetatio **91**: 3–13, 1991.
A. Henderson-Sellers and A. J. Pitman (eds).
Vegetation and climate interactions in semi-arid regions.
© 1991 *Kluwer Academic Publishers. Printed in Belgium.*

Desertification and global change

M. M. Verstraete[1] & S. A. Schwartz[2]

[1]*Institute for Remote Sensing Applications, CEC Joint Research Centre, Ispra Establishment, TP 440, I-21020 Ispra (Varese), Italy;* [2]*Department of Atmospheric, Oceanic and Space Sciences, The University of Michigan, Ann Arbor, MI 48109-2143, USA*

Accepted 24.8.1990

Abstract

Arid and semiarid regions cover one third of the continental areas on Earth. These regions are very sensitive to a variety of physical, chemical and biological degradation processes collectively called desertification. Although interest in desertification has varied widely in time, there is a renewed concern about the evolution of dryland ecosystems because (1) a significant fraction of existing drylands already suffers from miscellaneous degradation processes, (2) increasing populations will inevitably result in further over-utilization of the remaining productive areas, (3) climatic changes expected from the greenhouse warming might result in drier continental interiors, and (4) some of the desertification processes themselves may amplify local or regional climatic changes. This paper reviews some of the many aspects of this issue in the context of the Global Change research program.

Introduction

Humanity has had a long association with arid and semiarid regions: the first great civilizations (in Egypt and Mesopotamia) developed at the end of the climatic optimum some 3000 years B.C., at a time when the Sahara appears to have been vegetated as parts of the Sahel are today (Butzer 1966; Lamb 1977). By the time the Great Pyramids were erected in Egypt (around 2700 B.C.), the climate of Northeast Africa and the Middle East was in a drying phase that resulted in the arid landscapes we have known for much of the last 5000 years (El-Baz 1983). Superimposed on this long term climatic evolution, however, is the increasingly large impact of growing human populations. The expansion into new territories, and exploitation of the natural resources of these drylands beyond their carrying capacity has resulted in rapidly deteriorating environmental conditions.

This degradation now affects many or most ecosystems on this planet, and there is a progressive awareness by the public and the decision makers of the many aspects of this land abuse (destruction of tropical rain forests, various forms of pollution, including toxic and nuclear wastes, sewage, acid deposition, etc., the so-called 'ozone-hole', oils spills and chemical or nuclear accidents, to name but a few). The release into the atmosphere of large quantities of carbon dioxide, methane and other pollutants by industrial and agricultural activities is now so large that the composition of the atmosphere is affected, and that in turn is expected to affect the climate of the Earth (Henderson-Sellers & Blong 1989; Schneider 1989).

In response to this increased awareness, the scientific community has started to design and implement a coordinated research effort geared at documenting the current state and probable evolution of the global system. The International Geosphere Biosphere Program (IGBP) is an interdisciplinary research effort 'to describe and understand the interactive physical, chemical, and biological processes that regulate the total Earth system, the unique environment that it provides for life, the changes that are occurring in this system, and the manner in which they are influenced by human actions' (NAS 1988, p. 2). This ambitious project is coordinated by the International Council of Scientific Unions (ICSU). The expression 'Global Change' is often used as a synonym to IGBP research program, it also designates the US component of the IGBP. In this paper, we review the nature, extent and severity of desertification, and discuss possible interactions between this form of land degradation and the expected climate and environmental changes. The specific contributions of in situ observations and satellite remote sensing systems in the overall strategy to monitor global environmental degradation are described in a companion paper (Verstraete & Pinty 1990).

The concept of desertification

Much time and effort has been spent trying to define the concept of desertification. Such a task is difficult, if not impossible, because of the number and complexity of the issues involved, the interdisciplinary nature of the problem, and the range of spatial and temporal scales over which this concept is applied (Verstraete 1983, 1986). For the present purpose, we shall define desertification as the set of all environmental degradation processes in hot drylands (hyperarid, arid, semiarid and subhumid regions), as a result of either climatic stress or human mismanagement, or both. Desertification will also include the causes (to the extent these can be identified) and the impact of degradation on natural and managed ecosystems. Clearly, environmental degradation can occur in all biomes, but the fragile nature, harsh climate and expanding area of these drylands have made it a primary focus of attention. Desertification usually has severe long term consequences for the productivity of the land, and therefore for the populations that inhabit these regions.

The word 'desertification' was first introduced by the French forester Aubréville in his book

Table 1. Extent of drylands by geographical region.

[km², %]	Total	Hyperarid	Arid	Semiarid	Subhumid	Dryland
Africa	30,321,130	6,094,094 20.10	6,169,507 20.35	5,129,749 16.92	4,051,032 13.36	21,444,382 70.72
America	42,567,895	164,582 0.39	2,099,881 4.93	4,679,068 10.99	4,327,944 10.17	11,271,475 26.48
Middle East	6,139,098	1,125,997 18.34	3,052,996 49.73	985,490 16.05	769,062 12.53	5,933,545 96.65
Asia	38,120,322	382,439 1.00	4,008,463 10.52	5,312,779 13.94	3,925,704 10.30	13,629,385 35.75
Australia	7,686,884		3,766,572 49.00	1,537,377 20.00	1,229,900 16.00	6,533,850 85.00
Europe	10,507,630		10,096 0.10	239,179 2.28	267,503 2.55	516,778 4.92
Grand total	135,897,300	7,767,112 5.72	19,107,518 14.06	17,883,642 13.16	14,571,146 10.72	59,329,417 43.66

Sources of raw data: (Rogers 1981) and Hammond (1985).

Climats, Forêts, et Désertification de l'Afrique Tropicale (Aubréville 1949). He witnessed the degradation and disappearance of tropical forests in many humid and sub-humid parts of Africa, and attributed it to a large extent to the slash and burn agricultural practices of the local populations. This destruction of the forest lead to savannas with scattered trees, and then further to soil erosion and a general tendency towards more xeric environmental conditions (Dregne 1986, p. 6). Aubréville had identified climate change as a potential factor, but could not estimate its importance for lack of adequate data. It is only later on that the concept became commonly associated with arid and semiarid regions.

Hot drylands are characterized by high solar radiation, potential evapotranspiration rates, and diurnal ranges of temperature, and low precipitation and atmospheric humidity. The location and extent of these regions depends strongly on how they are defined, and many definitions have been proposed in the literature. Table 1 shows the extent of hyperarid, arid, semiarid and subhumid areas for six regions of the world. The data were compiled from information given by Rogers (1981), who digitized UNESCO's map showing the *World Distribution of Arid Regions* (UNESCO 1977). This map was constructed on the basis of hydrological data, using a water balance approach; it arguably represents the best map of aridity currently available.

It can be seen that almost 33% of the continents are hyperarid, arid or semiarid, and that almost 44% of all continental areas (excluding Antarctica) can be classified as drylands. While degradation in hyperarid regions is limited to oases (Meckelein 1980), much of the arid, semiarid and subhumid areas are generally considered to be a risk of desertification.

Table 2 shows the distribution of productive drylands, broken down by type of economic activity, as well as the percentage thought to be desertified as of the early eighties. These data are not directly comparable to those of Table 1, but it is interesting to note that rangelands are by far the largest areas affected by desertification.

In the late seventies, the United Nations estimated the rate of desertification at between 54 000 and 58 000 km^2 (UNCOD 1977, p. 9; United Nations, 1978, p. 2). By 1984, Norman Myers estimated that some 120 000 km^2 of agricultural and pastoral land were deteriorating beyond useful economic use per year (Myers 1984, p. 46). These numbers should be compared to estimates of deforestation worldwide, which range from 100 000 to 113 000 km^2 per year, with the bulk of the destruction occurring in the tropics (Myers 1984, p. 42; WRI 1986, p. 72).

Arid lands are dynamic regions, they have been evolving over thousands of years, mostly in response to climatic changes (Warren 1984). Throughout most of its history, humanity has been able to cope with such an evolution (and with its own degradation of the environment) by colonizing new and hitherto unaffected areas. Only during the last century has it become apparent that the land resource was finite; and individuals and societies are only now becoming aware of the consequences of unlimited growth and careless abuse of finite natural resources. Recurrent severe droughts in the Sahel and more recently in Ethiopia, have helped to periodically focus attention on this problem, although the degradation occurs in many other regions.

Drylands often experience significant interseasonal and inter-annual climate variability, and in particular periods of reduced water availability. Drought therefore constitutes a 'normal', expected (if largely unpredictable) component of their climate: it is a temporary situation where the amount of available water is smaller than what would be needed for the intented economic use, or for the normal growth and development of the

Table 2. Distribution of desertified areas by economic activity.

Economic use	Area [km^2]	% desertified
Rangeland	25,560,000	62
Rainfed agriculture	5,700,000	60
Irrigated agriculture	1,310,000	30
Total productive drylands	32,570,000	61

Source: WRI, 1986, p. 278.

plant cover (Gibbs 1975, p. 11). Droughts have been identified both on the basis of meteorological or hydrological parameters alone (see Landsberg 1975, for a review), or by reference to economic impacts, or even perceptions of risk (Heathcote 1969). Many authors view drought as a short-lived phenomenon responsible for recurrent but not permanent stress to the environment (e.g., Le Houérou 1977). This approach, although understandable *a posteriori*, may not be operational because it is impossible to know how long a dry spell is going to last. The Sahel is a case in point, where the relatively dry period that started around 1968 was long identified as a drought, even though is has not ended yet, more than 20 years later (Nicholson 1989). How long does a drought need to be before it is called a climatic shift or change?

Desertification is generally conceived as a much wider concept than drought, involving not only water availability issues, but also various forms of soil degradation, loss of biological productivity, and a host of human impacts. It is deemed to occur on 'long' time scales, causing a slow but cumulative decline in biological productivity and in the capability of the land to support its natural vegetation or agricultural exploitations. Desertification must be distinguished from drought, but since the end of a drought even cannot be predicted, its duration can only be known after the fact, and the distinction between desertification and drought in terms of their respective duration cannot be applied operationally. The combination of progressive desertification and droughts can be severely crippling to the environment, as the stress created by human overexploitation of the land becomes especially visible during severe droughts. Drought and desertification can amplify each other's impacts, and the resulting degradation of the environment can further affect adjacent areas, either directly (invading sand dunes, siltation of dams downstream), or indirectly (migration of populations, increased international tensions over increasingly scarce resources).

The intensity of desertification processes can range from slight to very severe in terms of the degradation of plant and soil resources (Dregne 1986). Desertification threatens many of the arid and semiarid regions of the world, and involves a large number of complex interactions between physical, topographical, edaphic, and biological parameters, as well as human components (land use and ownership, social structures, economic development, and health status) (Spooner 1982). Different processes and interactions take place in different ecosystems, but the general tendency towards degradation that results from the interplay of these processes and interactions is common to all situations. In that sense, desertification, which has been called the cancer of drylands, can affect any area which has been made vulnerable enough by climatic stress or human overexploitation, or both.

It is difficult to estimate the number of people affected by desertification because geographical, political and economic differences whithin each country result in some population classes being more affected or more at risk than others. Nevertheless, it is probable that around 100 million people are directly affected by desertification today, and that another 900 million may be at risk now or in the near future (UNCOD 1977, p. 8; United Nations 1978, p. 2; Paylore & Greenwell 1980, p. 14–18; Paylore 1984, p. 18–19; Al-Sudeary 1988, p. 13).

The processes of land degradation

Desertification often results from the degradation of the vegetation cover by overgrazing, over-trampling, wood collection, repeated burning, or inappropriate agricultural practices. It leads to a general decrease in productivity of the land and in the accelerated degradation of the soil resource due to soil erosion (both by wind and water), siltation, salinization and alkalinization of irrigated lands, or dry salting. The excessive loss of soil, nutrients, and sometimes even seeds from the ecosystem affects the capability of the vegetation to recover and constitutes the principal mechanism of irreversible damage to the environment.

The impact of grazing on pastoral rangelands

depends largely on its intensity and timing. Light grazing may increase the productivity of the range by stimulating new growth, while moderate to heavy grazing often results in the preferential removal of the more palatable species (Warren & Maizels 1977, p. 203–210). This facilitates the invasion of less palatable or inedible species (invaders) by modifying the ecological niches. Overgrazing during the dry season further reduces the vegetation cover and increases the risk of soil erosion. Perennial species are particularly important in stabilizing the soil, especially during the dry season, while ephemeral and annual species may help control soil erosion during the wet season.

The deterioration of the soil surface, which results from the removal of the vegetation cover, strongly affects the health, vigor and reproductive capacity of the remaining plants through the disruption of plant-water relations. Subsequently, there is an increase in runoff, sheet and gully erosion on sloping ground; the top soil is lost, together with its store of water, nutrients, and seeds. Compounding the problem is the higher bulk density of the soil caused by severe trampling, which decreases infiltration by water and also increases runoff. Ultimately, if erosion is not stopped, it will completely destroy the productive value of the land by soil stripping and gully extension.

Rainfed farming practices have their own distinctive signatures of desertification. The problem often originates on land cleared for cultivation or left fallow. As soon as the natural vegetation is cleared, the soil is made vulnerable to accelerated wind and water erosion. The introduction of agricultural equipment, designed for the deep soils of mid-latitude regions, into tropical regions with thin soils can further aggravate the soil loss (soil pulverization and removal of humus and nutrients). The removal of the plant cover during harvests results in bare soils and promotes the formation of a hard crust at the surface. This crusting of the soil reduces infiltration and increases runoff. Thus, as in pastoral rangelands, the fertile surface soil is stripped away, leaving behind infertile subsoils. Gullies may then form on lower parts of slopes, presenting similar physical obstacles to farming operations as they do to rangeland practices. In addition, runoff from the slopes causes sedimentation downhill; waterways and dams are filled, and flooding is intensified in low-lying areas (UNCOD 1977). The degradation processes act synergistically in both rainfed farming and pastoral systems.

Soil erosion by wind and water may take place in both arid and semiarid regions, and at various times during the year (Verstraete & van Ypersele 1986). The impact of wind erosion clearly dominates in the drier regions, while water erosion will affect mostly the wetter areas (Marshall 1973, p. 58). Furthermore, it is the extreme events which are the most destructive, especially if they occur in the dry season, a period of maximum sensitivity to perturbation. The risk of soil erosion depends on both the susceptibility of the soil to erosion, and on the probability of a meteorological event (wind gust or rainfall) of sufficient magnitude to displace the soil. For example, the erodibility of the soil may be highest during the dry season, when the protective vegetation cover is minimal, but the probability of intense rains is highest during the rainy season. The highest risk of water erosion therefore occurs at the end of the dry season, when the plant cover has not grown back yet, but the likelihood of an intense rainfall increases.

Wind erosion affects the soil structure and composition in different ways. First, the flow of particles is proportional to the cube of the friction wind speed above a threshold value (Bagnold 1943; Gillette et al. 1982; Skidmore 1986). Second, the coarser sandy materials drift over short distances (a few hundreds of meters to a few kilometers), while the finer materials are carried away in the wind, sometimes over continental distances: Dust from the Sahara has been observed to cross the Atlantic Ocean, and 'sand rains' have been reported from as far as Northern Europe and the Caribbeans (Carlson & Prospero 1972; Rapp 1974; MacLeod et al. 1977; Morales 1979).

The fine top soil which is lost to wind erosion is the most fertile portion of the soil complex and only sterile soil is left behind (UNCOD 1977). The lack of water in these shallow soils also prevents seedlings from surviving prolonged droughts, making permanent plant life even more difficult. When perennial seedlings do manage to emerge, they may easily be uprooted and blown away. With little protection left on the ground, the blasting impact of moving sand can destroy young crops (Le Houérou 1977).

Water erosion is intimately linked to the hydrological cycle of arid and semiarid lands. The primary physical agent of erosion is the transfer of kinetic energy from the raindrops to the soil particles. Once these particles have been dislodged, they can be transported away by surface runoff. The presence of vegetation greatly decreases the impact of raindrops because the leaves absorb much of their momentum: the drops break up and slow down considerably. The sparser the vegetation cover, the more rainfall reaches the ground, where it must either percolate into the ground or runoff.

Soils are extremely variable in their spatial distribution. The most important characteristics of soils are their texture, structure and chemical composition, as well as their depth. Soil texture is determined on the basis of the relative fractions of sand, silt, and clay. Runoff from a specific soil depends primarily on the intensity of precipitation, the infiltration rate and the local slope. The coarser soils (i.e., with an important sand fraction), have a high infiltration rate and can therefore absorb intense showers. Soils with a major clay fraction have a much reduced infiltration rate, and tend to generate more runoff, even under lower precipitation intensities (Hillel 1982). The breakup of soil aggregates by large raindrops may lead to the dispersal of the fine elements and the sealing of the surface when the soil dries up. This crusting contributes to the formation of an impermeable surface (Le Houérou 1977). Soil depth plays a crucial role because it directly affects the water storage capacity, and the amount of nutrients (both mineral and organic) available for plant growth and development. The hydraulic conductivity of soils is a non-linear function of the water content, with much lower values for drier

soils. The drying of the top soil impedes water movement and may therefore reduce the total evaporation. Deep soils have a better chance of conserving moisture than shallow soils.

Since water availability is the primary limiting factor for plant productivity in dry regions, irrigation has often been seen as the best way to boost production. This technique, however, presents its own set of problems and can also lead to the permanent degradation of the land. The productivity of the vegetation is directly linked to its transpiration rate; therefore the goal of irrigation is to provide plants with just enough water in the soil. Excess water should be drained. When this is not done, irrigation water infiltrates the soil and picks up soluble salts. The drying of the top soil creates a gradient of moisture and generates a slow but continuous upward flow of mildly salty water. These salts therefore accumulate at the surface as the water evaporates. Most plants cannot tolerate very high salt concentrations in the upper soil, as it affects their capability to absorb water in the root layer. Ultimately, crops must be switched to halophytes (salt tolerant plants) or the field must be abandoned. On the other hand, proper irrigation and drainage may be designed to flush the salts in the upper soil layers, but this requires high capital investments and constant maintenance of the drainage system.

Last but not least, fire has long been recognized as a potentially destructive agent. It is well known that a light fire may actually stimulate new plant growth, especially in grasslands, and accelerate the turnover of nutrients in the environment. Burning may also help control some bush or tree invasion, but there are many possible negative effects. Hot fires can sterilize the ground by destroying the nitrogen-fixing bacteria, thereby reducing the productivity of the land, especially if they are ill-timed. Hot fires can lead to the destruction of humus and a loss of fertility, a destabilization of the water relations in the topsoil, or even the loss of the seedstore (UNCOD 1977). Even light fires can take their toll by causing distillates of organic matter to coat particles of the soil surface, decreasing infiltration and increasing runoff or erosion vulnerability

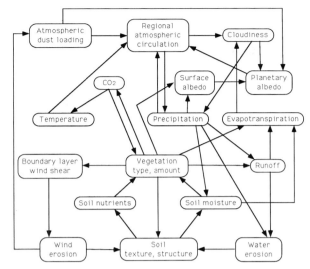

Fig. 1. Soil, vegetation and atmosphere interactions within the context of desertification.

(Warren & Maizels 1977). Fires can drastically modify the geophysical environment near the surface: the albedo may be affected, at least spectrally, the roughness of the surface may be changed in either direction, depending on the type and density of the cover, and hydrological relations at the surface may be modified.

The interactions between the components of the environment (soil, plants, climate) are shown shematically in Figure 1. This diagram is not intended to be exhaustive, or to describe the particular situation at a specific site; but it does attempt to show how various aspects of a dryland ecosystem depend on each other and are affected, directly or indirectly, by biogeophysical processes that take place in this environment. One of the major difficulties of modeling these interactions results from the fact that they take place at many different spatial and temporal scales (Avissar & Verstraete 1990).

This description of desertification would not be complete without a discussion of human aspects, both as causes of degradation and as victims of its impact. Societies directly affect the environment through such activities as pastoralism, agriculture, transportation, urbanization and industrialization. These activities are largely controlled by the interplay between economic constraints and the availability of natural resources, as well as by the expectations of individuals and other cultural factors. Progress in medicine and the provision of medical services has improved the quality of life and decreased mortality rates, at least for those segments of the population that have access to these services. On the other hand, environmental degradation and the breakdown in the production systems from desertification is often accompanied by a lowering of health standards, a decrease in nutritional intake, and an increase in morbidity. The health status of a population and the location and accessibility of natural resources, in turn, directly influence the rates of migration of populations. Hence, a region may be indirectly affected by desertification in a neighboring area. On a larger political scale, these problems can result in severe internal problems between social classes, or even in massive environmental migrations, possibly causing international tensions. There is an extensive literature on this subject. The interested reader should refer to Copans (1975), UNCOD (1977), Spooner (1982) and Glantz (1987) for introductions to this vast subject.

Climate and environmental change

Climate variability, and in particular extreme conditions, must be expected in drylands. What is more worrisome, however, is the possibility that human activities may be responsible for significant further climatic changes in the foreseeable future. The primary cause for concern lies with the observed change in the chemical composition of the atmosphere. Agricultural, industrial and urban activities, and specifically the heavy consumption of fossil fuels, result in the injection of large quantities of carbon dioxide and many other pollutants in the lower atmosphere. Some of these compounds find their way into the oceans or may be re-absorbed by the surface, but a large fraction remains airborne and contributes to the trapping of thermal infrared radiation emitted by the surface. This so-called greenhouse effect results in a substantial heating of the air near the ground.

It is currently estimated that the concentration of CO_2 in the atmosphere could reach twice its pre-industrial value in the first half of the next century. Many atmospheric scientists have attempted to evaluate the likely temperature change that would result, using general circulation models of the atmosphere. Different models predict somewhat different distributions of heating because of internal differences in their representation of the physics and dynamics of the atmosphere (Schlesinger 1989), but results from these numerical sensitivity studies show that the globally-averaged annual mean surface temperature could increase by 2 to 6 °C by 2050. Furthermore, these expected changes in temperature will likely be associated with major perturbations of the hydrological cycle. The actual impact on the water cycle is difficult to assess, but it is quite possible that the interior of large continents may experience further drying (Kellogg & Zhao 1988).

Furthermore, the impact of changes at the surface on the climate itself must also be considered. Focusing on land areas, the climate is affected by the underlying surface in three major ways: first, the optical properties of the surface, in particular albedo, control the rate of solar energy absorption by the climate system. Second, the vegetation cover affects the rates of transfer of energy, water and various chemical compounds between the surface and the atmosphere. And third, the roughness of the surface controls the exchange of momentum between the atmosphere and the planet, and strongly affects the divergence of the wind fields in the atmospheric boundary layer (Verstraete & Dickinson 1986; Nicholson 1988).

The concern for global change goes far beyond just climatic change, however. The study of the cycling of major biogeochemicals among the various components of the climate system has provided a new paradigm for global investigations. Arid and semiarid regions have long been considered wastelands, but they are now recognized as integral parts of the Earth system. Detailed investigations in the southwest United States showed that grazing can increase heterogeneities in the spatial and temporal distribution of water, nitrogen, and other soil resources. Biogeochemi-

cal cycles provide a new convenient framework to understand the invasion of desert shrubs and the progressive degradation of the environment, and to try to predict the likely future evolution of these areas as well as the interactions between arid lands and other ecosystems (Schlesinger *et al.* 1990).

It is important to appreciate the significance of the climatic changes described above. First of all, local and regional temperature or moisture changes may be much higher (or lower) than these global values. Some regions will likely be more affected than others. Second, a slight change in mean values also results in more frequent extreme conditions, even without a change in the statistical distribution of the climatic parameters. Third, a change of global mean temperature of this magnitude in such a short amount of time is essentially unprecedented: even the warming responsible for the deglaciation 15 000 years ago was ten times slower. As a result, there is some concern about the capability of the biosphere to cope with such a change.

Scientific issues

The bottom line is therefore relatively clear: the human population is growing at the rate of 95 million individuals per year (or about 11 000 per hour) (Ehrlich and Ehrlich, 1990, p. 9), and desertification alone accounts for the loss of 120 000 km^2 of productive land per year (about 1400 ha per hour). Since we also loose at least 120 000 km^2 of forests and 24 to 26 billion tonnes of top soil per year through erosion (Ehrlich & Ehrlich 1990, p. 28), it is not difficult to see that this situation cannot be sustained for much longer. The time is for action, not panic, and the scientific community has a definite role to play.

First of all, it is urgent to clearly assess the situation and establish the probable future evolution. In fact, the lack of precise information about the extent and severity of the population-environment-climate problem has consistently been recognized over the last 40 years. Here are a few selected quotes from the literature on this subject: Aubréville (1949), p. 330:

A few years ago, we studied this problem of desertification of West Africa along the Saharan border without reaching definite conclusions. We were not able to demonstrate a progressive drying of the climate because meteorological observations and statistics were not reliable enough and did not cover long enough periods. [Translated from French by M. Verstraete]

McGinnies *et al.* (1968), p. 3:
'To the extent that we have experienced widespread desert encroachment within historic times, we are far from unanimous as to its causes. [...] For the immediate future lack of basic data about desert environments remains our largest single category of arid-lands problem.'

Warren and Maizels (1977), p. 177 and 186:
'This chapter briefly reviews the evidence for desertification. It will soon become apparent that this is not an easy task. Statistics are seldom in the right form, and are hard to come by, and even harder to believe, let alone interpret. [...] The evidence for desertification is diffuse and almost impossible to quantify.'

Bie (1989), p. 2:
'Although the occurrence, if not the severity, of famines is well documented, there is a scarcity of data linking general dryland degradation to famine. There is little doubt, however, that local land degradation has taken place. There is therefore a need to develop better methods whereby land degradation can be assessed. Inaccurate methodologies appear in the past to have had major impact on donor attitudes on the occurrence, frequency and severity of dryland degradation.'

Monitoring the environment (in drylands as well as other ecosystems) plays a number of crucial roles and must be pursued to (1) establish a baseline against which future observations can be compared, (2) document the spatial and temporal variability of the relevant environmental parameters, (3) identify the regions at risk of further degradation, and the nature of the processes at work, (4) provide the data needed to build and validate the mathematical models of the environment that are needed to understand and predict the evolution of these ecosystems, (5) support policy decision making in such tasks as prioritizing the target areas for relief and conducting cost-benefit analyses of various remedial actions or feasibility studies, as well as support field activities geared towards minimizing further degradation or reclaiming affected areas, and (6) evaluate the effectiveness of these policies, plans and remedial actions.

It is unfortunately not clear where desertification is the most severe, or at what rate it is occurring. Preventive measures or remedial actions should be based on a accurate assessment of the situation, and that crucial element is still missing. While there are countless reports of impending disasters at the local scale, there is no coherent way to obtain a synthetic accurate and quantitative view of the global situation. Satellite remote sensing techniques appear to be the most appropriate for this task, but a lot more work must be done in order to retrieve quantitative information from these data, as opposed to qualititative estimates. A discussion of the potential contribution of space platforms, along with traditional in situ measurement techniques, can be found in a companion paper (Verstraete & Pinty 1990).

In parallel to this effort, the second contribution of the scientific community must be a better understanding of the processes involved and an analysis of the implications of the remedial actions that may be envisaged. The global models used to investigate the scenarios described above and their impacts represent the state of the art in climate modeling, even though they still suffer from significant shortcomings. For example, in most cases, the representation of the oceans and of air-sea interactions are rather crude, and the parameterization of cloud and surface processes cannot do justice to the number and complexity of the processes that actually take place. These models must be expanded to include the most significant biogeochemical cycles. In fact, a hierarchical set of models at various scales must be designed to study and understand better the interactions between physical, chemical, biological, and human processes. This will require further emphasis on interdisciplinary investigations, as well as a renewed effort to train generations of young scientists to deal effectively with this complex system. The Global Change Research Program is a major step forward in that direction.

Acknowledgments

The Department of Atmospheric, Oceanic and Space Sciences of the University of Michigan and the School of Earth Sciences of Macquarie University financed the participation of the first author to the 'Conference on Degradation of Vegetation in Semi-Arid Regions: Climate Impact and Implications', Macquarie University, January 29–31, 1990. The support and encouragements of Prof. W. Kuhn and A. Henderson-Sellers are gratefully acknowledged. Serena Schwartz is grateful to the Population-Environment Dynamics Program administred by the School of Public Health of the University of Michigan for financial support of her interdisciplinary studies.

References

Al-Sudeary, A. M. 1988. Alleviation of rural poverty in arid lands. In: Arid Lands: Today and Tomorrow. Edited by E. Whitehead, C. Hutchinson, B. Timmermann and R. Varady, Westview Press, Boulder, 13–20.

Aubréville, A. 1949. Climats, Forêts, et Désertification de l'Afrique Tropicale. Société d'Editions Géographiques, Maritimes et Tropicales, Paris, 351 pp.

Avissar, R. & Verstraete, M. M. 1990. The representation of continental surface processes in mesoscale atmospheric models. Reviews of Geophysics 28: 35–52.

Bagnold, R. A. 1943. The Physics of Blown Sand and Desert Dunes. William Morrow & Co., New York.

Bie, S. W. 1989. Dryland degradation assessment techniques. Outline of a presentation for Session 1 of the Professional Development Workshop on Dryland Management, The World Bank, Washington DC, May 10–11, 1989.

Butzer, K. W. 1966. Climatic changes in the arid zones of Africa during early to mid-Holocene times, in World Climate from 8000 to 0 BC. Proceedings of the International Symposium held at Imperial College, London, 18 and 19 April 1966. Royal Meteorological Society, 72–83.

Carlson, T. N. & Prospero, J. M. 1972. The large-scale movement of Saharan air outbreaks over the northern Equatorial Atlantic. Journal of Applied Meteorology 11: 283–297.

Copans, J. 1975. Sécheresses et Famines du Sahel, Volumes I and II. F. Maspéro, Paris, 150 pp. and 144 pp.

Dregne, H. 1986. Desertification of arid lands. In: Physics of Desertification, Edited by F. El-Baz and M. Hassan, Martinus Nijhoff Publishers, Dordrecht, 4–34.

Ehrlich, P. R. & Ehrlich, A. E. 1990. The Population Explosion. Simon and Schuster, New York, 320 pp.

El-Baz, F. 1983. A geological perspective of the desert. In: S. Wells and D. Haragan (Eds.), Origin and Evolution of Deserts. University of New Mexico Press, Albuquerque, 163–183.

Gibbs, W. J. 1975. Drought: Its definition, delineation and effects, in Drought: Lectures presented at the 26th session of the WMO Executive Committee, Special Environmental Report No. 5, WMO No. 403, Geneva, 1–39.

Gillette, D. A., Adams, J., Muhs, D. & Kihl, R. 1982. Threshold friction velocities and rupture moduli for crusted desert soils for the input of soil particles into the air. J. Geophys. Res. 87: 9003–9015.

Glantz, M. H. (Editor) 1987. Drought and Hunger in Africa: Denying Famine a Future. Cambridge University Press, Cambridge, 457 pp.

Hammond 1985. Ambassador World Atlas. Hammond Inc., 484 pp.

Heathcote, R. L. 1969. Drought in Australia: A problem of perception. The Geographical Review 59: 175–194.

Henderson-Sellers, A. & Blong, R. 1989. The Greenhouse Effect: Living in a Warmer Australia. New South Wales University Press, Kensington, 211 pp.

Hillel, D. 1982. Introduction to Soil Physics. Academic Press, New York, 365 pp.

Kellogg, W. W. & Zhao, Z. 1988. Sensitivity of soil moisture to doubling of carbon dioxide in climate model experiments. Part 1: North America, Journal of Climate 1: 348–366.

Lamb, H. H. 1977. Climate: Present, Past and Future, Vol. 2, Climatic History and the Future. Methuen and Co., London, 835 pp.

Landsberg, H. E. 1975. Drought: A recurrent element of climate. In: Drought: Lectures presented at the 26th session of the WMO Executive Committee, Special Environmental Report No. 5. WMO No. 403, Geneva, 41–90.

Le Houérou, H. N. 1977. The nature and causes of desertization. In: Desertification. Edited by M. Glantz, Westview Press, Boulder, 16–38.

MacLeod, N. H., Schubert, J. S. & Ananejionu, P. 1977. Report on the Sahel 4 African Drought and Arid Lands Experiment. In: Skylab Explores the Earth, NASA, Washington DC, 263–286.

Marshall, J. K. 1973. Drought, land-use and soil erosion. In: The Environmental, Economic and Social Significance of Drought. Edited by J. V. Lovett, Angus and Robertson Publishers, Sydney, 55–77.

McGinnies, W., Goldman, B. & Paylore, P. (Editors) 1968. Deserts of the World: An Appraisal of Research Into Their Physical and Biological Environments. University of Arizona Press, Tucson, 788 pp.

Meckelein, W. (Editor) 1980. Desertification in Extremely Arid Environments. Stuttgarter Geographische Studien, Band 95, Stuttgart University, 203 pp.

Morales, H. C. (Editor) 1979. Saharan Dust: Mobilization, Transport, Deposition. SCOPE Report 14, Wiley and Sons, Chichester.

Myers, N. (Editor) 1984. Gaia: An Atlas of Planet Management. Anchor Press, Garden City, 272 pp.

NAS 1988. Towards an Understanding of Global Change: Initial Priorities for U.S. Contributions to the IGBP. National Academy Press, Washington DC, 213 pp.

Nicholson, S. E. 1988. Land surface atmosphere interaction: Physical processes and surface changes and their impact. Progress in Physical Geography 12: 36–65.

Nicholson, S. E. 1989. Long-term changes in African rainfall. Weather 44: 46–56.

Paylore, P. 1984. One of us must be wrong: Fools rush in; Part 4. Arid Lands Newsletter 21: 18–19.

Paylore, P. & Greenwell, J. R. 1980. Fools rush in; Part 2: Selected arid lands population data. Arid Lands Newsletter 12: 14–18.

Rapp, A. 1974. A review of desertization in Africa: Water, vegetation and man, SIES Report No. 1, Secretariat for International Ecology, Stockholm, Sweden, 77 pp.

Rogers, J. A. 1981. Fools rush in, Part 3: Selected dryland areas of the world. Arid Lands Newsletter 14: 24–25.

Schlesinger, M. E. 1989. Model projections of the climatic changes induced by increased atmospheric CO_2. In: Climate and the Geo-Sciences: A Challenge for Science and Society in the 21st Century. Edited by A. Berger, S. Schneider and J. Cl. Duplessy. Kluwer Academic Publishers, 375–415.

Schlesinger, W. H., Reynolds, J. R., Cunningham, G. L., Huenneke, L. F., Jarrel, W. M., Virginia, R. A. & Withford, W. G. 1990. Biological feedbacks in global desertification. Science 247: 1043–1048.

Schneider, S. H. 1989. Global Warming. Sierra Club Books, 317 pp.

Skidmore, E. L. 1986. Wind erosion climatic erosivity. Climatic Change 9: 195–208.

Spooner, B. 1982. Rethinking desertification: the social dimension, in Desertification and Development: Dryland Ecology and Social Perspective. Edited by B. Spooner and H. S. Mann. Academic Press, New York, 1–24.

United Nations 1978. United Nations Conference on Desertification, Roundup, Plan of Action and Resolutions. United Nations, New York, 43 pp.

UNCOD 1977. Desertification: Its Causes and Consequences. Prepared by the Secretariat of the United Nations Conference on Desertification. Pergamon Press, Oxford, 448 pp.

UNESCO 1977. World Map of Arid Regions. United Nations Educational, Scientific and Cultural Organization, Paris.

Verstraete, M. M. 1983. Another look at the concept of desertification. In: S. Wells and D. Haragan (Eds.). Origin and Evolution of Deserts. University of New Mexico Press, Albuquerque, 213–228.

Verstraete, M. M. 1986. Defining desertification: A review. Climatic Change 9: 5–18.

Verstraete, M. M. & Dickinson, R. E. 1986. Modeling surface processes in atmospheric general circulation models. Annales Geophysicae 4: 357–364.

Verstraete, M. M. & van Ypersele, J. P. 1986. Wind versus water erosion in the context of desertification. In: Physics of Desertification. Edited by F. El-Baz and M. Hassan. Martinus Nijhoff Publishers, Dordrecht, 35–41.

Verstraete, M. M. & Pinty, B. 1990. The potential contribution of satellite remote sensing to the understanding of arid lands processes. In this volume.

Warren, A. 1984. The problem of desertification. In: Key Environments: Sahara Desert. Edited by J. L. Cloudsley-Thompson. Pergamon Press, Oxford, 335–342.

Warren, A. & Maizels, J. K. 1977. Ecological change and desertification, in Desertification: Its Causes and Consequences. Prepared by the Secretariat of the United Nations Conference on Desertification. Pergamon Press, Oxford, 169–260.

WRI 1986. World Resources 1986. World Resources Institute and International Institute for Environment and Development. Basic Books Inc., New York, 353 pp.

Vegetatio **91**: 15–21, 1991.
A. Henderson-Sellers and A. J. Pitman (eds).
Vegetation and climate interactions in semi-arid regions.
© 1991 *Kluwer Academic Publishers. Printed in Belgium.*

The role of land use planning in semi-arid areas

B. R. Roberts
University College of Southern Queensland

Accepted 7.9.1990

Abstract

Worldwide, the semi-arid grazing and cropping regions are being subjected to severe overuse. Australian semi-arid regions have been cleared on a vast scale for marginal cropping. Grazing land has been overused and shows widespread signs of vegetation and soil degradation. Ecological realism on land capability based on objective condition monitoring, is essential to land management if sustainable production is to be achieved. The Land Care Movement can markedly influence this achievement. The margins of semi-arid cropland and the application of ecologically-based animal stocking rates require attention in several States. Drought assistance policy requires amendment to safeguard natural pastures. Lease covenants have a potentially central role to play in land administration.

The global situation

The need for planning of land use according to capability in the world's semi-arid regions is emphasised by the large proportion of the total degraded land which occurs in the semi-arid climatic zone. Maps comparing the geographical distribution of rainfall and soil loss, indicate that the greatest losses are in the 500–1000 mm zone and not in the very low or very high rainfall zones as might be expected.

Estimates show that about 4 M ac. of productive land are lost to spreading deserts each year, while the loss of land to erosion is estimated at 7% of present productive land per decade (Anon 1988). When these losses are analysed, the semi-arid regions of Africa, South America and Asia are found to be most severely affected. It is worth noting that these three global regions also report the highest birth rates and population doubling times as low as 23 years (Anon 1988). This paper considers the reasons why the semi-arid areas are so susceptible to degradation and how planning principles can be applied to remedy the situation.

Land use planning principles

For several decades planners have based permanent land use systems on two simple principles:
1. Use each portion of land according to its potential to produce.
2. Protect each portion of land according to its susceptibility to degradation.

The simple principles enunciated in the 1940s by British colonial administrations take cognisance of the need to fully utilise and adequately protect the productive potential of all agricultural and pastoral lands irrespective of rainfall and soil type. These guidelines were widely used in the young developing countries of Australia, New Zealand, South Africa and India, at least as far as land settlement objectives were concerned. Unfortunately political expediency usually overrode ecological planning and, as a result, an unrealistic level of close settlement and animal stocking was allowed and even encouraged in many semi-arid regions.

In recent years the principles of land use planning have been expanded to take account of mining, tourism, recreation and other alternative land uses. In addition, the principle of keeping future options open has been added to planning criteria in an effort to minimise the end uses of land in a changing world in which predicting future needs is increasingly complex.

The Australian experience

The problems of permanent rural production from Australian semi-arid landscapes, are magnified by the unusually high variability of rainfall. Compared to other semi-arid homoclimates, Australia has a 25–35% less reliable rainfall, which makes planning of cropping land and of sheep and cattle stocking rates considerably more difficult.

Land use planning in the marginal cropping regions of South Australia and Western Australia, has relied heavily on estimates of the probability of reasonable harvests based on improved dryland farming techniques (Meinig 1962). The low evaporation in the winter-rainfall zone has allowed the production of wheat yields of the order of 1 tonne/ha under a rainfall of 270–330 mm (Australian Wheat Board 1989).

Earlier planners gave almost total attention to rainfall and soil but in more recent times additional factors such as salinity-proneness, wind erosion, structure decline and acidity have been given more attention. These criteria of land suitability for cropping have been emphasised, not as a result of scientific prediction, but rather as a response to vast and serious symptoms of degradation in several Soldier Settlement Schemes, irrigation projects and newly developed scrublands (Williams 1989). Land previously under Mallee in the south and Brigalow in the

north, is presently causing a re-think of 'the margins of the good earth', as a result of declining landscape stability and soil productivity problems.

The original Australian attempts to plan new cultivations were carried out without reliable information on the proneness of land to degradation under cultivation. With the wisdom of hindsight, it is clear that much of the Eastern Wheat belt of Western Australia apparently cannot sustain an annual cropping system. Thus modern planners must now recommend, not how far the new frontiers can be extended, but whether the present boundaries need to be pulled back.

Land use planning should always take the best technology into account when classing land as suitable for a particular use. As a result of improved tillage, plant breeding and weed control, Australia has led the world in dryland agricultural techniques, but the sustainability of the systems concerned are now being called into question.

The Australian pastoral zone poses very different problems for the land use planner. Vast areas of semi-arid grazing land have been severely degraded, in some cases apparently irreversibly. The extent and distribution of land degradation was reported in a nationwide survey (Anon 1978) and no follow-up survey has been published to date. Planning in the pastoral zone concentrates on stock carrying capacity estimates, the distribution of drinking water, the control of scrub regrowth and the possibilities of sowing improved pastures, especially legumes. In the case of Queensland, which has significantly higher cattle numbers than other States, 45% of the State's natural grazing land has been degraded by overgrazing (Weston & Harbison 1980). Advanced water erosion has permanently reduced the productive capacity of sections of this land to the extent that it is uneconomic to mechanically improve damaged areas. This is one case of future options being lost even where land has not been cleared for cropping. Similar constraints to rehabilitation occur on a large scale in the Gascoyne catchment in Western Australia and the Victoria River region of the Northern Territory. Scalded areas of the Western Division of

New South Wales are only restored by relatively expensive ripping, ponding and reseeding of saltbush (Roberts 1990b).

Past lessons in land use planning

Cocks & Parvey (1985) have provided one of the most useful overviews of how land use planning in inland Australia may be improved. They point out that land administration agencies have two main ways of directing land management: Reactive control and Planned control. The distinction is between a reactive and proactive approach, with the ability to use both in different circumstances and on different issues. Parvey *et al.'s* (1983) survey of land issues of major importance to local authorities, found that the two prime differences between arid and non-arid authority's problems were the importance of subdivision in non-arid areas and drought in arid areas.

Cocks & Parvey (1985) suggest that four types of land use control warrant attention in the arid zone:
1. Zoning of areas for predetermined land use.
2. Management plans for properties and allocated sites.
3. Performance standards, monitored by technically-based field techniques.
4. Impact Assessment of land use on particular projects.

While modern property planners and development advisers don't have all the information they desire, they cannot claim the absence of information which can be used to defend their predecessors' recommendations. Time has given us the opportunity to read the 'Writing on the Wall' if we care to. The symptoms of over-use are everywhere to be seen and lessons abound. These lessons can be learnt by keen observers willing to extrapolate to physically similar environments. However, the powers of scientists and geographers to implement their ecologically-based land suitability plans are very weak compared to the overriding political decision-making powers of the State legislatures. Identifying the ecological lessons to be learnt is considerably easier than

gaining political acceptance of constraints on land use. For several reasons, Environmental Impact Studies have not been applied to rural industries to any significant extent. This conflicts with the nationwide expansion of Environmental Impact Study requirements for other land users such as miners, developers, industrialists and tour operators. The reason for State governments' reluctance to insist on impact studies for agricultural and pastoral developments is apparently associated with both the cost of undertaking the large number of studies which would be required, and the fact that such action would be politically unpopular when it involves financially hard pressed individuals.

The writer has consistently pointed to the need to eliminate this double standard in rural land use. Some States have brought in legislation to control over-clearing and some forms of land degradation. Others either have only weak legislation or refuse to apply the law for political reasons. In yet other cases, such as Drought Assistance, legislation is administered in such a manner that the planned benefits to the land are not only nullified, but the legislation tacitly encourages over-use of pastoral land in some States. Similarly the so-called Pasture Protection Boards apparently have the opposite effect, to protection, despite the spirit of the legislation which they administer (Roberts 1984).

The fundamental lesson for Australia, is that we ignore Nature's laws at our peril, in other words the realities of land capability are as harsh as the economic realities which cause our short term blindness.

Planning for the greenhouse effect

Changes in the Australian situation may be both climatic and socio-economic. In recent years the traditional rights of rural production to have priority use of semi-arid land, have been challenged by alternative uses such as Aboriginal occupation, tourism and outdoor recreation. When these new pressures are added to the finding that rural industries have degraded 51% of

Australia's land, it may confidently be predicted that considerably more political credence will be given to alternative land uses in the future.

While some geographers are still asking 'What climatic change?' others are planning on what they believe to be irrefutable evidence of coming changes, notably an increase of 20% in the rainfall of northern Australia, a decrease of 10% in southern Australia and an overall temperature increase of 3–5 °C by the year 2030 (Graetz *et al.* 1988). For land use planners, the predicted changes in semi-arid climatic areas cannot form any dependable basis for planning of changed land use at this stage.

The writer (Roberts 1990a) has elsewhere analysed the possible changes in plant species and density, and thus carrying capacity, which could be expected from a 20% increase in rainfall. This change in productivity is likely to parallel the variability which presently exists between wetter and drier years in inland Australia. More important for sound land use in Australia, is the recent and emerging Land Care Movement. This is based on several hundred landholder committees which have the goal of testing, demonstrating and implementing on all rural lands, land use systems which will support sustainable production from crops and animals.

The role of Land Care Committees in applying on a vast scale what has been known about sound land use for decades, is an essential part of Australia's rural future. This unique movement, which epitomises the change in attitudes toward the land, has a greater potential to improve land use in Australia, than any other factor, including climate. The ultimate recognition of land degradation as the nation's prime environmental problem, has led to massive increases in federal funding of sound land use, and to the appointment of a significant corps of regional Land Care Officers in all States. For the first time in Australia's history, landholders, not urban conservationists, are taking the lead in planning and implementing sustainable land use (Roberts 1989).

Planning and the land care movement

In the past, the setting of conditions and guidelines for land use, including the determination of stocking rates, has been a function of government. The big change which has occurred with the Land Care Movement, is that landholders now have a framework or mechanism by which they can give the initiative. It remains to be seen how far landholders will use this opportunity to set their own guidelines, develop their own code of practice and produce peer pressure on their own members, to use the land within its limits.

Everyone agrees that legal controls should be avoided wherever voluntary action can be attained through education and incentives. However, two major factors are likely to hinder landholder-led changes:
(i) Determination of the margins of suitable cropland in semi-arid regions.
(ii) Control of animal numbers within the constraints of feed available.

It may reasonably be expected that the 115 committees now established in Western Australia for instance, would give special attention to where the north-eastern boundaries of the wheatbelt should be drawn for land use planning purposes. Similarly, in South Australia the embattled Eyre Peninsula community warrants a serious analysis of the sustainability of their cropping systems in terms of realistic rainfall estimates. Heathcote's (1988) analysis shows that for the 20 year period ending 1986, the longest drought-free period was three years. The most drought-affected areas and the cropping areas are shown in Fig. 1. This map indicates that drought is the norm in much of the inland, and as such, planning of carrying capacity estimates must be very conservative.

As Heathcote (1988) suggests, there must be a lesson for land use planners in the fact that in the 1980s and the 1930 depression, equivalent rates of rural unemployment, accelerated property amalgamations and a similar loss of young people from rural areas occurred. Acceptance of the natural limitations is the only basis on which to plan future land use if continual relief assistance is to be avoided. To quote Heathcote (1988), 'The

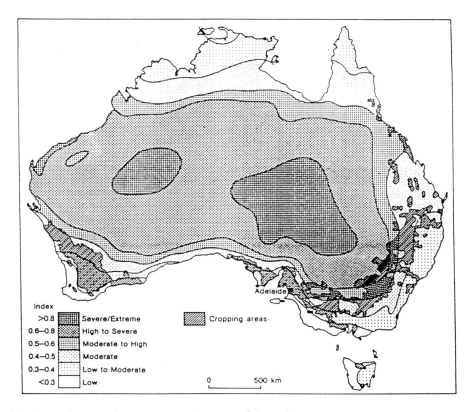

Fig. 1. Drought incidence relative to the cereal cropping areas of Australia.

Notes: Drought incidence is expressed as an index computed from rainfall percentiles (50, 30 and 10) in the formula:

$$\text{Index} - \frac{50\% - 10\%}{30\%}.$$

gamblers strategy on the semi-arid farming frontier – either a good crop or drought relief, may not have much of a future'.

The conscientious rural adviser is torn between the needs of the rural family for income and the needs of the land for ecological stability. Fisher (1986) who chaired a recent inquiry into land administration in the Western Division of New South Wales says, 'There still seems to be the greatest concern for the Western Land Commission's client – the poor battler – in the granting of cultivation licences. Despite a proclaimed embargo, "hardship of the lessee" is still the criterion. I wonder when the criterion will be otherwise – 'hardship of the land' which is, after all, the Crown's (and therefore the public's) resource.' This situation still exists to a large extent in 1990

in the region concerned. The Western Division is a particularly important case study in land use planning of semi-arid regions, because even though the Royal Commission of 1900 was gravely concerned about land condition, the Western Lands Act of 1901 was designed to enable closer settlement. For three quarters of a century, Australia's land laws seemed driven only by the political catchcry of 'Peopling the Inland'.

Young (1979) has pointed out that all the main wool-producing States carry similar total numbers of sheep, but that the number of pastoralists involved in managing the land concerned, is very different. This difference is almost entirely due to State land use policies. Table 1 (Young 1979) shows the effects of property size on enterprise viability in three States:

Table 1. Distribution of property size in southern pastoral Australia (%) (after Young 1979).

	South Australia	Western Australia	New South Wales
Substandard	16	5	51
Borderline	9	5	26
Adequate	27	27	23
Safe	48	63	0

A vital factor in semi-arid land use planning is determination of the elusive 'living area'. In New South Wales for instance, 68% of pastoralists lease a station that was less than the designated 'home maintenance area' in the mid-1970s. One of the consequences of this is that 12% of pastoralists left the Western Division in a five year period. More importantly, 44% of rural employees in this region lost their jobs and the number of jackeroos declined from 208 to 61 (Young 1979). Until more ecological realism penetrates the bureaucratic processes of land tenure and rates administration, the tragedy of pastoralists being 'birds of passage' will remain.

Planning and monitoring rangeland condition

Cambell (1966) maintains that the variation in vegetation and seasonal climate affecting individual properties in the pastoral regions make it '... difficult to see how any administrative arrangement could produce a superior result to that which would be achieved by a more flexible system which placed the responsibility squarely on the managers shoulders.' Cambell agrees that it might be necessary to penalise lessees who repeatedly overstock their land, 'provided they are not forced into such practices through external pressures such as inadequately-sized holdings or an ill-adapted system of rural finance'. He suggests that the evidence (in 1966) did not suggest that scientists or administrators have better managerial skills or technical information to warrant their taking over responsibility for land management decisions from experienced managers.

Cocks & Parvey (1985) suspect that this position is unchanged 20 years later in terms of practical improvements in advice based on well-tested models which can be adopted to individual properties. The CSIRO scientists responsible for the 'Rangepack' decision-making model would challenge this opinion (Foran & Stafford Smith 1988).

Most State governments now recognise the need to link the renewal of leases to trends in the condition of natural pasture. The science of range condition assessment is complex, long term and politically sensitive.

Young & Wilcox (1986) point out that the purpose of any monitoring system is to gain information which makes decision-making better. For this reason the data collected from monitoring must be understandable by property owners and administrators. Ideally the data can assist managers in their practical decisions, and be of use to administrators who are responsible for developing lease covenants which reward or penalise lessees for their record of stewardship. This implies that the monitoring techniques must not only be objective enough to identify changes in productivity. They must also include 'benchmark' sites with which grazed sites can be compared in a manner that allows climatic effects on productivity to be separated from grazing effects.

The writer has elsewhere (Roberts 1990b) reviewed the principles and practices of assessment for the purposes of both seasonal management by the pastoralist and long term administration by the Crown. Property planning and stock number determinations are rapidly becoming key activities of the Land Care Committees in the pastoral zone. For the first time in Australia's unimpressive history of sustainable land use, the pastoral States will this year attempt to reach agreement on a national basis for rangeland monitoring. This activity could have a significant effect on the planning and administration of our semi-arid lands and their vegetation – whichever school of thought wins the Greenhouse debate.

References

Anonymous. 1978. A Basis for Soil Conservation Policy in Australia. Interdepartmental Committee. Dept. Environment and Housing. AGPS, Canberra.

Anonymous. 1988. World Resources 1988–89. World Resources Institute, Basic Books, NY.

Australian Wheat Board. 1989. Fifty Years of Wheat Statistics. AWB, Canberra.

Cambell, K. O., 1966. Problems of adaptation of pastoral business in the arid zone. Aus. J. Agric. Econ. 10: 14–20.

Cocks, K. D. & Parvey, C. 1985. Prospects for land use planning in arid Australia, Aus. Rangel. J. 7(1): 47–50.

Fisher, M. P. 1986. Western Lands Administration in New South Wales and Possible Guidelines for Australia. Mulgalands Symposium Proc., Royal Society Queensland 155–159.

Foran, B. D. & Stafford Smith, D. M. 1988. Helping Pastoralists Make Better Decisions, Working Papers, 5th Biennial Rangeland Conf. Longreach, Aus. Rangel. Soc.

Graetz, R. D., Walker, B. H. & Walker, P. A. 1988. The consequences of climatic change in seventy percent of Australia. In: Greenhouse, planning for climate change, Ed. G.I. Pearman, CSIRO, Canberra.

Heathcote, R. L. 1988. Drought in Australia: Still a Problem of Perception, Geog. J. 16(4): 387–397.

Meinig, D. W. 1962. Margins of the Good Earth, Rigby, Adelaide.

Parvey, C. A., Blain, H. D. & Walker, P. A. 1983. A register and atlas of local government issues, CSIRO. Tech. Mem. 83/21. Div. Land and Water Res.

Roberts, B. R. 1984. Lessons from Past Land Development, In, Birth of Land Care. UCSQ Press, Toowoomba, Qld.

Roberts, B. R. 1989. The Land Care Movement: A New Challenge to Advisors. Soil and Water Conservation Assoc. J. 2(4): 4–8.

Roberts, B. R. 1990a. Predicting and Identifying Man–Made and Climatically-Induced Vegetation Changes in Australia's Pastoral Zone. ANZAAS Proc., February, Hobart, Tasmania.

Roberts, B. R. 1990b. Managing the Unmanageable: Building Environmental Feedback into Pastoral Systems. 4th Aus. Soil Conservation Conference Proceedings, Perth.

Weston, E. J. & Harbison. 1980. Assessment of the Agricultural and Pastoral Potential of Queensland. Tech. Rep. 27, Agric. Branch, QDPI, Brisbane.

Williams, J. 1989. Land Degradation: Evidence that Current Australian Farming Practice is not Sustainable. In: Management for Sustainable Farming, Ed. R. J. Hampson. Q. A. C., Lawes, Queensland.

Young, M. D. 1979. Pressures and Constraints on Arid Land Management, Aus. Rangel. Soc. Conference Proceedings: 1–10.

Young, M. D. & Wilcox, D. G. 1986. The Design and Use of Quantitative Monitoring Systems for Administration and Management. Proc. Second Int. Rangel. Congress, Adelaide. Aus. Academy of Science, Canberra.

Vegetatio **91**: 23–36, 1991.
A. Henderson-Sellers and A. J. Pitman (eds).
Vegetation and climate interactions in semi-arid regions.
© 1991 *Kluwer Academic Publishers. Printed in Belgium.*

The El Niño / Southern Oscillation and Australian vegetation

N. Nicholls
Bureau of Meteorology Research Centre, P.O. Box 1289K, Melbourne, Victoria, 3001, Australia

Accepted 7.9.1990

Abstract

The El Niño – Southern Oscillation (ENSO) phenomenon has a marked effect on Australia's rainfall. The tendency for major Australian droughts to coincide with ENSO "events" (i.e. anomalously warm sea surface temperatures in the east equatorial Pacific), and for extensive wet periods to accompany "anti-ENSO" events, is well documented. Also well-known is the partial predictability of Australian rainfall anomalies provided by ENSO.

Some other ENSO-related characteristics of interannual fluctuations of Australian rainfall are less-widely recognised, viz:
– rainfall variability is very large
– droughts and wet periods have time scales of about one year
– they exhibit very large (continental) spatial scales
– they tend to be phase-locked with the annual cycle
– they are often followed/preceded by the opposite rainfall anomaly.

The character of Australian rainfall fluctuations is thus very different from that of areas where the influence of ENSO is weak, Europe for instance. Rainfall in some other areas, notably southern Africa and India and parts of the Americas, is also strongly affected by ENSO and shares some of the above characteristics.

The relevance of these ENSO-related characteristics of Australian rainfall to its vegetation will be discussed. Australian native vegetation is adapted to these characteristics, especially in the semi-arid inland where ENSO's influence is strong. Most introduced plants are not adapted to ENSO and this has sometimes complicated their use here. The combination of ENSO-related rainfall fluctuations and European land-use strategies has resulted in some very rapid, unpredicted and undesirable changes in vegetation in the past two centuries. It has also increased the risk of soil erosion. Recognition of the real character of Australian rainfall fluctuations may help avoid further degradation of soil and vegetation.

Introduction – What is ENSO?

Variations in climate from year-to-year appear at first glance to be random. Examination of historical data, however, reveals a coherent global pattern of oceanic and atmospheric fluctuations called the Southern Oscillation. Extreme anomalies in this pattern involve dislocations of the rainfall distribution in the tropics, bringing drought to some regions and torrential rains to others (e.g. Ropelewski & Halpert 1987, 1989). These anomalies typically last about a year. Related anomalies of the atmospheric circulation extend high into the atmosphere and polewards into the temperate zones, especially in the Southern Hemisphere. Some major changes in the ocean currents and temperatures are also related to the Southern Oscillation. The best known of these is the El Niño, a marked temperature increase that occurs every few years in the eastern equatorial Pacific with catastrophic effects on the marine ecosystems along the west coast of the Americas. Because El Niño usually occurs with an extreme anomaly in the Southern Oscillation, the two phenomena are often referred to jointly as "ENSO". Periods with very warm sea surface temperatures (SSTs) in the east equatorial Pacific, and the global pattern of climate anomalies that usually accompany this warm water, are referred to as ENSO *events*. A major ENSO event occurred in 1982/83 with severe droughts in Australia, Indonesia, parts of Africa and India.

"Anti-ENSO" events, the other extreme of the Southern Oscillation, also occur. In these years the east equatorial Pacific is cold (a phenomenon now called "La Niña") and heavy rainfall and flooding is observed over the areas usually affected by drought during ENSO events. Heavy

rainfall in India, Africa and Australia during 1988 was associated with an "anti-ENSO" event. The dislocations in rainfall distribution associated with ENSO and anti-ENSO events mean that the areas the phenomena affects tend to have more variable rainfall than is the case elsewhere.

ENSO is the result of interactions between the tropical oceans (especially the Pacific) and the atmosphere. The detailed form of this interaction is yet to be determined. Major research programs aimed at modelling it are under way. Coupled models of the equatorial Pacific Ocean and the atmosphere, capable of crudely simulating some aspects of the phenomenon have been developed (e.g. Schopf & Suarez 1988). A model capable of a realistic simulation of the complete phenomenon, however, has yet to be developed.

This paper commences with a presentation of various aspects of Australian climate fluctuations that are caused by ENSO's influence. Some of these are quite well-known, but others, which may be more important to vegetation, are not. This is followed by a brief discussion of ways in which

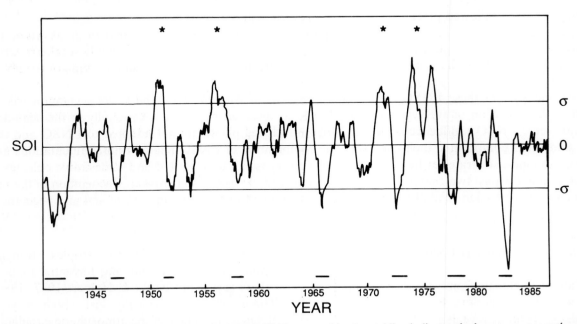

Fig. 1. The Southern Oscillation Index (or SOI) since 1940. The central horizontal line indicates the long term mean; the other two horizontal lines indicate one standard deviation above or below the mean. The horizontal bars along the bottom axis indicate the timing of major east Australian droughts, as determined from documentary records. The asterisks indicate epidemics of Murray Valley Encephalitis which accompany widespread, extended wet periods.

these ENSO-related aspects of Australia's climate have affected the vegetation, both on evolutionary time-scales, and since European settlement. Shorter sections on how ENSO can aggravate soil erosion, and the implications of all these aspects of ENSO for land management, conclude the paper.

There is often confusion about the terminology used in discussions of ENSO. ENSO "events" (i.e. El Niño events) are just one extreme of the quasi-cyclic ENSO phenomenon. The "anti-ENSO" events are the other extreme. The quasi-cyclic nature of ENSO is illustrated in Figure 1 which shows values of the Southern Oscillation Index (SOI). This index, which reflects the behaviour of ENSO, is the standardised difference in pressure between Tahiti and Darwin. ENSO events occur when the SOI is at large negative values; anti-ENSOs occur at large positive values of the SOI. The SOI fluctuates quasi-periodically. The nature of this quasi-periodicity, which has important effects on Australia's climate, will be discussed below.

Major Australian droughts are indicated in Figure 1. They tend to accompany large negative values of the SOI (i.e. ENSO events). The years when epidemics of Murray Valley Encephalitis (MVE) occurred in southeast Australia are also indicated. These epidemics occur in periods of widespread heavy rainfall and flooding and tend to occur when the SOI is large and positive (Nicholls 1986). The relationship between MVE and the SOI demonstrates the strong tendency for widespread wet periods (or pluvials) in Australia to accompany anti-ENSO events.

ENSO-related characteristics of Australian climate

Rainfall fluctuations

The best known characteristic of ENSO is the tendency for rainfall anomalies to appear in many areas at about the same time. Thus droughts in India, north China, Australia and parts of Africa and the Americas tend to occur approximately simultaneously (Williams et al., 1986; Ropelewski & Halpert 1987). At the same time, unusually heavy rainfall occurs in the central and east Pacific. These "teleconnections" are of no direct relevance to Australia's vegetation, except in the sense that all these areas experience some of the climate characteristics discussed below. So, vegetation in these areas might have attributes similar to Australian vegetation, as a result of adaptations to ENSO-related climate attributes. Some other features of Australian rainfall caused by ENSO *should* have important effects on vegetation.

High variability

One feature of Australian rainfall fluctuations caused by ENSO's influence is the large interannual variability. Conrad (1941) examined the dependence of interannual rainfall variability on the long-term mean annual rainfall, using data from 384 stations spread across the globe. He defined the relative variability of annual rainfall as the mean of the absolute deviations of annual rainfalls from the long-term mean, expressed as a percentage of the long-term mean. Conrad found that a function relating relative variability to mean precipitation fitted his data very well. The relative variability decreased, in general, as the mean precipitation increased. Over some large areas, however, the relative variability deviated in a consistent way from the global relationship with mean rainfall.

Some of these deviations were due to the influence of ENSO phenomenon on rainfall. Nicholls (1988), using Conrad's data, compared the relationship between relative variability and mean rainfall in areas affected by ENSO with the relationship elsewhere. The relative variability was typically one-third to one half higher for ENSO-affected stations compared with stations with the same mean rainfall in areas not affected by ENSO.

Nicholls and Wong (1990) confirmed, on recent data and using the coefficient of variation as a measure of relative variability, that ENSO does amplify rainfall variability in the areas it affects, relative to elsewhere. This effect was strongest at low latitudes and low rainfalls and so is especially relevant to semi-arid areas in the tropics and sub-tropics. In Figure 2 the variance of annual rainfall is plotted against the long-term mean annual rainfall for two groups of stations; those with little correlation with the SOI and those with relatively strong correlations with the SOI. Stations with intermediate strength correlations (absolute correlations between 0.05 and 0.50) are omitted for clarity. There is clear evidence in the figure that stations with annual rainfalls strongly correlated with the SOI tend to have more variable rainfall. So ENSO *amplifies* variability in the regions it affects rainfall, including Australia.

The amplification factor is substantial. The linear regressions between the logarithms of variance and mean for the two groups of locations are shown on Figure 2. The relationships between variance (V) and mean rainfall (M) derived from these regressions are:

$$V = 5.85M^{1.36} \text{ (locations with strong correlations with SOI)}$$
$$V = 1.01M^{1.52} \text{ (locations with weak correlations with SOI)}$$

In arid and semi-arid regions these expressions indicate that the variance of annual rainfall in an area strongly affected by the Southern Oscillation would be, typically, more than double that in an area with similar mean rainfall but not influenced

LOG OF VARIANCE VERSUS LOG OF MEAN RAINFALL

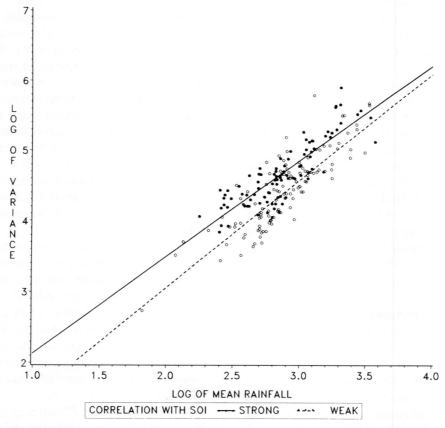

Fig. 2. Logarithm of variance of annual rainfall versus logarithm of mean annual rainfall. Open circles represent stations with weak correlations between their annual rainfall and the SOI (r < 0.05). Full circles indicate stations with strong correlations with the SOI (r < 0.5). Data from Nicholls and Wong (1989).

by the Southern Oscillation. Such a wide difference in variability should be reflected in the adaptations of indigenous vegetation.

The high variability of Australian rainfall was noticed soon after European settlement. The problems faced by farmers introducing crops was a major indicator of this higher variability. Rolls (1981) notes that "If a year was dry, the unsuitable English wheats set no seed. If a year was wet the fungus disease, rust, turned crops into a red mush as they headed".

Continental spatial scales

Australian rainfall fluctuations, as well as being more severe (i.e. more variable) because of ENSO's influence, also operate on very large spatial scales. The 1982/83 drought severely affected over half the continent, an area as large as Western Europe and 25 times the size of Wales and England.

The continental-scale of the 1982/83 drought is typical of many years, although it was more severe than most. Streten (1981) listed the percentage of Australia receiving above average rainfall, for each year from 1950 to 1969. In three of the 20 years more than 80% of the continent had above average rainfall; in three other years less than 20%. The chances of even one year deviating so far from the expected 50% of the continent receiving above average rainfall are negligible, unless strong spatial correlations exist. Gibbs & Maher (1967) found that in 22% of the years between 1885 and 1965 there was no part of Australia with annual rainfall in the lowest decile. Again, the chance of this happening even once in the 80 years is tiny, if Australian rainfall fluctuations were random, independent point processes.

Clearly Australian rainfall fluctuations do exhibit strong spatial correlations, and ENSO is one of the factors producing these. McBride & Nicholls (1983) correlated Australian seasonal rainfall with various indices of ENSO and found large areas with significant correlations. Rainfall over most of eastern Australia, about half the continent, is strongly correlated with ENSO. Ropelewski & Halpert (1987, 1989) also demonstrated that Australian rainfall fluctuations tend to coincide over large areas, during ENSO and anti-ENSO events.

The large spatial scale of the ENSO phenomenon is obviously the main reason for the large spatial scale of Australian rainfall fluctuations but the lack of major mountain ranges inland also contributes. The parts of eastern Australia where the influence of ENSO is weakest lie around the coastal fringe, where the Great Dividing Range can produce local rainfall effects thereby complicating the large scale influence of ENSO. Inland of the coastal mountains, however, there is little orography to differentiate the reaction to the broad-scale influence of ENSO, leading to widespread droughts and pluvials.

Long time scales

A "drought" in England is defined as a period of 15 days without rainfall. At the other extreme, the Sahel has been in drought for the past two decades. Australian droughts, and pluvials or wet periods, tend to last about twelve months or so (e.g. Ropelewski & Halpert 1987). ENSO and anti-ENSO "events" tend to last about twelve months (e.g Rasmusson & Carpenter 1982) and this sets the time scale of Australian rainfall fluctuations.

Figure 3 shows three-month averages of rainfalls for a district centred on Bourke, in western New South Wales. Two sets of years were used to produce this figure: a set of "ENSO" events (1918, 1940, 1957, 1965, 1972, 1976, 1982, 1986), and a set of "anti-ENSO" events (1916, 1938, 1950, 1955, 1970, 1973, 1975, 1988). The average rainfalls for the two sets of years are shown for a period of three years starting at the beginning of the years prior to those listed above.

There is a period of just over 12 months where the average rainfalls of the two sets differ in a consistent way, i.e. with each period in the anti-ENSO years receiving substantially more rainfall than the ENSO years. This period stretches from about February of year "0" to April of year "+1". The ENSO and anti-ENSO events are all aligned so that they occur in year "0". The difference in

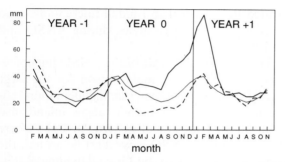

Fig. 3. Composite three-month running means of rainfall for the district around Bourke, New South Wales. The thin continuous line indicates the long term mean rainfall. The thick continuous line is a composite of eight "anti-ENSO" events; the thick broken line is for eight "ENSO" events. "Year 0" is the first year of both the ENSO and anti-ENSO events. "Year-1" is the year that precedes the events; "Year + 1" follows the events.

average rainfall received over this period is substantial. The average for the ENSO events is only 346 mm while the anti-ENSOs receive more than twice this, 720 mm.

A similar temporal pattern of rainfall anomalies occurs over much of eastern and northern Australia. Figure 4 shows the pattern of rainfall anomalies for a district around Alice Springs, for the same ENSO and anti-ENSO events used in Figure 3. As was the case around Bourke, the anomalies associated with anti-ENSOs last about

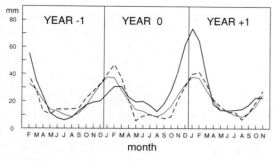

Fig. 4. Composite three-month running means of rainfall for the district around Alice Springs. The thin continuous line indicates the long term mean rainfall. The thick continuous line is a composite of eight "anti-ENSO" events; the thick broken line is for eight "ENSO" events. "Year 0" is the first year of both the ENSO and anti-ENSO events. "Year-1" is the year that precedes the events; "Year + 1" follows the events.

twelve months, starting late in one summer and lasting until late the next summer. The similarity of the temporal pattern of anomalies in Figures 3 and 4, for locations separated by over 1500 km, illustrates again the large spatial scales of rainfall anomalies produced by ENSO.

Phase-locking to annual cycle

Figures 3 and 4, and the results of Ropelewski & Halpert (1987) and many others, also demonstrate that these extended periods of drought or extensive rains do *not* occur randomly in time, in relation to the annual cycle. In fact, the ENSO phenomenon, and Australian rainfall fluctuations associated with it, are phase-locked with the annual cycle. Figure 3 and 4 demonstrate that the heavy rainfall of an anti-ENSO tends to start early in the calendar year and finish early in the following year. The dry periods associated with ENSO events tend to occupy a similar time period. This means that if an extensive drought or wet period is well-established by the middle of the calendar year it is unlikely to "break" until at least early the following year. The 1982/83 drought again provides a good example. The drought started about April 1982 and broke over much of the country in March and April 1983. Such phase-locking has been found in most other variables associated with ENSO (e.g. Rasmusson & Carpenter 1982).

Biennial cycle

This phase-locking is related to a biennial cycle which is a fundamental element of ENSO variability (Rasmusson *et al.*, 1989). There is also a lower frequency variation, but it is the biennial mode which captures the major features associated with ENSO "episodes". The biennial cycle is observed over the equatorial Pacific and Indian Oceans and is tightly phase-locked with the annual cycle. It varies in amplitude from cycle to cycle and sometimes changes phase. Nicholls (1979, 1984) discussed how ocean-atmosphere interaction around Indonesia, modulated by the seasonal cycle, could result in a biennial cycle phase-locked to the annual cycle.

The biennial mode means that ENSO events will often be preceded and/or followed by anti-ENSOs, and vice-versa. In terms of Australian rainfall, this means that year-to-year changes in rainfall can be extreme. The annual rainfall in the district around Bourke in 1950, an anti-ENSO year, was 841 mm (c.f. annual mean of 347 mm). The following year's rainfall was only 217 mm. In 1982 the district received only 162 mm; the following year 552 mm.

A tendency for a biennial cycle can even be detected in Figures 3 and 4. In both figures rainfall is higher in anti-ENSO years than in ENSO years (compare the "year 0" results). But their anomalies are reversed in the years preceding the events, i.e. the years before ENSO events tend to be wetter than average and the years before anti-ENSOs tend to be drier than average.

The change from ENSO-related drought to anti-ENSO and pluvial conditions can be rapid, and usually occurs early in the calendar year. An ENSO-related drought in 1888, for instance, was broken on New Years Day 1889 by heavy rains that caused severe flooding in many parts of eastern Australia (Nicholls 1990). The 1982/83 drought broke over much of Australia during March 1983, again with heavy rains.

The descent into drought can also be rapid. The Bourke district received 418 mm of rain in January and February 1976, at the end of the 1975 anti-ENSO event. This was five times the average for these two months. Only 98 mm fell in the next six months, about half the average for this period.

The long wet periods, followed rapidly by long dry periods, would enhance the likelihood of wildfires. Rapid growth during the pluvial would dry quickly during the ensuing drought and then "dry" thunderstorms could ignite the bush.

Winds and temperatures

Rainfall is not the only aspect of the climate affected by ENSO. Frosts will tend to be more common in inland Australia during ENSO events, because low rainfall is associated with decreased cloud cover, allowing increased radiative cooling at night. Maximum temperatures will usually be higher during ENSOs, also because of the decreased cloud cover. Both of these features may add to the damage to vegetation caused by the low rainfall.

There are also strong variations in wind between ENSO and anti-ENSO events, although these variations are different in different parts of the country. North of about 25°S low-level winds in *winter* during ENSO events tend to be about 3-4 times stronger than during anti-ENSO events (Drosdowsky 1988). In the southeast the winds in summer tend to be 2-3 times stronger in ENSO events. These relationships with ENSO suggest that wind variability may also be relatively high in Australia, as is the case with rainfall. No comparative studies have yet been made to check this.

The stronger winds associated with ENSO related droughts would further increase the likelihood of wildfire. Any fire ignited by a "dry" thunderstorm during an ENSO event could spread further and faster because of the stronger winds. The latitudinal variation in the season in which these ENSO-amplified winds occur (winter in the north, summer in the south) means that the strong winds occur at the time of year when fires are most likely anyway.

Predictability of climate fluctuations

The biennial cycle underlying the ENSO phenomenon, and the phase locking of this cycle to the annual cycle, provide some regularity to the phenomenon, and to climate variables associated with it. This regularity provides a degree of predictability. The phase-locking means that ENSO, or an index of the phenomenon (e.g. the SOI), will tend to change phase around March-May and only rarely at other times of the year. Thus if the SOI is strongly positive (anti-ENSO) during the Australian winter, it will probably stay in that phase until early the following year. So climate variations normally associated with this phase of ENSO, and which occur towards the end of the calendar year, may be predictable simply by monitoring the SOI earlier in the year. This observation underlies the *Seasonal Outlooks* prepared by the Australian Bureau of Meteorology. These outlooks use observed values of the SOI to predict seasonal rainfall fluctuations over eastern and northern Australia, from early winter until the end of the year.

Effects of ENSO on Australian vegetation

Characteristics of Australian vegetation attributable to ENSO

There is reason to believe that ENSO has been operating for at least many thousands of years (e.g. Nicholls 1989). Westoby (1980) noted that "climates with the same general level of aridity can offer very different mixtures of growth opportunities, because of the patterning of rainfall in time; accordingly different mixtures of growth-forms are found". So Australian vegetation should be somewhat different from vegetation in other arid and semi-arid areas where ENSO is not an active control on the climate, because of the different "patterning of rainfall in time" that ENSO produces in Australia. Australian vegetation should be suited to an environment of highly variable rainfall, with frequent severe droughts or pluvials extending, typically, for about 12 months. It should also be able to survive frequent fires since, as noted above, the long alternating wet and dry periods, and the coincidence of strong winds with the dry periods are conducive to wildfire.

The following is a list of just some of the characteristics of Australian vegetation that may be, at least in part, attributable to ENSO's influence on the climate. The list is not comprehensive and is provided just to illustrate that the "patterning of the climate" caused by ENSO can affect Australia's vegetation. Many of the characteristics listed below appear to be so well-known to ecologists that citation of individual works

seems inappropriate. Most of the information that follows was taken from Recher *et al.* (1979, 1986) and Stafford Smith and Morton (1989).

Absence of succulents

Cacti, although adapted to arid climates and requiring little moisture, need regular rain. Such plants would be poorly-equipped to survive twelve-month droughts and to take advantage of extensive wet periods when they do occur, i.e they seem unsuited to the high rainfall variability ENSO produces over much of Australia. Succulents are almost totally absent from the Australian arid and semi-arid regions. Other such regions elsewhere often display a large variety of such plants. The few species found in Australia occur in particular locations where runoff after even very small rains can be collected.

Establishment dependent on extended wet periods

Successful establishment of bladder saltbush and mulga requires extended periods of wet weather. Mulga requires heavy summer rains to produce large numbers of flowers, and subsequent heavy winter rains to set seed. The seeds then germinate in summer and further heavy rains are needed to allow the young seedlings to establish. As noted earlier, anti-ENSO events produce heavy rains starting in late summer and extending well into the next summer, as required for the establishment of mulga seedlings.

Drought tolerance/avoidance

Many Australian plants are remarkably tolerant of severe, extensive droughts. Well-developed tolerance or avoidance strategies are essential because of the frequent severe droughts caused by ENSO. Mulga, once established can withstand all but the most severe droughts and can survive more than 50 years. Such longevity is a useful characteristic in a country of highly-variable rainfall where the conditions for successful establishment will occur relatively rarely. Some other species will survive perhaps 90-95% of years, but rely on seed reserves to re-establish after the more exceptional events (Westoby 1980). Many Australian plants have large root systems, relative to their above ground structures. Plants with large roots are more tolerant of drought because they can gather soil moisture long after it has disappeared from near the surface. Spinifex, for example, have very long root systems. A variety of drought avoidance mechanisms are used by Australian plants. Surviving droughts as seeds is a favourite avoidance technique. Some Australian plants have life cycles tailored to take advantage of the relatively consistent aspects of the climate to set seed, rather than relying on the more variable aspects. For instance some tropical annuals mature at the earliest time at which the rainy season might end, thus ensuring they set seed every year (Andrew & Mott 1983). If they aimed to mature at the "average" time for the end of the rains, there would be many years in which they would fail to set seed. "Resurrection" plants can survive almost total desiccation for years and then recover within 24 hours of a rainfall to flower and set seed quickly before the water disappears again (Gaff 1981). Ephemerals, which complete their life cycle and set seed rapidly before the soil dries, are common over much of the continent and produce spectacular massed flower displays.

Diverse life history strategies

In a predictable climate, there are usually only a few optimal strategies in the balance between drought tolerance and establishment vigour. Where conditions differ more from year-to-year, different establishment strategies are preferred in different years and a wider range of strategies can co-exist (e.g. persistence as seed, dormant tuber or growing adult). The "extreme irregularity of the Australian arid zone climate thus contributes to a relatively high diversity of life history strategies" (Stafford Smith & Morton, 1989).

Vegetation height

Australia has more trees at a given level of aridity than elsewhere. Milewski (1981) attributes this to larger rainfalls producing deeper water penetration. Larger trees, with large root structures able to remove water from far below the surface, would be favoured by such rainfalls, relative to an area where rainfall was lighter and more frequent. The heavy rainfall events usually accompany anti-ENSOs and persist for about a year or so. This persistance would also favour establishment of larger trees, relative to an area of totally unpredictable rains.

Fire resistance/dependence

Much of the Australian flora is fire-resistant or even dependent on fire for successful reproduction. Some plants, e.g. spinifex, actively promote burning by producing flammable oils in their leaves. The fire releases the nutrients and clears away the old plants, allowing young ones to germinate. Fire usually induces substantial seed release from eucalypts. An enhanced "seed rain" after fire temporarily exceeds losses to seed predation by harvester ants, thus improving the chances of successful establishment (Noble et al., 1986). In the least fertile environments (e.g. spinifex-dominated communities) numerous species persist only as seeds or dormant root-stock between fires which release scarce nutrients from perennial tissue, and make space amongst perennials.

'Fluctuating climax'

The high interannual variability of annual rainfall results in such dramatic changes in vegetation from year-to-year in the arid and semi-arid parts of Australia affected by ENSO that the concept of a true static climax vegetation may be inappropriate. Rather, the vegetation appears adapted to the climate in such a way that demographic composition is in a state of unstable equilibrium. This difference in Australian vegetation from the conventional concept of "climax" has been recognised in scientific circles since the late 1930s (Heathcote 1965).

The above are just some examples of the numerous and diverse characteristics of Australian native vegetation attributable to the climate rhythms imposed by ENSO. A more detailed study of the vegetation, guided by knowledge of these ENSO-induced rhythms, may reveal others. Understanding of how ENSO has affected Australian native vegetation would seem to be important if we are to restrict the accidental degradation of the Australian environment. Comparative studies of vegetation in the Enso-affected parts of Australia with vegetation in areas less strongly affected by ENSO (e.g. southwest Australia) and in other countries affected by ENSO (e.g. southern Africa) may extend our understanding.

ENSO and vegetation changes

Since the native Australian vegetation was adapted to the climate rhythms induced by ENSO, and Europeans were unaware of these rhythms, it is not surprising that the effects of introduced plants and animals not adapted to ENSO did lead to rapid and undesirable changes in vegetation. The best known of these changes is probably the area now known as the Pilliga Scrub (Rolls 1981; Austin & Williams 1988). Much of this area of 400,000 ha was open grassy country with only about eight large trees per hectare when Europeans arrived in the 1830s. Frequent burning by Aboriginals, and grazing by indigenous marsupials, restricted the opportunities for trees and shrubs to establish. Fire germinated the seed of the trees and shrubs but rat kangaroos ate many of the resulting seedlings before they could establish.

The introduction of sheep reduced the numbers of rat kangaroos, by destroying their cover and their food. A severe drought during the major ENSO event of 1877/78 further reduced the numbers of indigenous marsupials. The following year, a major anti-ENSO event (Wright 1975), was very wet. The few large *Callitris glaucophylla* trees seeded well and when stock owners burnt to destroy grasses with seeds that got into their sheep's wool, *Callitris* seedlings came up thickly, unhindered by the grasses which would usually compete with them for space. This time there were

no rat kangaroos to eat the seedlings either and the trees grew unchecked.

Over the next decade there were several further periods of establishment, again synchronised with ENSO-anti ENSO oscillations. The European rabbit, also an enthusiastic eater of seedlings, arrived in the area in the late 1880s and prevented further establishment until myxomatosis in 1951 reduced the rabbit population. The first successful release of myxomatosis occurred in 1950. Earlier releases of the disease had not led to widespread establishment. The extensive rains and flooding in 1950, associated with a major anti-ENSO, contributed to the successful establishment of the disease by providing ideal breeding conditions for the insects that spread it.

In 1917 the Forestry Commission stopped burning in the Pilliga and by 1950 large amounts of forest litter had accumulated. So had decades of seed production. The forest dried in the ENSO event of 1951, following good growth during the "anti-ENSO" of 1950, and a major fire started in November 1951. In the absence of rat kangaroos and rabbits, the new growth induced by the fire had nothing to stop it. From now on regeneration and thickening of the forest would occur in every suitable wet year.

In less than a century Europeans unintentionally transformed the area from grazing land into the dense Pilliga Scrub supporting sustained timber harvesting. The ENSO phenomenon played a critical role in this transformation. The biennial nature of ENSO, and of the climate fluctuations associated with it, appears to have been especially important in determining these changes. In the establishment of the Pilliga Scrub, for example, the temporal pairing of ENSO and anti-ENSO events (1877 and 1878; 1950 and 1951) was a major factor, not just extreme rainfalls falling randomly in single years. The very large variability of rainfall caused by the ENSO phenomenon was also important. Heathcote (1965) notes that there was considerable concern early this century that the combination of grazing and the high variability of Australian rainfall was leading to progressive degradation of the native vegetation in semi-arid areas.

Austin and Williams (1988) and McKeon *et al.* (1990) cite other examples where the extreme climate events associated with both extremes of ENSO resulted in major long-term vegetation degradation. In western Queensland there was a rapid increase in sheep during the above-average rainfall years of the early 1890s. Major ENSO events between 1899 and 1902 resulted in very low rainfall and a rapid drop in animal numbers. Heavy utilisation of edible grasses and shrubs during this drought led to a spread of inedible plants and carrying capacities seem to have been permanently impaired (Heathcote 1965; McKeon *et al.* 1990). In the subtropical grasslands of southern coastal Queensland, rapid change in species composition to bunch spear grass (*Heterpogon contortus*) appears to have resulted from overgrazing with sheep during the ENSO-related drought of 1881-82 (McKeon *et al.* 1990). More recently, low beef prices in the mid-1970s led to increased stocking rates in Queensland. These years were wet, the result of the 1973-75 anti-ENSO, but attempts to maintain the high stocking rates into the 1980s with their drier, ENSO-related conditions has led to pasture degradation, species changes and soil erosion (McKeon *et al.* 1990).

ENSO and erosion

Many of the climate characteristics caused by ENSO would appear to lead to an increased risk of soil erosion, by both flood and wind. The extreme rains associated with anti-ENSOs cause large flows of water which are major creators of landscape structure. McMahon *et al.* (1987) have shown that streamflow tends to be more variable in Australia than is the case in similar areas in the rest of the world. Because of the biennial nature of ENSO, heavy rains will often follow soon after a major drought, which would have reduced vegetation cover, leaving the soil bare and vulnerable. So the biennial rhythm, and the high variability, of Australian rainfall, both the result of ENSO, increases the risk of erosion.

Wind erosion is also favoured by ENSO.

Strong winds often accompany ENSO events. Again, with the vegetative cover removed by the drought the strong winds can result in major erosion. The tendency for the strong winds to occur in the hotter and drier seasons (in the winter in the north; in the summer in the south) would aggravate this since the high evaporation in these seasons would also tend to increase the vulnerability of the soil to wind erosion. Southeastern Australia has been subjected to immense dust storms towards the end of ENSO events, as a result of the high winds and desiccation of the soil during these events. A spectacular example occurred in February 1983 but frequent dust storms were also observed late last century. Occasionally rain accompanied these storms, leading to references to "Red Rains" or "Rains of Blood". Roads, edges of towns, and railways were all threatened by drifting sands in mallee country.

The vulnerability of Australian soils to erosion by wind and water is at least partly due to the severe extremes of climate associated with ENSO, and to its inherent biennial cycle. So severe erosion events must have occurred before European settlement. Agricultural practices, especially last century and in the early decades of this century, aggravated the situation. The extension of cropping into arid areas, with dry-land farming "techniques" such as deep cultivation and bare cultivated fallowing provided an open invitation to erosion. In the grazing areas, overstocking and retention of high stocking rates well into droughts led to the destruction of ground cover exacerbating the potential for erosion.

Implications for land management

The use of some European land management practices, designed for a less-variable climate, has led to land and vegetation degradation in Australia, either through exacerbating the already high potential for soil erosion, or by favouring the establishment of "woody weeds" and other changes in species composition. This degradation is the direct result of employing land management practices appropriate to a benign and predictable climate in an area where ENSO produces a climate of extremes. Friedel *et al.* (1989) argue that lack of knowledge about the Australian climate, on the part of early settlers and governments, led to misuse of Australia's arid lands. If further degradation is to be prevented then a thorough understanding of the *real* Australian climate is needed, along with appropriate land management practices. McKeon *et al.* (1990) note that the influence of the Southern Oscillation on northern Australian savannas is "strong enough to suggest that the Southern Oscillation should be included in any prescription for savanna management".

Some possible changes to land management (mainly from Griffin & Friedel 1985; and Friedel *et al.* 1989), to avoid degradation through interaction between land management strategies and the extreme, ENSO related Australian climate variations, are listed below.

Many graziers make management decisions largely on the basis of their stock, not recognising that damage to pasture can occur well before stock condition declines. Rapid de-stocking at the start of droughts would help the vegetation to survive. Such a policy is justified because of ENSO's tendency to produce long droughts. For example, if a drought is underway by winter, it is unlikely that it will be broken before mid-summer. So de-stocking, rather than "holding on" in the hope of an early break, would often be the best strategy.

Large-scale regeneration of vegetation is infrequent and associated with the extended high rainfall events (associated with major anti-ENSOs). Cattle populations respond to the same pluvial phases but their peak populations are delayed and they can therefore place great pressure on vegetation during the post-pluvial phase, when the plants are already experiencing drought stress. The biennial tendency of ENSO means that a very dry period, rather than "average" rainfall, may follow the pluvial phase. Thus, a conservative re-stocking strategy is needed, if this pressure is not to lead to degradation and aborted regeneration.

Excessive numbers of woody plants will establish in some areas unless they are controlled by strategic fires lit soon after major rainfall periods. Wildfire will inevitably follow, unless management burns are initiated within a critical time span. Such management burns replace the effect of Aboriginal burning and small marsupials which, before European settlement, would have restricted the establishment of woody weeds.

Westoby (1980) suggested that setting equilibrium stocking rates, even conservative ones, was an inappropriate range management strategy in areas where grazing pressures in conjunction with high rainfall variability and patterning of rainfall in time can produce radical vegetation changes. He proposed that grazing pressure should be used as a management "tool" to help or obstruct transitions in the state of vegetation produced by specific rainfall sequences. So, in an area where ENSO produces occasional very wet periods and these are the only periods which allow re-establishment of desired perennials, de-stocking at these times may be the best strategy. Westoby *et al.* (1989) have extended this approach and suggest that a state-and-transition model provides a better basis for rangeland management than the conventional range succession model in areas where episodic events are important and influences of grazing and vegetation change act intermittently. Under this model, range management would not aim at establishing a permanent equilibrium, but would be opportunistic and seize opportunities and evade hazards as they arose. Areas where ENSO influences the climate, such as the semi-arid areas of Australia, seem prime candidates for such an approach.

These are just a few management strategies which would seem appropriate in a land where ENSO dominates. There should be others. For instance, recognition that increased winds during ENSOs and heavy rains at the end of ENSOs (both of which could exacerbate soil erosion) might lead to changes in land management which could offset the increased erosion risk. Re-examination of land management strategies in the light of our relatively-new knowledge of the climate rhythms imposed on Australia's climate by ENSO may be fruitful, both in terms of avoiding degradation and improving the economic use of the land.

The predictability of climate variations arising from ENSO, in particular, should lead to many strategies to improve the economic return to graziers while reducing the likelihood of degradation of Australia's arid and semi-arid lands and their vegetation. McKeon *et al.* (1990) demonstrate how ENSO-related seasonal predictions can help grazing management in the subtropical grasslands where spring burning is used to improve diet quality. If dry conditions follow burning there is economic loss due to feed restrictions, reductions in pasture growth, and increased risk of soil erosion. McKeon *et al.* used a pasture growth model to simulate growth for each of the past 114 years and then compared pasture growth in years when the SOI was negative in winter (i.e. ENSO events) and years when it was positive in winter (anti-ENSO events). The percentage of years when spring growth was insufficient to prevent the adverse results listed above was 53% when the winter SOI was more than half a standard deviation below the mean, and 23% when it was more than half a standard deviation above the mean. They conclude that seasonal forecasts based just on the SOI "could be used to make better decisions to reduce area or frequency of burning under conditions... when production losses and erosion risks are greatest".

Since European settlement there has been substantial degradation of Australia's land and vegetation, due to ignorance of the effects of ENSO on the climate, and of how European land management practices will interact with the ENSO-related climate fluctuations. This has particularly been the case in the arid and semi-arid areas, where ENSO's influence is strongest. Our increasing knowledge of ENSO, its effects on Australia's climate, and how native and exotic biota interact with climate, should provide the basis for more profitable use of these regions while, at the same time, reducing the likelihood of further exploitation and degradation.

It must also be noted, however, that the variability of rainfall caused by ENSO complicates the management of arid and semi-arid rangelands.

Harrington *et al.* (1984) observed that "in a reliable climate... it is possible to plan a grazing and/or fire management which is responsive to the rainfall regime... and thereby obtain a measure of control over the composition of the plant community. In unreliable climates the acquisition of such knowledge is much more difficult because any particular run of rainfall events is infrequently repeated and may give an effectively unique vegetation response". Wilson and Harrington (1984) note that "it is the variability of rainfall and the relatively fixed nature of flocks and herds that causes the greatest management problems in arid regions. While some change in the composition and productivity of pasture communities following droughts seems inevitable, it is the challenge of range management to identify and avoid the critical points where serious long-term damage occurs: that is where the ecosystem is pushed beyond the bounds of its resilience". The patterning in time of rainfall produced by ENSO events may provide the regularity needed to manage these areas better.

References

Andrew, M.H. & Mott, J.J. 1983. Annuals with transient seed banks: the population biology of indigenous Sorghum species of tropical north-west Australia. Aust. J. Ecol. 8: 265–276.

Austin, M.P. & Williams, O.B. 1988. Influence of climate and community composition on the population demography of pasture species in semi-arid Australia, Vegetatio 77: 43–49.

Conrad, V. 1941. The variability of precipitation. Mon. Weath. Rev. 69: 5–11.

Drosdowsky, W. 1988. Lag relations between the Southern Oscillation and the troposphere over Australia. BMRC Research Report No. 13, Bureau of Meteorology Research Centre, Melbourne, Australia, 201 pp.

Friedel, M.H. Foran, B.D. & Stafford Smith, D.M. 1989. Where the creeks run dry or ten feet high: Pastoral management in arid Australia. Proc. Ecol. Soc. Aust. 16: in press.

Gaff, D.F. 1981. The biology of resurrection plants, in Pate, J.S. and McComb, A.J. (eds), The biology of Australian plants. Univ. of West Australia Press, Nedlands, 412 pp.

Gibbs, W.J. & Maher, J.V. 1967. Rainfall deciles as drought indicators, Bulletin No. 48, Australian Bureau of Meteorology, Melbourne Australia, 118 pp.

Griffin, G.F. & Friedel, M.H. 1985. Discontinuous change in central Australia: some implications of major ecological events for land management. J. Arid Environments 9: 63–80.

Harrington, G.N. Friedel, M.H., Hodgkinson, K.C. & Noble, J.C. 1984. Vegetation ecology and management. In: Harrington, G.N., Wilson, A.D. and Young, M.D. (eds), Management of Australia's rangelands, CSIRO, Melbourne, 354 pp.

Heathcote, R.L. 1965, Back of Bourke. Melbourne Univ. Press, Melbourne.

McBride, J.L. & Nicholls, N. 1983. Seasonal relationships between Australian rainfall and the Southern Oscillation. Mon. Wea. Rev. 111: 1998–2004.

McKeon, G.M., Day, K.A., Howden, S.M., Mott, J.J., Orr, D.M., Scattini, W.J. & Weston, E.J. 1990. Management for pastoral production in northern Australian savannas. (in preparation).

McMahon, T.A., Finlayson, B.L., Haines, A. & Srikanthan, R. 1987. Runoff variability: A global perspective, in, The influence of climate change and climatic variability on the hydrologic Regime and water resources. IAHS Publication No. 168.

Milewski, A.V. 1981. A comparison of vegetation height in relation to the effectiveness of rainfall in the mediterranean and adjacent arid parts of Australia and South Africa. J. Biogeography 8: 107–116.

Nicholls, N. 1979. A simple air-sea interaction model. Quart. J. Roy. Meteor. Soc. 105: 93–105.

Nicholls, N. 1984. The Southern Oscillation and Indonesian sea surface temperature. Mon. Weath. Rev. 112: 424–432.

Nicholls, N. 1986. A method for predicting Murray Valley Encephalitis in southeast Australia using the Southern Oscillation. Aust. J. Exper. Biol. Med Sci. 64: 587–594.

Nicholls, N. 1988. El Niño – Southern Oscillation and rainfall variability. J. Climate 1: 418–421.

Nicholls, N. 1989. How old is ENSO? Climatic Change 14: 111–115.

Nicholls, N. 1990. The Centennial Drought, in Webb, E. (ed), Windows on Australian Meteorology, Australian Meteorological and Oceanographic Society, (in press).

Nicholls, N. & Wong, K. 1990. Dependence of rainfall variability on mean rainfall, latitude, and the Southern Oscillation. J. Climate 3: 163–170.

Noble, J.C., Harrington, G.N. & Hodgkinson, K.C. 1986. The ecological significance of irregular fire in Australian rangelands. In: Joss, P.J., Lynch, P.W. and Williams, O.B. (eds.), Rangelands: A resource under seige. Australian Academy of Science, Canberra, 634 pp.

Rasmusson, E.M., Xueliang Wang & Ropelewski, C.F. 1989. The biennial component of ENSO variability. J. Marine Systems (submitted).

Rasmusson, E.M. & Carpenter, T.H., 1982, Variations in tropical sea surface temperature and surface wind fields associated with the Southern Oscillation / El Niño. Mon. Weath. Rev. 110: 354–384.

Recher, H.F., Lunney, D. & Dunn, I. 1979 & 1986. A natural legacy: Ecology in Australia (1st and 2nd editions). Pergammon, 276pp & 443 pp.

Rolls, E.C. 1981. A million wild acres, Nelson, Melbourne.

Ropelewski, C.F. & Halpert, M.S. 1987. Global and regional scale precipitation patterns associated with the El Niño / Southern Oscillation. Mon. Weath. Rev. 115: 1606–1626.

Ropelewski, C.F. & Halpert, M.S. 1989. Precipitation patterns associated with the high index phase of the Southern Oscillation. J. Climate 2: 268–284.

Schopf, P.S. & Suarez, M.J. 1988. Vacillations in a coupled ocean-atmosphere model. J. Atmos. Sci. 45: 549–566.

Stafford Smith, D.M. & Morton, S.R. 1989. A framework for the ecology of arid Australia. J. Arid Environments, (in press).

Streten, N.A. 1981. Southern hemisphere sea surface temperature variability and apparent associations with Australian rainfall. J. Geophys. Res. 86: 485–497.

Westoby, M. 1980. Elements of a theory of vegetation dynamics in arid rangelands. Israel J Botany 28: 169–194.

Westoby, M., Walker, B. & Noy-Meir, I. 1989. Opportunistic management for rangelands not at equilibrium. J. Range Management 42 (in press).

Williams, M.A.J., Adamson, D.A. & Baxter, J.T. 1986. Late Quaternary environments in the Nile and Darling basins. Aus. Geo. Studies. 24: 128–144.

Wilson, A.D. & Harrington, G.N. 1984. Grazing ecology and animal production. In: Harrington, G.N., Wilson, A.D. and Young, M.D. (eds.), Management of Australia's rangelands. CSIRO, Melbourne, 354 pp.

Wright, P.B. 1975. An index of the Southern Oscillation. Climatic Research Unit, University of East Anglia, 22 pp.

Measurement

Introduction

Identifying land degradation and seeking to understand why it is occurring is of considerable concern to semi-arid land managers. Dunin discusses the spatial significance of evaporation measurements and comments on how point measurements might be extrapolated to larger spatial scales. Leys reviews and comments on models of soil erodability and suggests an improved methodology for measuring and monitoring which is supported by field measurements of soil erosion and the effects of increased ground cover on this process. Verstraete and Pinty discuss the value of satellite remote sensing to land degradation studies and show that, while satellites offer considerable potential, they are not yet a fully satisfactory source of data. Finally, in this section, Denmead discusses the sources and sinks of trace gases especially 'greenhouse gases'. Knowledge of these sources and sinks is a pre-requisite for understanding the predicted warming due to these trace gases. Current opinions are mixed, but the consensus appears to suggest that greenhouse warming with the associated intensification of the global hydrological cycle and poleward shifts in the 'climate belts' might lead to many semi-arid regions becoming even more water-stressed. Thus the implications of Denmead's chapter need to be considered in the planning and management of semi-arid regions.

Measurement

Introduction

Vegetatio **91**: 39–47, 1991.
A. Henderson-Sellers and A. J. Pitman (eds).
Vegetation and climate interactions in semi-arid regions.
© *1991 Kluwer Academic Publishers. Printed in Belgium.*

Extrapolation of 'point' measurements of evaporation: some issues of scale

F.X. Dunin
CSIRO Division of Plant Industry, Canberra, Australia

Accepted 24.8.1990

Abstract

The paper focuses on extrapolation of observed values as a means to determine regional evaporation. Evaporation data from diverse plant communities in southern Australia were drawn together to assess the magnitude and causes of areal variation in the process over the landscape. Soil water effects on daily evaporation rates were responsible for pronounced variability over a few hectares of uniform vegetation, variability being comparable to that encountered at an extended scale subject to combined influences of soil water and vegetation effects. In the longer term, local effects of soil water on evaporation were apparent, albeit with attenuated areal variability. Short term differences between extensive plant communities did not necessarily persist, sometimes resulting in a reversal of differences in evaporation rate. Estimating regional evaporation at time scales ranging from daily to monthly calls for an understanding of spatial and temporal variation in factors imposing control at the surface, especially those dealing with biological response. Improving this understanding to achieve accuracy of estimates means that demands for finer temporal resolution in descriptions of regional evaporation must be accompanied by greater density of measurement points to resolve areal variability in flux rate.

Further complexities were identified with advective effects caused by discontinuities in the landscape, notably for irrigated regions. Accounting for an advective enhancement of 20% in evaporation rate in an irrigation region involved a description of recurrent developing boundary layers at downwind scales of several hundreds of metres. Scale considerations for the developing characteristic boundary layer were also an issue in extrapolating data from a ventilated chamber study to predict evaporation response in a future environment enriched in atmospheric carbon dioxide. A potential change in evaporation rate by 50% accompanying the doubling of atmospheric carbon dioxide in a ventilated chamber was suggested as being moderated to a 10-20% reduction at the extended scale.

Introduction

Increasing the accuracy of estimates of areal evaporation is demanded in endeavours to manage natural resources without irreversible detriment to the environment. Much of the hydrologic evidence in Australia points to subtle changes in annual evaporation, less than 10%, accompanying land-use changes that have had devastating consequences with erosion, salinity and more recently with soil acidification. It would appear that the severalfold enhancement in plant productivity, as has characterised agricultural management, has been achieved with changes in water use that are proportionately less by about two orders of magnitude. Given that major disruptions in land-use induce only perturbations on the evaporation regime, it is not surprising that existing models to describe vapour loss have proved deficient. We can note a retreat from model approaches to land and water management to be replaced with increasing emphasis on evaporation measurement in research proposals. Supply of evaporation data can ultimately lead to improved model definition but in the first instance, it serves as a crucial inventory item to address pressing issues of resource conservation.

Simple extrapolation of a data set of measurements at a single location can be misleading at an extended scale of landscape which is typified by heterogeneity in attributes which influence the evaporation process. This paper reviews some case studies of evaporation by plant communities to identify important factors contributing to areal variability with a view to defining the time and space resolution necessary in extrapolation to reduce uncertainty to 10% or less for regional estimates of evaporation. Topographic effects on evaporation response are examined to illustrate an example of variable control by soil water supply, causing modified catchment vapour loss. Evaporation regimes of contrasting categories of plant community are compared to indicate differences in magnitude and their persistence over time. Advective influences arising from abrupt changes in factors influencing evaporation rate are explored to gauge their significance in determining areal evaporation. Their implications are also examined for evaluating scale effects of elevated levels of atmospheric carbon dioxide on evaporation response.

Topographic effects on evaporation response

A variable evaporation response can occur within a primary grassland catchment as small as 5 ha (Fig. 1) in which daily rates differed between topographic domains by as much as a factor of 2 (21 December, Fig. 2) using point measurements deployed within the catchment. A progressive summer drying of the upper reaches comprising more than 50% of the catchment was associated with a corresponding fall (after December 8) in evaporation rate relative to atmospheric demand,

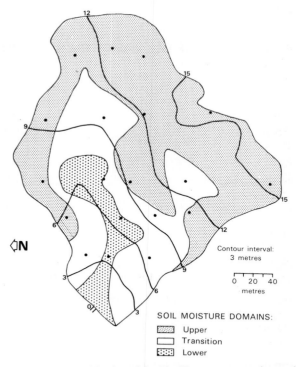

Fig. 1. Topographic plan of the 5 ha Krawarree experimental catchment showing soil moisture measurement locations, contours and delineated domains of soil moisture content (after Dunin & Aston 1981).

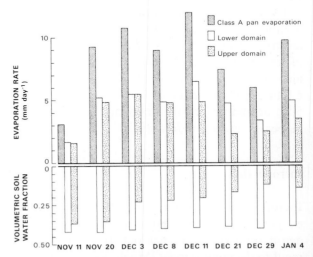

Fig. 2. Progression in evaporation rate and volumetric soil water content of the upper and lower domains of the Krawarree catchment during a drying cycle in 1975/76 (after Dunin & Aston 1981).

reflected by Class A pan evaporation. By contrast, the lower reach occupying 10% of catchment area, maintained constant relative evaporation rate at about 0.5 of Class A pan rate due to sustained high levels of soil water content (> 0.40).

Continuing soil water drainage transferred water down slope to the lower domain throughout the period for a near constant supply that was apparently non-limiting to plant evaporation. During the earlier part of the observed period, evaporation rates were comparable between domains whilst the volumetric soil water fraction exceeded 0.20 throughout the catchment.

The significance of systematic effects of topography in this catchment on soil water supply for evaporation is indicated in a comparison (Fig. 3) of total catchment evaporation (determined approximately monthly) with corresponding independent lysimeter determinations but representative of the upper domain (Dunin & Aston 1981,

1984). Lysimeter values accounted for catchment vapour loss for all but a four month period between September 1973 and January 1974, during the two years of observation. Lysimeter underestimates in this period were attributed to differential drying within the catchment, akin to the independent observations of Figure 2 (made in 1975/76). An adjustment to lysimeter values for soil water effects on evaporation response (Dunin et al. 1978) notably for the lower domain, produced a description of each domain of the time course of evaporation rate which, when integrated over the catchment on a monthly basis, corresponded closely to values of total catchment evaporation. This study of catchment response incorporating independent determinations of evaporation loss at single locations within the catchment clearly demonstrated substantial variability in evaporation losses at times within 5 ha of apparently homogeneous natural grassland. Reducing error consistently below 10% for areal

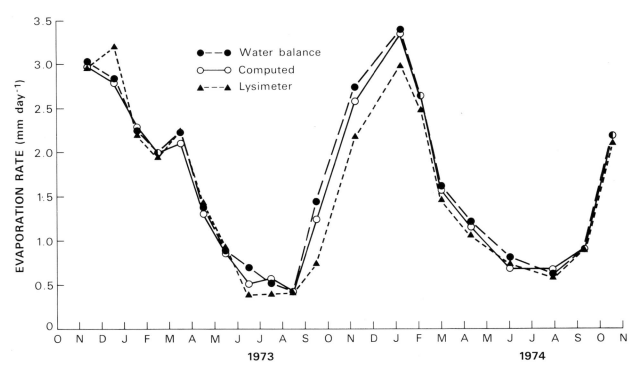

Fig. 3. Comparison of time averaged values of evaporation rate as determined by catchment water balance solution (●) with lysimeter measurement (Δ) and lysimeter determinations adjusted for spatial variability of soil water content (o) within the Krawarree catchment (after Dunin & Aston 1981).

estimates of vapour loss at this scale calls for spatial resolution of the order of 1 ha and time resolution of one day or even less.

Variation between plant communities

A comparison experiment at Krawarree for pasture evaporation involved a natural grassland (Themeda-carrying capacity 2 sheep ha^{-1}) and a fertilised exotic grassland (Demeter-carrying capacity 15 sheep ha^{-1}) with observations in fields < 5 ha between November 4 and December 31, 1975. The adjoining fields initially contrasted in green leaf area index from 1.5 (Themeda) to 3.5 (Demeter). Grazing pressure by domestic stock was eliminated during the period of observation.

Rainfall was 65 mm for the 58 day period but was insufficient to prevent the drying of 20 cm depth of topsoil, initially quite moist, to almost wilting point in both cases (Fig. 4a). Evaporation rates from Demeter exceeded those from Themeda by about 50% initially, but with the more rapid depletion of soil water under Demeter, evaporation rates fell below those by Themeda by early December (Fig. 4b). A loss of green leaf area in Demeter accompanied the progression towards wilting point, further depressing its evaporation rate to extend the difference in evaporation rates between communities (c.f. December 3 and December 12). Towards the end of observation, rates of evaporation were comparable between pastures, associated with the convergence of soil water status at dry levels.

The progressions of cumulative evaporation (Fig. 4c) diverged initially in response to the greater rates by Demeter and then tended to converge when the ranking in evaporation rate was reversed. Evaporation totals were comparable for the observation period (hence equivalent average evaporation rates) but clearly limitations on evaporation imposed by biological factors of evaporation differed greatly between communities.

A need to account for biological control, variable in time, of grassland evaporation is evident from this comparison both for characterising

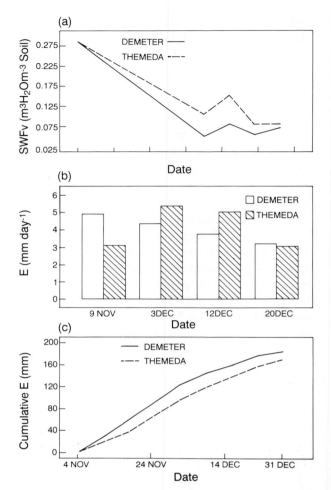

Fig. 4. Comparison of soil water content and evaporation rates from two pastures at Krawarree during a drying cycle in 1975 showing; a) dynamics of volumetric soil water fraction of the upper 20cm of topsoil. b) selected daily rates of evaporation determined by the energy balance technique concurrently in both fields. c) progressions in cumulative evaporation based on both micrometeorological measurements and complemented by soil water measurement in the profile to 1.5m soil depth.

response and for extrapolation. The shallow rooting habit of grassland, a determinant of soil water supply, exerts a profound influence in contributing to rapid fluctuations in evaporation rate. Superimposed is the impact of foliar development in pasture which can arise from grazing, growth and phasic development and from soil water stress. A complex array of interacting influences on grassland evaporation thus poses problems of reproducibility for an observed evaporation regime. The call for detail in time and spatial resolution for extrapolation is strengthened with this case study which reflects an overriding importance of grassland status for areal determinations of evaporation.

The dynamically variable evaporation regime of fertilised pasture is also evident from a 6 month comparison with a eucalypt forest (Fig. 5), using weighing lysimeters that were representative of the respective surrounding communities. The sites were separated by 100 km but were similar in terms of radiation inputs and soil conditions in that soil water deficits were not so great as to induce major changes in leaf area; pasture grazing was only minor so that both communities were not subject to variable leaf area index with values of 3 or greater for most of the period. Peak daily pasture rates each month were either comparable to or exceeded forest evaporation rates by as much as 20%. Monthly pasture rates were systematically less than forest values; in February the discrepancy was as much as 50%, significantly when the soils were at their driest for the period. The forest evaporated more water in the long term due to greater rooting depth and potentially greater availability of soil water by a factor of almost 2. The more rapid pasture evaporation

rate, reflected in peak daily rates, could not be sustained in the long term due to a shortfall in rainfall to maintain its limited soil water reservoir at high levels. In addition, greater interception losses from forest relative to pasture contributed to long term enhancement of forest evaporation.

The less perturbed regime of forest evaporation compared to pasture can be attributed to a large extent to morphological attributes related to vegetation height. Height induces aerodynamically rough conditions which were responsible for a near doubling of annual interception loss by the forest over pasture (Dunin 1987; Dunin et al. 1988). A greater rooting depth of forest, apparently related to its height, buffered the eucalypt canopy against heat stress and leaf loss during periods of rainfall deficiency with sustained transpiration by comparison with the more opportunistic pasture. This role of vegetation morphology calls for vertical resolution to discriminate between vegetation characteristics which introduce major differences in evaporation in space between communities. A reversal of this example for greater annual pasture loss of vapour in humid zones over forest evaporation, as has been suggested by Jarvis and Stewart (1979), merely emphasises the critical importance of three dimensional resolution for vegetation and water supply

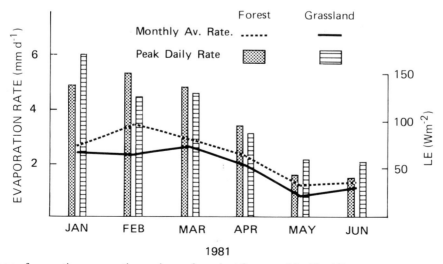

Fig. 5. Comparison of respective evaporation regimes of eucalypt forest and fertilised Demeter pasture over a 6 month period showing averaged monthly rates and peak daily rates for each month (after Dunin et al. 1985).

44

characteristics. This approach accommodates a variable impact of vegetation type on evaporation in accounting for transient effects that become systematic over time with changed circumstances.

Advection

Advection accompanies spatial variability in evaporation caused by a change in one or more factors influencing the rate. Immediately downwind of a boundary, non-uniform conditions apply horizontally both in air properties and in the surface turbulent fluxes. Further downwind, the equilibrium condition is characterised by horizontal uniformity associated with a well developed zone of constant vertical flux in the near surface layer. Over that downwind distance with progressive adjustment in gradients of temperature and humidity causing flux divergence (horizontal and vertical), advective influences perturb the evaporation rate relative to the equilibrium condition.

Concern for advective effects varies between studies of plant evaporation. Micrometeorological determinations of evaporation loss usually involve instruments deployed with adequate fetch, often hundreds of metres based on Dyer (1963), to ensure near equilibrium conditions with one-dimensional transfer. By contrast, many studies in agronomy and plant physiology dealing with plant water relations are focussed on differences in evaporation response but employ experimental design based on plants in isolation or as small plots under field conditions. There are acknowledged problems in relating measurements between these scales of determination (Jarvis & McNaughton 1986). Resolving the issue of scale of measurement depends to a large extent on accounting for advective effects of spatial variability on evaporation response. The following example of irrigated soyabeans at Griffith, NSW, indicates the magnitude and extent of advective problems experienced in commercial irrigated agriculture.

The diurnal course of latent heat flux diverged after 1000 hr between lysimeter values (λE_o) and

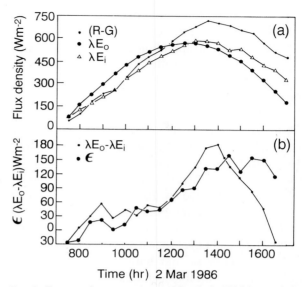

Fig. 6. Evaporation response on March 2, 1986 by an irrigated soyabean crop of 2 ha; a) diurnal course of available radiant energy (R-G), lysimeter latent heat flux, λE_o, and micrometeorological determinations at 0.75 m above the crop 1.0 m high. b) comparison of the discrepancy $\lambda E_o - \lambda E_i$ with ε, Lang's (1973) correction for vertical flux divergence between the micrometeorological determination and lysimeter measurements of latent heat flux.

adjacent atmospheric determinations involving the Bowen ratio (λE_i) (Fig. 6a). The discrepancy became pronounced once lysimeter rates exceeded the available radiant energy as determined from measurements of net radiation (R) and ground heat flux (G). The wind was a persistent SSE for the period shown and blew over a fetch of 35 m of turgid crop consistent with the lysimeter community. Upwind of this fetch was a mixture of mature crops of soyabeans and sorghum recently withdrawn from irrigation and under obvious stress, having a canopy temperature estimated to be 3°C higher than the irrigated crop around noon. Fuller details of the micrometeorological apparatus and the centrally located lysimeter in the 2 ha experimental field with representative behaviour are described elsewhere (Meyer *et al.* 1987, Dunin *et al.* 1989). The departure in patterns of latent heat flux was explained in terms of local advection (Fig. 6b) with the discrepancy between λE_o and λE_i corresponding fairly well with a calculated correction to account for flux divergence between the lysimeter crop and the height of Bowen ratio determination at 0.75m above the crop (Lang 1973); this correction used both windspeed and horizontal gradients of temperature and humidity derived from profiles with aspirated psychrometers separated by 30 m upwind of the lysimeter.

General agreement, within an acceptable experimental error of 40 Wm^{-2} occurring up till 1100 hr coincided with the period of atmospheric instability $((R-G) > \lambda E_o)$ when turbulent mixing tended to offset the advective effect of the upwind discontinuity between crops; it was also possible that discrepancies in canopy temperature between crops was not pronounced during this stage due to overnight hydration of the stressed crops. Following 1100 hr the discrepancy enlarged with the onset of a temperature inversion indicated by $R-G < \lambda E_i$; the discrepancy reached 80 Wm^{-2} by noon and persisted in excess of 100 Wm^{-2} throughout the afternoon. Stable conditions were a likely cause of inversion advection prevalent during the afternoon; a lack of turbulent mixing associated with atmospheric stability may have suppressed the development of the boundary layer. A progressive shift in boundary conditions with canopy temperature differences increasing may have contributed to the advective condition. Irrespective of the basic causes, the irrigated field was subjected to changing fetch requirement causing significant advective enhancement of evaporation up till evening (c.f. λE_o and $(R-G)$, Fig. 6a).

The foregoing example provides a basis for determining an average enhancement in evaporation for the field subject to advective influences for most of the day. Enhancement may be approximated by the discrepancy $(\lambda E_o-\lambda E_i)$ for the lysimeter and its proportionate effect is given by $(\lambda E_o/\lambda E_i)$-1. Enhancement at other points in the field can be deduced using the analytic solution of Philip (1987) for describing changes in canopy temperature and surface turbulent fluxes with distance downwind from the boundary. The upwind canopy temperature is assumed as being 3°C greater than ambient temperature measured at

2 m height above the lysimeter site. Furthermore, enhancement is taken as occurring during inversion conditions. The analytic solution using measured values of windspeed (3.5m s^{-1}) and water vapour density (8.7 × 10^{-3} kg m^{-3}) indicates evaporation rates in excess of available radiant energy (inversion condition), extending over a downwind distance of 250 m or greater than the extent of the irrigated field of soyabeans. Enhancement observed at the lysimeter, 35 m downwind, is deduced as 20% and is approximately equivalent to an averaged value determined from the map of advection inversion across the field.

This example justifies concern for advection in causing areal variability and enhancing vapour loss significantly but which is difficult to determine by normal procedures. In irrigation areas typified by discontinuities at scales of a few hundreds of metres, recurrent development of boundary layers poses problems of measurement and interpretation as to areal losses of evaporation from those intensively managed regions.

Evaporation from 'Greenhouse' environments

Experimental studies of elevated levels of atmospheric carbon dioxide on evaporation response are advective in that isolated plant communities are exposed for short periods to air enriched in carbon dioxide concentration. The characteristic boundary layer appropriate to a modified evaporation response has failed to develop in an environment characterised by properties of air consistent with external evaporation rate.

A forest study of doubling the CO_2 concentration at Kioloa State Forest, NSW involved a ventilated chamber on a weighing lysimeter in a eucalypt forest. A 30% reduction in evaporation rate was deduced by comparing periods of enriched CO_2 with periods of ambient concentrations of CO_2. This effect was attributed to a corresponding lowering of canopy conductance during the periods of CO_2 enrichment on the lysimeter (Wong & Dunin 1988). The observation for a

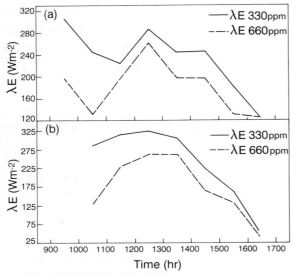

Fig. 7. Diurnal progressions of latent heat flux to compare forest evaporation response at normal concentrations of atmospheric CO_2 (determined micrometeorologically) with that subject to doubled concentrations of atmospheric CO_2 (determined with a lysimeter enclosed with a ventilated chamber, (a) March 17, 1984, and (b) April 3, 1984

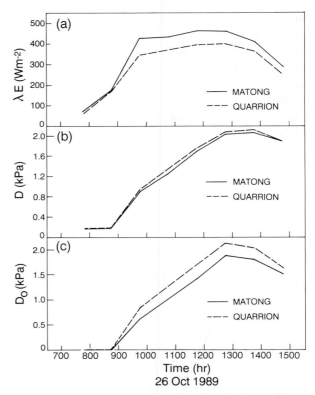

Fig. 8. Comparison on October 26, 1990 of wheat cultivars at Wagga, Matong having 50% greater stomatal conductance than Quarrion: a) diurnal progression in latent heat flux. b) vapour pressure deficits (D) measured at 2.0 m height, each centrally located in a 5 ha field. c) vapour pressure deficits at the bottom of the turbulent boundary layer and at the top of the canopy laminar boundary layer (D_o).

30% reduction in evaporation rate is confirmed in Figure 7, but using a comparison involving the enriched lysimeter environment and that concurrently for normal CO_2 concentrations of the surrounding forest. This comparison also suggests a comparable lowering by 30% of canopy conductance with a doubling of atmospheric CO_2.

A complementary approach to understand changed evaporation response accompanying doubled CO_2 concentrations is to draw comparisons of field communities which differ in stomatal control. Figure 8 shows this type of comparison for two cultivars of wheat in adjoining fields each of 5 ha; these cultivars differed in leaf conductance by 50% but had comparable leaf area indices at 2.0 for this time of measurement thereby experiencing a 50% difference in canopy stomatal control. Latent heat fluxes differed by 10-20% in their diurnal progression after the early morning period of canopy saturation caused by overnight dew (Fig. 8a). A difference in boundary layer conditions was reflected in vapour pressure

deficits (D) varying by as much as 0.1 kPa at a height of 2.0 m above the soil surface with both crop heights being 1.0 ± 0.1 m. The cultivar with lower conductance, Quarrion, had its canopy temperature warmer by as much as $1.5°C$ with an accompanying vapour pressure deficit at canopy level (D_o) being greater by as much as 0.3 kPa (Fig. 8c).

This example reflects a moderating influence of boundary layer effects to compensate for contrasts in stomatal effects on evaporation response; a potential difference of 50% in evaporation rate due to stomatal effects had been attenuated to approximately 15% with warming of the air at and above the crop with lower conductance. Greater vapour pressure deficits being more pronounced near the canopy were generated in response to the warming causing effects of stomatal control to be offset to a large extent. A similar moderation of stomatal effects might be expected with increased concentration of ambient CO_2.

Concluding remarks

A range of effects, relating to soil water availability and plant growth response, was identified as making important contributions to localised areal variability in the evaporation rate. The impact of these effects was likely to be propagated in both space and time to complicate estimates of the evaporation regime at an extended scale of some kilometres. Estimates of evaporation loss at a single location were unlikely to provide satisfactory values for a region at time scales from daily up to monthly. Increased deployment of evaporation measurements maintained for a range of seasonal conditions can facilitate determinations of regional evaporation.

The presence of discontinuities in the landscape has a two-fold effect to induce spatial variability in evaporation response. Variation in attributes such as soil depth, rainfall and vegetation endow elements of the landscape with differences in evaporation rate. These differences are complicated further by horizontal variability in evaporation along downwind reaches of a discontinuity. The extent of this effect can be considerable in irrigation areas but is generally unknown in rain fed conditions. Furthering the understanding of interactions between boundary layer development and evaporation response can benefit climatic predictions in a future environment enriched in atmospheric carbon dioxide.

Acknowledgments

I am indebted to Mr Wybe Reyenga for his continuing association and competence to compile evaporation data over a wide range of conditions. Repeated consultations with Drs J.R. Philip & O.T. Denmead have been invaluable in the interpretation of data as well as in the design and conduct of field expeditions.

References

Dunin, F.X. 1987. Run-off and drainage from grassland catchments. In: Managed Grasslands, B. Analytical Studies, pp. 205–13, (Ed.) Snaydon, R.W., (Elsevier Science Publishers B.V., Amsterdam).

Dunin, F.X., Aston, A.R. & Reyenga, W. 1978. Evaporation from a Themeda grassland, 2. Resistance model of plant evaporation. J. Appl. Ecol. 15: 847–58.

Dunin, F.X. & Aston, A.R. 1981. Spatial variability in the water balance of an experimental catchment. Aust. J. Soil Res. 19: 113–20.

Dunin, F.X. & Aston, A.R. 1984. The development and proving of models of large scale evapotranspiration: An Australian study. Agric. Water Manage. 8: 305–23.

Dunin, F.X., McIllroy, I.C. & O'Loughlin, E.M. 1985. A lysimeter characterization of evaporation by eucalypt forest and its representativeness for the local environment, in Hutchison, B.A. and Hicks, B.B. (eds) The Forest Atmosphere Interaction, pp. 271–91, Proc. Forest Environmental Measurements Conf., Oak Ridge, Tennessee, 1983, (D. Reidel Publishing Company, Dordrecht.)

Dunin, F.X., O'Loughlin, E.M. & Reyenga, W. 1988. Interception loss from eucalypt forest: Lysimeter determination of hourly rates for long term evaluation. Hydrol. Processes, 2: 315–329.

Dunin, F.X., Meyer, W.A., Wong, S.C. & Reyenga, W. 1989. Seasonal change in water use and carbon assimilation of irrigated water. Agric. For. Meteorol., 45: 231–250.

Dyer, A.J. 1963. The adjustment of profiles and eddy fluxes. Q. J. R. Meteorol. Soc. 89: 276–280.

Jarvis, P.G. & McNaughton, K.G. 1986. Stomatal control of transpiration: scaling up from leaf to region. Advances in Ecological Research 15: 1–49.

Jarvis, P.G. & Stewart, J.B. 1979. Evaporation of water from plantation forest, in Ford, G.D., Malcolm, D.C. and Atterson, J. (eds.), The Ecology of Even-Aged Forest Plantations, Inst. Terrestrial Ecology, pp. 327–350, NERC, Cambridge.

Lang, A.R.G. 1973. Measurement of evapotranspiration in the presence of advection by means of a modified energy balance procedure. Agric. Meteorol. 12: 75–81.

Meyer, W.S., Dunin, F.X., Smith, R.C.G., Shell, G.S.G. & White, N.S. 1987. Characterising water use by irrigated wheat at Griffith. New South Wales, Aust. J. Soil Res. 25: 499–515.

Philip, J.R. 1987. Advection, evaporation and surface resistance. Irrig. Sci. 8: 101–114.

Wong, S.C. & Dunin, F.X. 1988. Photosynthesis and transpiration of trees in a eucalypt forest stand. Aust. J. Plant Physiol. 14: 619–32.

Vegetatio **91**: 49–58, 1991.
A. Henderson-Sellers and A. J. Pitman (eds).
Vegetation and climate interactions in semi-arid regions.
© 1991 *Kluwer Academic Publishers. Printed in Belgium.*

Towards a better model of the effect of prostrate vegetation cover on wind erosion

J. F. Leys

Soil Conservation Service of New South Wales, P.O. Box 7, Buronga, N.S.W. 2648, Australia

Accepted 24.08.1990

Abstract

Vegetation cover is the key to controlling wind erosion. A brief review of wind erosion/cover models is outlined. Fryrear's (1985) soil cover (wheat stubble) model was evaluated against field wind tunnel results from far south-west N.S.W. Fryrear's equation over estimated the soil loss compared to field wind tunnel results.

Fryrear's model failed to provide meaningful results at low cover levels with the soil loss ratio, $SLR > 1$ for percent soil cover, $\% SC < 6$. A single parameter exponential model was fitted to the wind tunnel data which ensured that SLR did not exceed 1 for $100\% SC$. Even with this improvement, the exponential model has drawbacks.

Results suggest that the SLR is sensitive to wind velocity and that SLR goes to 0 well before $\% SC = 100$. A method for approximating the threshold wind velocity required to initiate erosion for various cover levels is described. Using the recurrence interval for a prescribed wind velocity, the probability of erosion hazard for a field can be determined.

It is the authors belief that the wind tunnel is underestimating the occurrence of wind erosion events in this study. Three reasons why the wind tunnel may be underestimating erosion events are given.

Introduction

The main source areas of windblown mineral dust are the arid and seasonally-arid regions of the world. (From Pye 1987, p. 5).

Wind erosion is likely to occur when the vegetation layer is diminished below the level required to protect a dry, loose, surface soil. This condition often occurs in semi-arid regions of the world due to climatic variations and inappropriate land management methods. Vegetation can protect the soil by directly absorbing the force of the surface wind and hence reducing the wind velocity at ground level.

The significance of wind erosion is both environmentally and economically important. In Australia some of the costs are evident through the forms of soil removal, nutrient and organic matter loss from productive soils (Carter & Findlater 1989), sandblasting of crops, undermining of structures, sedimentation of transport links and capital improvements (Leys 1986), air pollution (Lourensz & Abe 1983), contamination of food and water supplies, increased maintenance and cleaning costs and increased incidence of respiratory disease Burnley (1987). Vegetation is the key to controlling wind erosion in semi-arid and arid lands and is the major natural resource that land managers manipulate in their day to day management.

There are two major land uses in semi-arid and arid Australia, extensive grazing on native pastures, (ie, the rangelands) and dryland cropping at the margin of the semi-arid lands Carter (1986). The short and long term production from both these land uses relies on the management of the vegetation layer. In the case of the rangelands this management is through stocking rates and in the case of the croplands it is through cultivation.

With both land uses it is critical that vegetation, or its residue, be maintained on the surface if land degradation caused by wind erosion is to be controlled. Graziers aim to utilise as much cover as possible to promote production without running the risk of soil degradation. Similarly, farmers

have to manage cover levels to ensure they do not have too much crop residue which can hinder cultivation and sowing operations or too little which results in erosion.

The Land Degradation Survey of N.S.W. 1987–1988 (Owen *et al.* 1989, Fig. 1) indicated that the most severe wind erosion occurs on the cropping lands that have soils with sandy surface textures and are used for the production of cereal grains by the practice of long bare-fallowing (i.e. the agricultural practice of maintaining a weed free paddock by the use of aggressive cultivation equipment for 10 months prior to the crop). Erosion was also evident in the north-west of N.S.W. where native woody shrubs (woody weeds) have increased their distribution and density in recent years, thereby restricting pasture

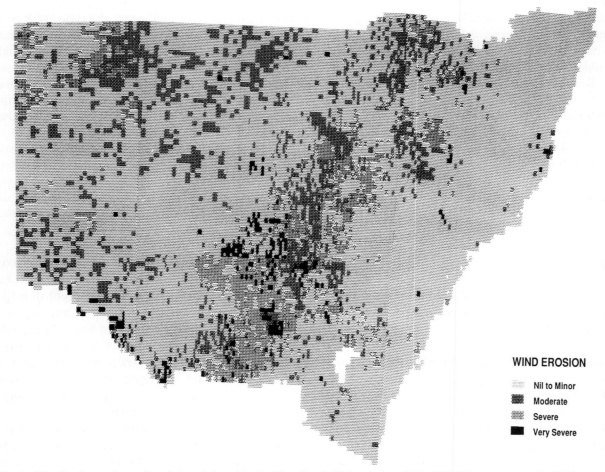

WIND EROSION

- Nil to Minor
- Moderate
- Severe
- Very Severe

Fig. 1. Distribution and severity of the wind erosion in New South Wales during 1987–1988 (After Owen *et al.* 1989).

growth and increasing the incidence of bare ground.

The case of woody weeds demonstrates that simply having vegetation cover is not enough to control erosion. The shape and density of the vegetation cover is also very important, as is its management.

The management of vegetation cover is therefore critical to controlling erosion. The question asked by land managers is how much cover is required to control erosion. Many authors have attempted to answer this very important question through a series of controlled wind tunnel and field experiments.

The aim of this paper is to briefly review wind erosion/cover models published in the literature and to evaluate one of these models that deals with the effect of prostrate cover on wind erosion. It also suggests possible improvements to the modelling of the relationship between vegetation cover and wind erosion and provides a case study on the important aspects of a proposed model.

Previous work

Most early work was undertaken in wind tunnels within laboratories using highly disturbed soils and simulated residue cover. The aim of the work was to identify how erosion rates were affected by changing cover levels. The work can be divided into two groups, those who considered prostrate cover and those who addressed the affect of surface roughness.

Prostrate cover was first addressed by Chepil (1944) who undertook a laboratory wind tunnel experiment to determine the influence of straw residue on erosion. He found that erosiveness of the soil q, for any given wind velocity u, varies exponentially with the quantity, expressed as weight of surface residue, for different soils. Siddoway et al. (1965) and Lyles and Allison (1981) found similar exponential relationships between various types, orientations and amounts (expressed as weights) of stubble and q.

Not all cover is prostrate and to model the more upright vegetation, Chepil (1950) introduced the concept of a *critical surface constant* to describe the importance of surface roughness elements. This concept applies to soils as well as vegetation. The experiments used mixed beds of erodible and non-erodible elements exposed to the wind until erosion ceased. He found that the relationship between the height of the roughness element (H) divided by the distance between them (d) was constant when erosion ceased. Chepil and Woodruff (1963) used the reciprocal of this ratio and termed it the *critical surface barrier ratio* (CSBR). On cultivated land the ratio was found to vary between 4 and 20, depending on the friction velocity and the threshold velocity of the erodible particles.

Marshall (1970) used a drag partition theory to approach the problem of how vegetation of different shapes and densities effects the potential for wind erosion. His paper describes the partitioning of the force exerted on the vegetation elements and the intervening ground for a range of different artificial roughness arrays. He found that total shearing stress did not appear to vary appreciably with the type of distribution when arrays are at the same density. He further stated that a 'critical condition' occurs when the average surface force becomes negligible. This implies that if there is negligible force to the ground then there is no erosion. The condition is computed by describing the roughness array (e.g. frontal area per unit ground area) and by characterising the roughness elements (e.g. unobstructed drag coefficient).

Lyles et al. (1974) further developed Chepil and Woodruff's (1963) CSBR. They found that CSBR was not constant for a given *friction velocity, u.*, (a measure of the shear stress on the ground), and as roughness increased so did the total surface stress with the greater proportion of the stress being absorbed by the non-erodible elements. This led to the development of the *critical friction velocity ratio*, CFVR. Using regression equations Lyles et al. (1974) related the CFVR, (i.e. u_*/u_{*t} where u_{*t} is the *threshold friction velocity*) to the dimensionless parameters H, L_x and A_c (where H is the average roughness height; L_x is the spacing between roughness elements in the flow direction;

and A_c = the proportion of the surface area covered by nonerodible elements).

Lyles *et al.* (1976) further developed the CFVR, to determine when erosion would begin. They found that CFVR was a two parameter power function of height of the elements divided by the distance between the elements H/L_x and the proportion of the surface area covered by nonerodible elements A_c.

Gregory (1984) derived a model to predict relative soil loss as a function of canopy and residue cover. Unlike earlier empirical models, Gregory provides a model based on the reduction of kinetic energy delivered to the soil by canopy and residue. This is similar to the approach of Marshall (1970) and Lyles *et al.* (1974; 1976).

Fryrear (1985) derived a two parameter exponential relationship between the soil loss ratio, SLR (soil loss from covered soil/soil loss from bare flat soil) and simulated percent soil cover %SC. The development of the SLR allows the comparison of erosion rates of different soils and allowed Fryrear to compare his results with those of other field experiments. Similarly, the use of %SC allows comparisons of different crop types and is considerably easier to collect than weights per unit area, that have to be collected, washed, dried and weighed. This model will be evaluated later in this paper.

Van de Ven *et al.* (1989) in a laboratory wind tunnel followed up Fryrear's 1985 work and determined the relationship between soil loss, wind velocity and the number, diameter and height of simulated plant stalks (wooden dowels). They found that the height of the stalks is more important than number or density of stalks and that soil loss SL = $-3.396 + 298\ (U_h - U_t)/(NDH)$, where U_h is free-stream velocity, U_t is threshold condition, N is density; D is diameter and H is height of the stalks.

Evaluation of Fryrear's soil cover and wind erosion model

The Soil Conservation Service's primary role in arid lands is to advise land holders on land management systems that are both sustainable and prevent land degradation. As stated earlier, the best way to do this is by vegetation management. Land holders must be able to predict how much cover they need to retain to prevent erosion. The need for a model that can be readily used in the field, to indicate the cover required to prevent erosion, is very important.

Of the above published models, the majority aimed to gain a better understanding of the processes and relationships between cover and wind erosion. Very few models use parameters that are easily measured in the field or are readily understood by extension staff and land holders. Of the above models, Fryrear's (1985) model that relates prostrate simulated percent soil cover (%SC) to the soil loss ratio (SLR) is the most applicable to wind erosion problems on cultivated land.

Budd and Leys (1989) showed that in the semi-arid lands of south-east Australia, long fallowing (the practice of keeping the paddock weed free for 10–12 months prior the sowing the crop) is a primary cause of wind erosion. Stubble residue from crops is invariably laid flat by the time of the erosion period due to previous grazing and cultivation under this farming system. Therefore a prostrate cover model was chosen for evaluation. The use of models that predict erosion with standing stubble (Lyles & Allison 1976; Van de Ven 1989) are questionable, as these conditions rarely exist on long fallow cultivation paddocks during the erosion period.

Methods

A portable field wind tunnel (Raupach & Leys 1990) was used to assess the erosion rates from plots with a range of stubble covers. The soil was a siliceous sand with a sandy surface texture, classified under the Northcote (1979) system as a Uc5.11. The research was undertaken on 'Milpara' Station 25 km north of Wentworth in far south-west N.S.W. The paddock was harvested for wheat in December 1987 and the wind tunnel tests undertaken during the late fallow period in February 1989.

Table 1. Actual stubble weights, soil flux and soil loss ratio (*SLR*) data from 'Milpara', calculated percentage cover (from Gregory, 1982) and predicted *SLR* from Fryrear's wheat stubble Equation 1.

Group	Sample no. per group	Stubble weight (kg/ha)	Cover (%)	Soil flux (g/m/s)	'Milpara' actual (*SLR*)	Fryrear's predicted (*SLR*)
1	5	0	0	90.0	1.00	1.63
2	4	380	18	16.9	0.189	0.43
3	8	730	30	7.64	0.085	0.18
4	5	1110	42	0.56	0.010	0.07
5	6	1530	53	0.49	0.005	0.03

Twenty eight stubble levels (determined by cutting and drying using the method of Leys & Semple 1984) were tested for their erosion rate (expressed as soil flux, *Q*, which is the mass of soil transported through an imaginary 1 metre wide door of an arbitrarily large height and perpendicular to the wind direction). The 28 stubble levels were divided into 5 groups, based on stubble weight, and the results averaged (Table 1). Stubble weights were converted to percent ground cover using the area cover-probability methods developed by Gregory (1982).

Results and discussion

The aim was to determine if Fryrear's (1985, p. 782) 'wheat stubble equation' could accurately predict the *SLR* for five different cover levels when compared to actual *SLR*s calculated from wind tunnel data from 'Milpara'.
Fryrear's equation is:

$$SLR = 1.63 \, e^{-0.074 \, \%SC} \qquad (1)$$

Where *SLR* is the soil loss ratio (soil loss from covered soil/soil loss from bare flat soil and %*SC* is the percent soil cover.

As shown by Table 1, Fryrear's predicted *SLR*s are higher for all cover levels compared to the 'Milpara' actual *SLR*s thus overestimating the erosion rate.
The 'Milpara' data was fitted using a two para-meter exponential regression model (SAS, 1988) giving:

$$SLR = 1.14 e^{-0.107 \, \%SC} \qquad (2)$$

The model is significant at the 1% level (P < 1%) and has a regression coefficient of $r^2 = 0.95$. This shows a similar relationship to Fryrear's model (1) (Fig. 2).

While there is good agreement between the two studies, there is one major limitation to this approach. The two parameter model fails to provide meaningful results at low cover levels, for when applying the model to the condition when %*SC* = 0 then *SLR* = 1.63. This is an unrealistic answer as *SLR* has a maximum value of 1.

To overcome this deficiency, a single parameter exponential model with no intercept would ensure that when %*SC* = 0 then *SLR* = 1. The data was refitted to a model where the regression line was forced through the origin (SAS 1988, Fig. 2). This gave Equation (3) with P < 1% and $r^2 = 0.98$.

$$SLR = e^{-0.103 \, \%SC} \qquad (3)$$

The single parameter model is a reasonable empirical approach to the problem of predicting the effects of changing vegetation cover on wind erosion, however, there are two major limitations to the approach. Firstly, the model *SLR* never falls to 0 no matter how high the %*SC* because of the exponential relationship. Secondly, the model gives the same *SLR* for all wind velocities.

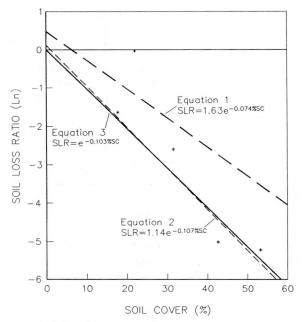

Fig. 2. Predicted soil loss ratios (*SLR*s) using Fryrear's (1985) wheat stubble equation were consistently higher for all percentage cover levels (*%SC*) compared to the actual *SLR*s from 'Milpara'. The single parameter model is preferred because it provides meaningful results at low *%SC* and predicts *SLR* = 1 at *%SC* = 0 unlike the other two parameter models.

Needs of a new model

Experience indicates that with increasing cover, *SLR* reaches 0, and that this generally happens well before 100% soil cover. While the relationship between *SLR* and *%SC* is exponential through most of its range there is a need for the model to account for a cover level below 100% where *SLR* = 0.

Similarly, experience indicates that the same *%SC* that controls erosion at low wind velocities is less effective at high velocities. This is very important to the soil conservationist because erosion control is expressed to land managers in terms of erosion hazard (i.e. the susceptibility of a parcel of land to the prevailing agents of erosion). For example, 40% cover may be enough to reduce erosion to negligible levels for the 1 in 1 year wind event, however, a 1 in 5 year event will

result in significant erosion for the same *%SC* level.

A model that reflects not only the relationship between *SLR* and *%SC*, but also its variation with wind velocity *u*, is highly desirable in order to highlight the probability of erosion occurring for different *%SC*.

For the eight different wind velocities simulated in the wind tunnel, soil flux *Q*, was detected for the four highest *u* values at all cover levels at 'Milpara'. Figure 3 shows that *SLR* is sensitive to changes in mean u for constant *%SC* levels. For example, comparing the differences in *SLR* for the highest ($u = 25$ m s^{-1}) and lowest ($u = 16.8$ m s^{-1}) velocities when *%SC* = 30, *SLR* = 0.09 and 0.005 respectively.

A case study of erosion hazard for 'Milpara'

To make comparisons between the meteorological and the wind tunnel data, and thereby assess the erosion hazard, *u* values measured in the wind

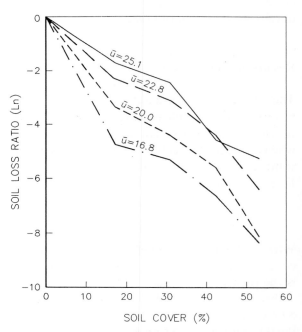

Fig. 3. Actual data from 'Milpara' demonstrates that the soil loss ratio is sensitive to changes in velocity (u in m s^{-1}) at constant percent soil covers.

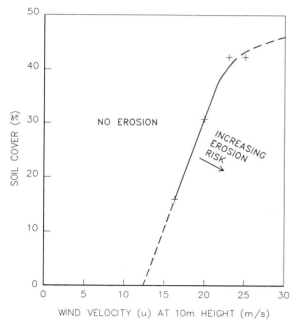

Fig. 4. The relationship delineates what percentage soil cover (% *SC*) is required to reduce erosion to negligible levels (Soil loss ratio = 0.01) for a range of wind velocities m s $^{-1}$ (at 10 m height). For the conditions above the line there is no erosion hazard, below the line there is increasing erosion hazard with decreasing % *SC* and increasing wind velocity.

tunnel at 0.02 m were converted to an equivalent *u* at 10 m (the normal meteorological wind recording height). As wind erosion occurs in strong winds (estimated as > 12 m s $^{-1}$ for Milpara, see Fig. 4) conditions are close to neutral and diabatic effects are insignificant.

Conversion was done by using the log law, which is written as:

$$u_z = \frac{u_*}{k} \ln \frac{z}{z_0}$$
(4)

where *k* is the Von Karmen's constant, taken as 0.4; *z* is the height above the ground (m) and z_0 is the surface roughness length (m).

The u_* and z_0 data derived from the velocity profiles in the wind tunnel were substituted into Equation 4 for the conversions of *u* at 0.02 m to *u* at 10 m height. The percentage occurrence of wind events were then determined from the

Mildura meteorological records for the condition when *SLR* = 1 and are presented in Table 2.

Erosion hazard is reduced as wind velocity decreases and % *SC* increases. Therefore, once the threshold velocity u_t of one % *SC* level is exceeded a greater % *SC* is required to control erosion. To determine the threshold velocities of the 'Milpara' site for a range of cover levels, % *SC* and u were plotted for the condition when *SLR* = 0.01 (Fig. 4). It is assumed, in the case of the 'Milpara' site, that *SLR* = 0.01 equates to a negligible erosion rate.

Figure 4 indicates that the relationship between % *SC* and *u* is approximately linear for low % *SC* and *u* values. Assuming this, extrapolation of the relationship to the intercept of the x axis gives u_t of 12 m s $^{-1}$ for a bare soil at 'Milpara'. At high % *SC* and *u* values the relationship shows a slower rate of change indicating a maximum % *SC* that controls erosion regardless of increasing *u*. This equates to the observation made earlier that *SLR* reaches 0 well before 100% soil cover.

The line in Figure 4 represents the threshold % *SC* required to contain erosion to negligible levels for a range of *u*. For the conditions above the line, there is no erosion hazard, while below the line there is.

According to the wind recurrence interval data (Table 2) there are 3% of summer days (2.7 days) that have wind velocities greater than the u_t of 12 m s $^{-1}$. Experience from the Mildura area suggests that erosion occurs more frequently than this. McTainsh *et al.* (1990) report that Mildura had an average of 5.3 dust storms per year between 1960–1984. The difference between the number of observed (based on dust storm frequency) and predicted (based on recurrence intervals) wind erosion events may be due to three factors:

i. The flow in the wind tunnel is at a constant velocity and lacks large the scale turbulence of natural winds, such as gusts. Therefore, the mean wind velocities of the wind tunnel are less erosive than the meteorological mean wind from Mildura.

The effect of the gusts on *Q* is that they raise the wind velocity for short periods thus

Table 2. Measured velocities, $\bar{u}$, calculated friction velocity, u_*, and surface roughness length z_0, from the wind tunnel's velocity profiles. Conversion of wind velocities from 0.02 m height to 10 m height using the log law (Equation 4), and the percentage occurrence of those wind events for Mildura airport.

$\bar{u}$		u_* (m/s)	z_0 (mm)	$\bar{u}$ at 10 m		# Percentage occurrence
50 mm (m/s)	200 mm (m/s)			(m/s)	(k/hr)	
2.412	3.290	0.25	1.10	5.76	20.7	42
3.100	4.129	0.30	0.77	7.03	25.3	22
4.378	5.8720	0.43	0.86	10.08	36.3	7
5.808	7.6960	0.55	0.70	13.02	46.9	2
6.886	9.2000	0.67	0.81	15.73	56.6	1
7.990	10.886	0.84	1.09	19.06	68.6	<1
8.916	12.960	1.17	2.35	24.37	87.7	<1
9.878	14.052	1.20	1.88	25.83	93.0	<1

Percentage occurrence of mean summer (Dec–Feb) wind velocities.

increasing Q. Q does not return to the pre-gust level because additional saltation material mobilised by the gust continues to drive the erosion process when velocity drops.

For example, for an average wind $Q = 10$ g m^{-1} s^{-1}, a gust then raises the wind velocity and Q increases to 15 g m^{-1} s^{-1} for the period of the gust. After the gust has passed velocities drop back to the level of the average wind, however, Q is now 12 g m^{-1} s^{-1}. This phenomena was described by Bagnold (1941) as the difference between the fluid (wind initiated) and impact (saltation sustained) threshold velocities. Therefore the mean wind velocities recorded in the tunnel lack the gust events of the meteorological winds that promote higher Q values compared to those of the wind tunnel.

ii. Meteorological records are taken over long period and high velocities that may have occurred during the recording period are obscured by the long averaging periods. Average wind velocities underestimate the erosivity of the natural wind. Therefore, the mean wind velocities recorded in the wind tunnel have less erosive power than the meteorological mean wind from Mildura.

iii. The wind tunnel is 5 m long and only simulates what erosion is likely to occur across the first 5 m of a paddock. Chepil (1957) reports that it takes 30 m for a sand to reach maximum soil flow. As such the Q values from the wind tunnel are not reaching the maximum Q for the soil. Therefore, Q values from the wind tunnel are an under estimate of the potential Q value.

Based on these three factors, it is reasonable to assume that the wind tunnel underestimates Q for a given u. This explains the difference between the number of observed (based on dust storm frequency) and predicted (based on recurrence intervals) wind erosion events.

Conclusions

Vegetation cover is the key to wind erosion control. Fryrear's (1985) soil cover (wheat stubble) model was evaluated against field wind tunnel results in far south-west N.S.W. Fryrear's equation over estimated the soil loss indicating that with 18% soil cover, soil loss was 43% of the soil loss of a bare flat soil, while wind tunnel results indicated that is was 18%.

Fryrear's model failed to provide meaningful results at low cover levels with the $SLR > 1$ for %SC value < 6. To overcome this, a single parameter exponential model was fitted to the wind tunnel data which ensured that *SLR* did not exceed 1 for 0 *%SC*. Even with this improvement, the exponential model has two drawbacks. Firstly, *SLR* will never reach zero, which according to experience is highly likely, and secondly, the model takes no account of the affect of wind velocity on *SLR* despite the evidence presented in this paper.

Results from 'Milpara' suggest that the *SLR* is sensitive to wind velocity and that *SLR* goes to 0 well before *%SC* = 100. A technique for converting wind velocity data within the wind tunnel to an equivalent meteorological wind (10 m height) by using the log law is described. This allows comparison of the winds in the tunnel with the meteorological records and indicates the recurrence interval of various erosion levels.

A method for approximating the threshold wind velocity required to initiate erosion for various cover levels is described. Using the recurrence interval for a prescribed wind velocity, the probability of erosion hazard for a paddock can be determined.

It is the authors belief that the wind tunnel is underestimating the occurrence of wind erosion events in this study. Three reasons why the wind tunnel may be underestimating erosion events are given.

To overcome these draw backs, the wind tunnel is being calibrated against wind and soil loss data recorded at a paddock scale, and a new model, based on the requirements outlined in this paper, is under development that better describes the relationship between soil erosion and prostrate vegetation cover.

Acknowledgments

The author acknowledges the Soil Conservation Service for funding this research; Dr M. Raupach for assistance in developing this paper and helpful criticism of the draft; Mr Bill Grace of 'Milpara' who has offered his time and help during the project; and my fellow soil conservationists Messrs G. Budd and R. Bath for help with data collection and continual critical assessment of this research.

References

Bagnold, R. A. 1941. The physics of blown sand and desert dunes, Chapman and Hall, London.

Budd, G. R. & Leys, J. F. 1989. A farming system for sandy soils in far south-western New South Wales, Symposium proceedings of Soil Management '88, 377–390, Darling Downs Institute of Advanced Education, Toowoomba, Queensland, 19–21 September, 1988.

Burnley, I. H. 1987. Health indicators. In: Atlas of N.S.W.: portrait of a state, CMA Dept. Lands, N.S.W. (Eds) Harriman, R. J. and Clifford, E. S.

Carter, D. 1986. Wind erosion research techniques workshop. Conducted under the auspices of the Standing Committee of Soil Conservation, Esperance, W. A., February 1985.

Carter, D. & Findlater, P. 1989. Erosion potential of phomopsis-resistant lupin stubbles. W.A. J. Agric. 30: 11–14.

Chepil, W. S. 1944. Utilization of crop residue for wind erosion control. Sci. Agric. 24: 307–19.

Chepil, W. S. 1950. Properties of soil that influence wind erosion. I. The governing principle of surface roughness. Soil Sci. 69: 149–162.

Chepil, W. S. 1957. Width of field strips to control wind erosion. Kansas Ag. Exp. Stn. Tech Bull., 92, United States Department of Agriculture.

Chepil, W. S. & Woodruff, N. P. 1963. The physics of wind erosion and its control. Adv. Agron. 15: 211–302.

Fryrear, D. W. 1985. Soil cover and wind erosion. Trans. of ASAE 28: 781–784.

Gregory, J. M. 1982. Soil cover prediction with various amount and types of crop residue. Trans. of the ASAE 27: 1333–1337.

Gregory, J. M. 1984. Prediction of soil erosion by water and wind for various fractions of cover. Trans. of the ASAE 29: 1345–1350, 1354.

Leys, J. F. 1986. Situation statement on wind erosion in N.S.W., in Wind erosion research techniques workshop. D. Carter (ed), conducted under the auspices of the Standing Comittee of Soil Conservation, Esperance, W. A., February 1985.

Leys, J. F. & Semple, W. S. 1984. Estimating the weight of crop residues for wind erosion protection. Western area Tech. Bul. 24, Soil Conservation Service of N.S.W.

Lourensz, R. S. & Abe, K. 1983. A dust storm over Melbourne. Weather 38: 272–75.

Lyles, L. & Allison, B. E. 1976. Wind erosion: The protective role of simulated standing stubble. Trans. of the ASAE 19: 61–64.

Lyles, L. & Allison, B. E. 1981. Equivalent wind erosion protection from selected crop residues. Trans. of the ASAE 24: 405–408.

Lyles, L., Schrandt, R. L. & Schneidler, N. F. 1974. How aerodynamic roughness elements control sand movement. Trans. of the ASAE 17: 134–139.

McTainsh, G. H., Lynch, A. W. & Burgess, R. C. 1990. Wind

erosion in eastern Australia. Aust J. Soil Res. 28: 323–39.

Marshall, J. K. 1970. Drag measurements in roughness arrays of varying density and distribution. Agr. Meteorol. 8: 269–292.

Marshall, J. K. 1972. Principles of soil erosion and its prevention. In: The use of trees and shrubs in the dry country of Australia. Deparment of National Development, Forestry and Timber Bureau. Australia Government Printing Service, Canberra.

Northcote, K. H. 1979. A Factual Key to the Recognition of Australian Soils. Rellim. Adelaide.

Owen, O. P., Emery, K. A., Abraham, N. A., Johnston, D., Pattemore, V. J. & Cunningham, G. M. 1989. Land Degradation Survey of New South Wales: 1987–88. Soil Conservation Service of N.S.W.

Pye, K. 1987. Aeolian Dust and Dust Deposits. Academic Press, London.

Raupach, M. R. & Leys, J. F. 1990. Aerodynamics of a portable wind erosion tunnel for measuring soil erodibility by wind. Aust. J. Soil Res. 28: 177–92.

SAS Institute Inc., 1988, SAS/STAT™ User's Guide. Release 6.03 Edition, Cary, NC, SAS Institute Inc. 1028 pp.

Siddoway, F. H., Chepil, W. S. & Armbrust, D. V. 1965. Effect of kind, and placement of residue on wind erosion control. Trans. of ASAE 8: 327–31.

Van de Ven, T. A. M., Fryrear, D. W. & Spann, W. P. 1989, Vegetation characteristics and soil loss by wind. J. Soil and Water Cons. 44: 347–349.

Vegetatio **91**: 59–72, 1991.
A. Henderson-Sellers and A. J. Pitman (eds).
Vegetation and climate interactions in semi-arid regions.
© 1991 *Kluwer Academic Publishers. Printed in Belgium.*

The potential contribution of satellite remote sensing to the understanding of arid lands processes

M. M. Verstraete[1] & B. Pinty[2]
[1]*Institute for Remote Sensing Applications, CEC Joint Research Centre, Ispra Establishment, TP 440,
I-21020 Ispra (Varese), Italy*; [2]*Laboratoire d'Etudes et de Recherches en Télédétection Spatiale, 18 Av.
Edouard Belin, F-31055 Toulouse, France*

Accepted 24.8.1990

Abstract

This paper discusses the need for monitoring the state and evolution of arid and semi-arid environments, and compares the specific contributions of *in situ* and satellite-based techniques. The role of physically-based models in the quantitative interpretation of the measurement is stressed, and a strategy is proposed for the systematic exploitation of space technologies.

Introduction

The natural resources of an area can be classified as renewable, reusable, and non-renewable. Renewable resources typically include those hydrological and biological resources that are regenerated through the cycling of matter and energy within these ecosystems (rivers and lakes, grasslands, agriculture). Reusable resources include those resources that exist and can be reused indefinitely if they are properly managed, but that are not regenerated on the space and time scales of human activities: the top soil is a case in point. This mismanagement of a reusable resource may lead to its destruction or disappearance (*e.g.*, soil erosion), and therefore to the irreversible degradation of the natural resource base of the area. Non-renewable resources include those resources that can only be used once, they are typically acquired by mining activities (oil or minerals, but also fossil water).

Arid and semi-arid regions, which offer all three types of natural resources, are also considered fragile environments, in the sense that the rate of exploitation of their renewable resources can easily exceed the rate at which these regions can deliver them, or in the sense that their exploitation may easily lead to the loss of some or most of their reusable resources. This occurs because the carrying capacity of these lands is quite variable in space and time, and because of the inability of human systems to adapt fast enough to these changes. An accurate evaluation of the status of the resource base of arid and semi-arid regions, and the continuous monitoring of its evolution, therefore constitute, in principle, necessary but not sufficient bases for the sustained development of these regions.

Until recently, the monitoring of natural resources and environmental conditions in arid and semi-arid regions has been carried out exclusively with *in situ* measurements. All of what we know of the past history of these areas come from direct field observations, or from reconstructions based on indirect evidence (proxy data). Over the last decades, however, the emergence of satellite remote sensing platforms has allowed the repetitive observation of these environments from space. These new monitoring techniques present definite advantages, but also specific difficulties, and it is the objective of this paper to discuss the respective roles and complementarity of *in situ* and remote sensing approaches to environmental monitoring.

Overview of the physical characteristics of dry lands

Arid and semi-arid areas are characterised by a generally low level of moisture availability, compared to other types of ecosystems. The location and extent of these regions is a matter of convention, and different authors have adopted different criteria in this respect. Arid and semi-arid regions cover about one third of the Earth's continental surface (Paylore & Greenwell 1979), and may support as much as one fifth of the total human population of this planet. Various processes of desertification result in both the expansion of these areas and the progressive degradation of their natural resources (*e.g.*, UNCOD 1977). Furthermore, climate scenarios produced by various climate models suggest the possibility of a drier phase as a result of the intensification of the greenhouse effect, especially during the summer in the interior of the continents (*e.g.*, Schlesinger 1989). The synergistic interaction between mismanagement and overexploitation on one hand, and climatic change on the other hand, may result in irreversible damage to the environment and in a worsening of the living conditions for the populations concerned.

The general properties and dynamical behaviour of arid and semi-arid regions in the context of global changes are reviewed in Verstraete and Schwartz (1990). The primary ecological features of hot dry regions are the low biomass and limited species diversity; they result from severe environmental conditions, including a lack of water to support plant life and a large diurnal temperature range.

Plants synthesise organic materials by combining water and minerals obtained from the soil with carbon dioxide extracted from the atmosphere, using solar radiation as a source of energy. Since the latter two are widely available, primary pro-

ductivity is mainly limited by the availability of water and nutrients. Soils are extremely variable in space. They result from the slow decomposition of the parent rock materials by physical and biochemical processes. Once formed, soils can easily be displaced by wind and water if they are not protected by vegetation. For its part, the input and disposition of water in the environment is quite variable both in space and time. A clear understanding of the implications of these scale considerations is crucial to appreciate the constraints under which plants must operate.

Locally, the availability of water is conditioned by the difference between the input of water through precipitation and its disposal through evapotranspiration, infiltration in the soil profile, and runoff. This can be expressed as follows:

$$\frac{\Delta W}{\Delta t} = P - E - I - R \tag{1}$$

where $\Delta W/\Delta t$ is the time rate of change in soil moisture content, P is the rate of precipitation, E is the rate of evapotranspiration, I is the rate of infiltration in the deeper soil layers, and R is the runoff. Only a fraction of the soil moisture content W is useful for plant growth and development, because of the inability of the roots to extract water from the soil beyond a limit which depends on the nature and texture of the soil.

The space and time distribution of precipitation (P) is determined by the state and evolution of the atmosphere at a variety of scales, from local to regional. The exchange (E) of water between the surface and the atmosphere depends on the amount of radiation absorbed at the surface, the speed, temperature, and humidity of the air in the boundary layer, and the efficiency of soil and plant processes to evaporate water. Infiltration (I) into the soil is controlled by the nature and structure of the soil profile, as well as its moisture. At any given place and time, there is an upper limit to the rate at which water can infiltrate the soil. If the rate of precipitation of liquid water exceeds this maximum infiltration rate, water accumulates at the surface (ponding), or runs off (on sloping terrain). The timing of rainfall events is also crucial: a given amount of water may be damaging on bare soil at the end of the dry season, and beneficial at the end of the rainy season. Similarly, the rate of delivery of precipitation is important: short intense rainfalls may result in strong erosion, while the same amount of water delivered over a longer period of time may infiltrate the soil and remain available for plant growth. Although the response of the biosphere to the atmospheric forcing is damped and delayed, some processes are quite dramatic, such as the 'green flash' observable in certain regions of Africa. Clearly, edaphic, biologic, and atmospheric processes interact over a wide range of spatial and temporal scales. This greatly complicates the task of modelling the interactions between the biosphere and the atmosphere, or of computing meaningful spatial averages (Avissar & Verstraete 1990).

The biosphere and the atmosphere form a closely coupled dynamical system, where variations in one component affect all other components. For example, a reduction of the vegetation cover increases surface albedo, runoff, and the risk of soil erosion, while modifying the roughness of the surface and also decreasing the availability of seeds. Bare rocks are more susceptible to fragmentation, while barren soils tend to form crusts, and severe weather events may result in dust storms. On the other hand, the selective removal of certain plants may provide new niches for invaders and reduce the competition for water between the remaining individuals.

On a regional scale, the same water balance equation is still applicable, but the loss of water from one ecosystem may correspond to a source of water for an adjacent one, since the water that runs off from one location may contribute to the water balance downstream. Similarly, the local export of water in the deep aquifers or through evapotranspiration may enhance the recycling of water at these larger scales. The interaction between the atmospheric branches of the hydrological cycles in adjacent regions has been studied with mesoscale numerical models (See, for instance, Anthes 1984; Pinty *et al.* 1989; and Segal *et al.* 1988). These investigations have documented the nature and extent of circulations induced by strong gradients in vegetation cover similar to land-sea breezes.

Recent and expected changes in dry regions

Arid and semi-arid regions have evolved in time, and will continue to do so, under the influence of natural forces. The observed accelerated degradation of these areas, however, results from superimposed perturbations of human origins. Overgrazing, inappropriate agricultural practices, and many other processes have resulted in the destruction of the vegetation cover and the degradation of the soil resource in many parts of the world. This problem, although local in essence, repeats itself globally and must be studied with the appropriate conceptual and mathematical tools. Regional and general circulation models (GCM) provide the natural context to approach these questions.

Several experiments have been performed with GCMs to evaluate the sensitivity of the regional and global circulation of the atmosphere to continental surface conditions (Mintz 1984). These studies compare the changes in climate of the numerical model between a reference case and a perturbed case where the surface is drastically modified, for example by forcing the evapotranspiration over continents to vanish. Other experiments have imposed a change in surface roughness, albedo, or soil moisture content. Sud and Smith (1985) and Sud *et al.* (1988) observed that a change in surface roughness of dry lands may affect the convergence patterns in the lower atmospheric layers, and thereby modify the precipitation regime of these regions. All these effects constitute positive feedback mechanisms, which have the potential for maintaining or enhancing desertification.

Local events can have global consequences. For example, chlorofluorocarbons, which are held responsible for the 'ozone hole' over a large fraction of the Southern Hemisphere, are primarily produced in the United States, Europe and Japan. Similarly, processes that originate in dry lands may have implications in temperate regions.

Arid lands export dust sometimes across entire oceans (Carlson & Prospero 1972), while water erosion in semi-arid regions often results in the siltation of dams in other countries. There are also global biogeochemical processes that seem to affect all regions of the planet simultaneously. For instance, the increase in methane concentration currently observed in the atmosphere may result from either or both enhanced production rates from the biosphere, including in the semi-arid regions, and a reduction of the oxidising capacity of the atmosphere (Brasseur & Verstraete 1989). This, in turn, has profound effects on the chemistry of the troposphere and the life-time of many other pollutants. Arid regions also appear to be active contributors to the nitrogen and calcium cycles (Schlesinger *et al.* 1990). A better understanding of the dynamics of the climate system, including its chemical and biological components is therefore crucial if these issues are to be addressed.

Monitoring arid lands phenomena

Environmental degradation has been observed in dry lands, and is projected to continue for the foreseeable future. It is therefore imperative to inventory the extent and condition of the natural resources of arid and semi-arid regions in order to support the drafting and implementation of rational development plans for their sustained use. Repeated monitoring is required to establish the space and time variability of the relevant environmental parameters, and to identify the regions at risk of severe degradation and the processes at work. The efficiency of these development plans and policies must be periodically evaluated in the light of the intended socio-economic goals and the actual effect of these plans on the environment. The monitoring of environmental resources and their depletion or destruction is therefore germane to the entire development process. Last but not least, there are intimate relations between the monitoring and modelling activities: the environmental data bases created by monitoring provide essential initial and boundary conditions for the models needed to

predict to evolution and impact of environmental and policy changes; and these models, in turn, help identify the sensitive parameters and specify the accuracy and scales at which these parameters should be monitored.

Two complementary approaches can be followed in the interpretation and use of data, especially those acquired from remote sensing. A qualitative approach may be suitable when the desired information is apparent by visual inspection. For example, since man-made buildings have quite different spectral reflectance characteristics than natural surfaces, it is easy to detect the expansion of urban areas into adjacent agricultural or other natural regions. This phenomenon can be observed by applying specific techniques of image processing, including contrast enhancement, image differencing, and mapping. Statistical analyses such as pattern recognition and clustering can also be applied to identify targets of interest, or the occurrence of certain events. This qualitative approach does not require a deep understanding of the physical processes involved, and only makes use of advanced image processing techniques. The main shortcoming is the inherent inability to estimate quantitative parameters such as the temperature of the surface, the quantity of water in the ecosystem, or the amount and productivity of the biosphere.

A quantitative approach to data interpretation requires the development of mathematical models. This is necessary because the various factors that affect the measurements must be taken into account if the desired signal is to be isolated from instrument noise and contamination by other processes. For example, radiances measured by satellite platforms outside of the atmosphere are affected by the transfer of this radiation through the atmosphere, and by the nature and structure of the surface. Climate and environmental models require data to specify initial and boundary conditions, and their validation is also accomplished by comparing their predictions to actual measurements. Finally, numerical models can be used to identify the most sensitive parameters, and to specify limits on the accuracy that can be expected from such measurements.

As explained in the previous section, the availability of water is the most important factor controlling the primary productivity of arid and semi-arid regions. Since it is difficult to observe directly available water, it is necessary to observe other components of the hydrological cycle, as described in Equation 1. Studying the evolution of dry lands therefore requires the systematic collection of relevant data and their analysis and interpretation at the space and time scales of interest.

In this section, we will compare the advantages and disadvantages of *in situ* versus satellite data, in terms of the geographical coverage, spatial and temporal frequency of information, the type of variables measured, and the scale of representativity of these measurements.

In situ *measurements*

In situ measurements are obtained by placing a sensor directly into the environment under study. Meteorological and climatological information is typically generated by one or more complementary networks of surface and upper air observing stations. The World Meteorological Organization defines the standards and ensures the intercomparability of the measurements. Typically, climatological stations measure only temperature and precipitation at the surface, on a daily or weekly basis. The network of synoptic stations is much sparser and includes mostly airports and major cities, but these stations measure additional parameters such as wind speed and direction, humidity, pressure, cloudiness, etc, at the surface and, in some cases, in the free atmosphere. These observations are often taken at least twice a day. Finally, there is an even sparser network of radiation stations, where calibrated instruments measure various aspects of the solar and terrestrial radiation budget, such as incoming direct, diffuse, global or net radiation, and duration of sunshine. These measurements may be available continuously or with a high temporal resolution. Hydrological stations, which measure river level or soil moisture, are usually set up and maintained at the watershed level. In a few selected regions, specialised networks may have been installed for specific purposes, such as agrometeorological uses or pollution control. No similar systematic collection of data has ever been organised internationally to document the dynamics of the biosphere.

In situ instruments typically need to be maintained and supplied in power, paper, tapes, and other materials. As a result, they require the periodic visits of observers and are operated only (or mostly) in populated areas. Since the geographical distribution of habitat is very uneven (a large fraction of humanity lives on the coast of the major continents), the data from these stations does not adequately cover the globe. The almost total lack of regular oceanic stations is a case in point.

In principle, the density of the observation network should depend upon the natural variability of the variable of interest: denser networks are needed to resolve strong gradients, while a few scattered stations may be sufficient in areas homogeneous with respect to the variable of interest. Since different variables display variability at different spatial and temporal scales, and since most variables are measured at a finite number of stations for operational reasons, it is clear that no single network of *in situ* stations can provide the optimum coverage for all aspects of environmental monitoring.

From the previous sections, it is clear that the hydrological cycle is the prime controlling factor of arid regions. Within the cycle, soil moisture is arguably the most interesting factor to monitor, since it directly controls the primary and secondary productivity of the environment. It turns out, however, that measuring soil moisture profiles directly is either expensive (neutron probes) or inconvenient (gypsum blocks and weighting methods). Soil moisture monitoring also requires, in principle, a dense network of stations, because of the intrinsic characteristics of soil, and, in particular, their high spatial variability. Since this is not affordable, indirect techniques are used to estimate soil moisture as a residual of all other components of the water balance equation. Evapotranspiration, one of the primary components of this equation, needs to be estimated either from humidity and wind measurements at a minimum of two levels in the atmosphere, or on the basis of an energy balance model and measurements such as solar radiation, temperature and stomatal resistance. Consequently, it is important to measure as many of the parameters that affect the water, radiation and energy balances as possible, to constrain the estimation of the derived parameters.

One major issue raised by environmental measurements is the representativity of these measurements. *In situ* measurements are local: they are only representative for a small portion of the 4-dimensional space (3 spatial and 1 temporal) centred around the location and time of observation. The scale of representativity is then defined as the maximum distance (in space or time) over which the difference between the true value of the variable and the *in situ* measurement is acceptable for the particular application at hand. The scale of representativity of a measurement is therefore related to the size of the error that can be tolerated on the estimate, and the lower the requirement in accuracy, the larger the volume and the longer the time period over which the measurement may be considered an acceptable estimate.

The scale of representativity is not the same for all measurements made at the same location, or for all components of the water and energy balance equations. For example, at any given location, the local temperature during the day, measured in a screen at 2 m above the ground, is primarily driven by the surface global radiation balance. Far from sources of pollution or in the absence of topographic effects, the global radiation balance depends on atmospheric conditions, such as cloudiness, which is itself controlled by the circulation of the atmosphere at a much larger scale (synoptic wave patterns). Local temperature, however, is also controlled by a series of local processes whose intensity depend on the characteristics of the local environment (*e.g.*, nature and extent of the vegetation cover, soil type and moisture content, etc). For these reasons, temperature measurements may be representative

of much smaller scales than some of the driving mechanisms (such as radiation). Since the synoptic and seasonal components can be identified even in local measurement time series, it is clear that these measurements include information at all scales of the processes that affect the measurement, but the large spatial scales may be only clearly isolated from the long time behaviour, while the impact of smaller scale processes may dominate the variability of the signal for smaller periods.

Rainfall events are often very localized, especially if they result from convective clouds. On the other hand, the level of a river represents the cumulative precipitation that has occurred in the part of the hydrological basin upstream from the station over some period of time. It may therefore be difficult to combine such measurements as precipitation and runoff in the same water balance equation, unless great care is taken to combine sets of measurements that are representative of the same scale. This discussion leads naturally to the problem of interpolation, which is intrinsically linked to the discrete nature of the measuring station network. Since the scale of representativity is a function of the nature of the parameters as well as their space and time variability, and since all variables are measured on a network of finite resolution, the interpolation of values may be more applicable or accurate for some variables than for others. In particular, if the scale of representativity is smaller than the typical distance between two stations in the network, or shorter than the time period between two consecutive measurements, numerical interpolation schemes may lead to significant biases (over- or under-estimation). This problem may be partially addressed by designing physically-based interpolation schemes that may compensate for the lack of resolution by imposing physical relations on the likely changes of the value of the variable in space or time. The joint use of mesoscale and global circulation models is an example in which a higher resolution model is used to describe the regional variations of various variables on the basis of fields given by the large scale and low resolution model (Giorgi & Bates 1989).

Remote sensing measurements from space

Remote sensing measurements are obtained by locating the sensor away from the environment to be observed. This implies that the information relative to the variables of interest must be transferred between the environment and the sensor over distances much greater than the corresponding scale of representativity. In some cases, this may involve material fluxes, but in the current context we will restrict our discussion to the remote sensing of radiation. Satellite remote sensing platforms are orbiting around the Earth at altitudes ranging from 600 km to as much as 36 000 km, depending on the characteristics of the orbit and the desired trade off between longevity (the lower the orbit the higher the friction in the upper atmosphere and the shorter the lifetime of the satellite) and spatial resolution (low orbit satellites typically have a higher spatial resolution). In terms of monitoring surface environmental conditions, radiation is the only parameter that can be measured on board of a satellite. Typically, space platforms are equipped with radiation sensors such as television cameras and spectrometers. A mechanical device allows the sensor to scan the planetary surface in a direction perpendicular to the direction of motion, and an optical device may provide spectral information. Images can be reconstructed from successive measurements with these instruments by juxtaposing the various picture elements or 'pixels'. Pixels vary in size between a few meters and a few kilometers, depending on the design of the instrument and the characteristics of the orbit. The orientation of the sensor and the timing of the measurement are known, and can be used to compute the geographical position of each pixel.

Three types of information can be retrieved from radiation measurements: the intensity of the radiation, its spectral composition, and its polarisation. These physical parameters are entirely controlled by the way the radiation was emitted or scattered in the atmosphere and at the surface of the Earth. If we knew the composition, structure and temperature of all materials at the surface and in the atmosphere, we could use our understanding of the transfer of radiation in these media to predict the intensities, spectral distribution and polarisation that should be observed on board the satellite. In reality, however, we observe the radiation and we would like to know what combination of composition, structure and temperature may have lead to these measurements. This is known as an inverse problem. Since the signal measured by the instrument on the satellite is affected by both the surface and the atmosphere, it is possible, in principle, to retrieve information on either components of the system with varying degrees of accuracy. Insofar as we are concerned with surface processes, atmospheric effects are contaminating the signal and mask or dilute the information content of the observed signal.

The orbits of satellite remote sensing platforms can be calculated to provide a wide range of conditions of observations in space and time, or, in the case of solar radiation, in illumination and observation angles. A geostationary satellite, for example, can be located in such a way that its angular speed of rotation around the Earth equals that of the Earth. In this case, the satellite always looks as the planet from the same relative view point, and therefore observes the same side of the planet. Geostationary satellites are necessarily located above the Equator, and can typically scan the planetary surface in 20 minutes or so. By contrast, a polar-orbiting satellite will complete a revolution around the planet in a few hours or less, and therefore be able to observe all locations within a finite amount of time. The time period between two observations of the same location under the same geometry can be as long as a couple of weeks, however.

The density of observations available from satellites is therefore set by the design and engineering of the platforms, not by the intrinsic variations of the parameters of interest. This is not usually a problem, however, because the size of the individual pixels is so small compared to the scale of variability of many of the interesting parameters that the space variations can be observed with great resolution. There remains, however, the issue of inhomogeneities within the pixel. For the issues that are discussed in this paper (water balance, regional to global phenomena), small pixels such as those observed by low orbiting satellites (LANDSAT, SPOT) may be considered small enough to be homogeneous. However, large pixels of 1 km or more on the side may contain a variety of surfaces and a diversity of cloud conditions. In the case of non-geostationary satellites, there may be a mismatch between the spatial and temporal density of observations: in first approximation, the higher the spatial resolution, the longer the period of time between two consecutive observations of the same target in the same geometrical configuration. Such satellites intrinsically observe the spatial variability, i.e., resolve spatial scales, much better than the temporal variability.

Three main regions can be identified in the range of electromagnetic radiation that is observable from space: the solar range, which includes visible and near-infrared radiation, the thermal infrared range, and the microwave range. All the radiation in the first spectral range originates from the Sun and has been scattered towards the instrument by the surface or the atmosphere. In the case of the thermal and passive microwave ranges, the radiation is emitted by the surface and the atmosphere. Active mircowave systems include a source of microwave radiation on board the satellite itself. The variety of sources of radiation and the differential interactions between this radiation and materials provide the basis for retrieving different information from multispectral observations of the same environment. However, the information obtained at the level of the satellite platform in the form of a radiation measurement can, in principle, only be interpreted in terms of the radiative properties of the surface and the atmosphere that affected the transfer of this radiation.

Any remote sensing measurement of radiation is a volume average of the parameter of interest, over the volume 'seen' by the sensor. The scale representativity in space of an individual pixel measurement is therefore the size of the pixel itself. The time scale representativity of the information, however, remains dependent on the nature and time rate of change of the variable

characteristic of the process being observed, as is the case for *in situ* measurements. If the sampling of the variable provided by the frequency of observation is shorter than the typical time period over which the variable changes by a significant amount (relative to what is desired), then these variations will be resolved.

In situ *versus remote sensing data*

Satellite remote sensing data are therefore of a different type than *in situ* measurements: They represent volume-averaged measurements by opposition to point measurements, and their interpretation in terms of either surface or atmospheric processes raises issues of contamination of the signal by the other medium. Furthermore, the nature of the variable actually measured is different between remote sensing and *in situ* measurements. For example, consider a temperature estimate deduced from a radiation measurement on board a satellite. If the contaminating effect of the atmosphere has been correctly removed, this temperature (obtained by inversion) represents the average skin temperature of the object at the surface; it is the temperature used in the radiation balance equation.

On the other hand, temperature measurements from the meteorological networks are either soil or air temperatures at finite heights or depths from the surface, and neither of them are skin temperatures. These temperatures actually trace the evolution of the energy balance equation for the thermometer taken as an object. They may represent the environmental temperature only to the extent that the thermometer is in thermodynamic equilibrium with its environment, and then only at the scales of representativity discussed above. Because of the location of the thermometer inside the medium (air or soil), the energy balance equation for the thermometer may in fact be controlled by a different set of physical processes than the equation of balance for radiation at the interface between the solid surface and the gaseous atmosphere above it. To retrieve the skin temperature of the solid surface from such local measurements would require the inversion of a physically-based model that relates the skin temperature to these bulk measurements, through energy transfers. The direct comparison of data obtained in such different ways and resulting from entirely different physical processes cannot be done easily.

It should be clear that *in situ* and remote sensing measurements actually sense different physical processes. This is both an advantage and an inconvenience: on the one hand, they cannot be compared directly without a detailed discussion of the physical processes by which they are linked, but on the other hand, they actually provide complementary information and therefore a different view of the same environment. Since satellite sensors measure radiation fluxes conditioned by either surface reflectances or skin temperatures, they do not provide directly all the physical parameters that must be included in the energy or water balance equations at the surface. Some of these parameters, however, may be directly obtained from local *in situ* observations.

The main advantage of satellite observations results therefore from their global and repetitive coverage, and the acquisition of area- or volume-averaged values at the scales of interest for regional or global climate monitoring. This is particularly advantageous for regional or global climate modelling studies, since the desired variables must be representative of the grid scales of the models. Since all observations are obtained with the same sensor, there is the further advantage of compatibility between these measurements. Anybody who has dealt with problems of comparing *in situ* measurements obtained with instruments from different manufacturers will appreciate the benefits of not needing inter-comparisons and inter-calibration campaigns. Apart from technical issues such as calibration and drift of satellite-based instruments, the price to pay for these advantages is the necessary use of physically-based models to interpret these measurements in terms of the variables that are of interest to determine the various terms of the water and energy balance equations.

The retrieval of relevant information from remote sensing data

As indicated above, only reflectances and temperatures can be observed directly from radiation measurements. Consequently, only those physical parameters that directly affect the observed radiation flux can be retrieved by inversion. Typically, the reflectance of a medium in the solar spectral range depends only on the optical properties and relative geometrical arrangements of the scatterers (Verstraete *et al.* 1990). Once these physical parameters have been estimated on the basis of a set of measurements, this information can be inverted a second time, with the help of process models, to yield another set of variables of more direct interest for the specific applications at hand. For example, leaf optical properties such as the single scattering albedo and the phase function may be linked explicitly to the chlorophyll or liquid water content of the leaves, which are directly relevant to the study of the primary productivity of the biosphere or the water cycle near the surface of the Earth.

Inversion procedures

In all generality, a measurement F_0 taken by remote sensing can be modelled as

$$F_t = F_t (X_1, X_2, \ldots, X_n; p_1, p_2, \ldots, p_m) \quad (2)$$

where F_t represents the theoretical value of the observation at the level of the satellite, X_i, $i = 1, 2, \ldots, n$ are n physical variables that condition the measurement, and p_j, $j = 1, 2, \ldots, m$ are m known parameters. If the functional dependency of F_t on these variables and parameters is known, and if the values of these variables and parameters were given, the theoretical value of F_t could be calculated and compared to the actual observation F_0. This is called a direct problem.

In most cases of practical interest, however, we have one or more measurements F_0, and we would like to retrieve the values of the physical parameters X_i that best account for these observations. This is called an inverse problem. Such a problem is difficult to solve for $n > 1$, since there is only one equation. Instead, we seek to obtain l observations of the same surface under different conditions (i.e., for different sets of parameters p_j), and we apply an optimisation procedure that seeks iteratively the best set of values for X_i, in some statistical sense (Pinty *et al.* 1990). If multiple observations are not available, the problem is ill-posed, and no explicit solution can be found because the mathematical system does not contain enough information to assign a unique value to each variable.

Another important issue is the accuracy with which the variables X_i can be retrieved. Indeed, much of the variability of F_t may be due to the predominant role of a subset of the controlling variables X_i. Clearly, those variables that have less effect on the measurements cannot be retrieved reliably from observed variations of F_t. The problem is said to be ill-conditioned with respect to these variables: the inversion procedure may assign a value to all variables, but the reliability of the estimates may be so low as to be useless for some of these variables. These issues have been discussed by Isaacs *et al.* (1986) in the context of numerical weather prediction.

In practice, it is necessary to estimate first, and a priori, the accuracy with which the values of physical variables must be known for a particular application. Such requirements could derive, for example, from sensitivity studies of climate models. The level of detail and complexity necessary in the processing of observational data and in the modelling of F_t is determined by this required accuracy.

Accuracy of retrieval procedures

There are at least two general strategies to retrieve the value of a parameter X_i from an equation such as (2). The first one consists in assigning the values of all variables X_k, with $k \neq i$, so that we are left with a single equation in a single unknown. This requires additional information from one of three alternative sources: *in situ* measurements,

results from objective analyses, and other satellite remote sensing data.

The corrections that need to be made to account for the effects of the atmosphere on satellite data are a case in point: observed profiles of atmospheric composition, temperature, etc, could be used directly, but this approach has the disadvantages of *in situ* observations mentioned above (scale of representativity). A second and better source of information would be to use the results of objective analyses produced by numerical weather prediction models: this has the advantage of using balanced fields, but the drawback is that these values are known only at a finite number of locations, and for the times for which the analyses were made. These locations and times may or may not correspond to those of the satellite observations. In either case, we are left with a problem of interpolation. The third source of information consists in using additional remote sensing data from different satellites or different spectral bands: this seems a priori more coherent, but often results in even larger sets of unknown variables and enhanced needs for mathematical and physical models. In all three cases, the issue of data assimilation must be addressed, if the retrieval of physical variables from satellites must be achieved repeatedly and on an operational basis.

A particular alternative approach can be followed when the scale of variability of the variable X_i is significantly smaller than that for all other variables X_j. In this case, it is possible to take advantage of the dense coverage provided by satellite observing systems, since the typical size of the pixel is smaller than the characteristic scale of representativity of the physical variables X_j, $j \neq i$ that also control the signal. Pinty and Ramond (1986) applied such a strategy to the study of spatial and seasonal variations of surface albedo in West Africa. To the extent that this approach requires the input of auxiliary data from independent sources, it is important to estimate the sensitivity of the results of the inversion to the potential errors of these data.

The second strategy consists in solving the inverse problem by applying an iterative numerical procedure to determine the best values of all variables X_i simultaneously. Root mean square or maximum likelihood statistical methods can be used in conjunction with a non-linear numerical optimisation technique, as described in Pinty *et al.* (1989). In this approach, the main difficulty comes from the highly non-linear nature of the function F_t. The systematic use of this approach requires at detailed discussion of the sensitivity of the results with respect to errors in the measurements. The problem is to document the maximum level of noise that can be tolerated in the measurements in order to retrieve the physical parameters X_i with a given level of accuracy, as well as the stability of the retrieved results with respect to the formulation of F_t as it relates to the input noise. Finally, it is also important to document how errors of measurements are distributed on the physical variables, in other words, how the variance and higher moments of these variables depend on the variance of the measurements (Pinty & Verstraete 1990). In this respect, it would be advantageous to standardise inversion procedures on a small set of well-documented and tested optimisation schemes.

The two approaches described above are not exclusive and can be combined. If too many parameters need to be retrieved simultaneously, it may prove advantageous to constrain the inversion process with a partial input of independent measurements. In this case, the specification of the values of the less sensitive variables provides the best course of action, because a given relative error on these parameters will have less impact on the final result than the same error on the most sensitive parameters.

Interpretation of retrieved values

In some cases, the physical parameters obtained from the first inversion maybe of direct interest. For example, microwave observations of natural surfaces may result in a direct estimation of the roughness of the surface. In many cases, however, the information that is really desirable is a complex function of these primary parameters.

These new variables must then be estimated on the basis of the information obtained from the first inversion, and this is achieved by performing a second inversion, using the newly derived parameters as input to a new generation of process models. This approach is rarely followed explicitly, however. Instead, many researchers establish direct statistical correlations between the observed radiances at the level of the satellite and whatever variables of interest they are investigating. The high spatial variability of satellite-derived pictures may indeed suggest the presence of more information than is objectively contained in the data. A statistical approach can provide qualitative information for specific purposes, but cannot lead to a deeper understanding of the processes involved. These approaches may not extend to other locations or times, and may therefore be of limited or no value beyond the limits of the data sets from which they were derived. The most important contribution of qualitative statistically-based approaches is to explore new potential relations, and to provide a strong motivation for the derivation of physically-based approaches to pursue research in these directions.

Arid and semi-arid regions have been extensively monitored from satellite platforms, especially with the NOAA AVHRR sensor and, more recently, the SMMR instrument on board the Nimbus-7 satellite. In particular, raw data from channels 1 and 2 of the AVHRR instrument have been used to produce maps of the Normalised Difference Vegetation Index (NDVI). Such an index can be produced at regular intervals to monitor the evolution of the vegetation cover during the growing season, and cumulated values over long periods have been statistically related to other integrated variables such as yield, productivity, or seasonal precipitation. While such correlations may suggest avenues for research, a closer look at the physical processes that control the radiation signal at the level of the satellite underlines the number and complexity of the technical and scientific issues that need to be addressed. For instance, both atmospheric corrections and bi-directional effects can affect the NDVI by as much as 10% of its value.

One example of the difficulty of interpreting statistical relations is provided by the observed correlations between the Microwave Polarisation Difference Temperature (MPDT), derived from the SMMR instrument, and the NDVI (Townshend 1989). Over fully covering canopies, the roughness of the surface may be assumed constant, or to have a minimal contribution to the microwave signal. The MPDT is then principally controlled by the liquid water content of the canopy. However, because of the particular location of the channel 1 of AVHRR, these measurements are, in principle, essentially insensitive to the presence of liquid water. This would also be the case of channel 2 measurements, unless the leaf structure is strongly affected by its water content. The signals obtained from SMMR and AVHRR are therefore not measuring the same primary physical characteristics of the canopy, and any observed correlation must result from the internal behaviour of the plant system: when water is present, the plant cover is able to grow, and therefore also displays a high chlorophyll content. Focusing exclusively on these correlations removes any possibility to extract a different information from the two signals, and in particular the opportunity to gain a better understanding of the evolution of the canopy.

Another case that requires careful consideration is provided by the interpretation of the relation between the observed global values of NDVI and the leaf area index, or the concentration of atmospheric carbon dioxide. Over a vegetated area, channel 1 of the AVHRR instrument is directly sensitive to the absorption band of chlorophyll because the latter fundamentally controls the single scattering albedo of leaves. For this channel, a relatively shallow canopy (small value of leaf area index) may be optically thick and absorb the bulk of the radiation: the reflectance is not affected by the addition of more leaf layers, hence it is not possible to determine the leaf area index for dense canopies. On the other hand, channel 2 of AVHRR measures the radiance of the surface in a region which is characterised by very little leaf absorption. This implies the predominant role of the multiple scattering of the radiation, itself controlled by the internal structure of leaves, as well as the structure of the canopy. A canopy may therefore be optically thick for channel 1 and optically thin for channel 2. In the latter case, the near-infrared signal is sensitive to the leaf area index, but it is also contaminated by the contribution of the underlying soil under the vegetation. It is therefore questionable, in principle, to try to relate the observed reflectances to the leaf area index, outside of the limited range $1 < LAI < 3$. The interpretation of correlations between NDVI and other environmental parameters, such as stage of development, rate of growth, or the fixation of carbon dioxide, is therefore very difficult, since the radiative properties that control the reflectances in these two channels are different and do not depend explicitly on these parameters.

In the case of a sparse canopy cover, the interpretation of the signal is even more difficult, since the NDVI is more strongly controlled by changes in vegetation cover than by changes in the optical thickness of the canopy. If r_1 and r_2 are the observed reflectances of a partially vegetated area in the spectral bands of channels 1 and 2 of AVHRR, σ is the vegetation cover, and ρ_{vi} and ρ_{si} are the reflectances of the vegetation and the soil in channels $i = 1,2$, respectively, the following relations hold:

$$r_1 = \sigma\rho_{v1} + (1 - \sigma)\rho_{s1}$$
$$r_2 = \sigma\rho_{v2} + (1 - \sigma)\rho_{s2} \tag{3}$$

These two equations can be combined to approximate the NDVI of a partially covered region as

$$NDVI = \frac{r_2 - r_1}{r_2 + r_1}$$

$$= \left[\frac{\rho_{v2} - \rho_{v1}}{\rho_{v2} + \rho_{v1}} + \frac{1 - \sigma}{\sigma}\frac{\rho_{s2} - \rho_{s1}}{\rho_{v2} + \rho_{v1}}\right]\Bigg/$$

$$\left[1 + \frac{1 - \sigma}{\sigma}\frac{\rho_{s2} + \rho_{s1}}{\rho_{v2} + \rho_{v1}}\right] \tag{4}$$

In first approximation, this equation can be simplified by assuming that $\rho_{s2} \approx \rho_{s1}$, and that $\rho_{s2} + \rho_{s1} \approx \rho_{v2} + \rho_{v1}$. These assumptions are roughly verified for typical soils and plant covers. This leads to the simple expression

$$NDVI = \frac{\rho_{v2} - \rho_{v1}}{\rho_{v2} + \rho_{v1}} \sigma \qquad (4)$$

Despite the rough assumptions, this equation shows clearly why the global NDVI values have such a strong dependency on the vegetation cover and display so vividly the seasonal evolution of the canopy cover.

Concluding remarks

In summary, arid and semi-arid lands are very dynamic regions, and they can only be monitored systematically and regularly from satellite remote sensing platforms. As indicated above, the interpretation of these satellite-derived data raises a number of scientific issues, and requires the design and implementation of a well-conceived strategy to extract the information content from these data. This strategy can be outlined in three consecutive phases, namely the pre-processing of the data to remove the various sources of contamination of the signal by the atmosphere or the idiosyncrasies of the sensor and the orbit of the satellite, then the inversion of the data against a physically-based model to retrieve the values of the primary parameters that control the radiation transfer at the surface, and finally the retrieval of secondary environmental parameters from another inversion against biological or other process models that can use the radiation parameters as input.

The design and launching of a new generation of satellites over the next decade will provide new opportunities to observe the surface of this planet with higher resolution sensors. It is essential that the major drawbacks to the interpretation of satellite-derived data be removed, because of the sheer size of the data sets that will be generated. This implies that such problems must be addressed immediately, and the existing data bases of satellite data can serve as test cases for these new approaches to their interpretation.

Acknowledgments

The Department of Atmospheric, Oceanic and Space Sciences of the University of Michigan and the School of Earth Sciences of Macquarie University financed the participation of the first author to the 'Conference on Degradation of Vegetation in Semi-Arid Regions: Climate Impact and Implications', Macquarie University, January 29–31, 1990. The support and encouragement of Prof. W. Kuhn and A. Henderson-Sellers are gratefully acknowledged. Serena Schwartz contributed some of the background materials on remote sensing in arid lands. This cooperative paper would not have been possible without the financial support of the EROS Data Centre for the second author.

References

Anthes, R. A. 1984. Enhancement of convective precipitation by mesoscale variations in vegetative covering in semi arid regions. J. Clim. Appl. Meteorol. 23: 541–554.

Avissar, R. & Verstraete, M. M. 1990. The representation of continental surface processes in atmospheric models. Rev. Geophys. 28: 35–52.

Brasseur, G. & Verstraete, M. M. 1989. Atmospheric chemistry-climate interactions, in Climate and the Geo-Sciences: A Challenge for Science and Society in the 21st Century. Edited by A. Berger, S. Schneider and J. Cl. Duplessy. Kluwer Academic Publishers, 279–302.

Carlson, T. N. & Prospero, J. M. 1972. The large-scale movement of Saharan air outbreaks over the northern Equatorial Atlantic. J. Appl. Meteorol. 11: 283–297.

Giorgi, F. & Bates, G. T. 1989. Modelling the climate of the western U.S. with a limited area model coupled to a general circulation model. In: Proceedings of the Sixth Conference on Applied Climatology. American Meteorological Society, Boston, 201–208.

Isaacs, R. G., Hoffman R. N. & Kaplan L. D. 1986. Satellite remote sensing of meteorological parameters for global numerical weather prediction. Rev. Geophys. 24: 701–743.

Mintz, Y. 1984. The sensitivity of numerically simulated climate to land-surface conditions. In: The Global Climate.

Edited by J. Houghton. Cambridge University Press, New York, 79–105.

Paylore, P. & Greenwell, J. R. 1979. Fools rush in: Pinpointing the Arid Zones. Arid Lands Newsletter 10: 17–18.

Pinty, J. P., Mascart, P., Richard, E. & Rosset, R. 1989. An investigation of mesoscale flows induced by vegetation inhomogeneities using an evapotranspiration model calibrated against HAPEX-MOBILHY data. J. Appl. Met. 28: 976–992.

Pinty, B. & Ramond, D. 1986. A Simple Bidirectional Reflectance Model for Terrestrial Surfaces. J. Geophys. Res. 91: 7803–7808.

Pinty, B. & Verstraete, M. M. 1990. Extracting information on surface properties from bidirectional reflectance measurements. Accepted for publication by J. Geophys. Res.

Pinty, B., Verstraete M. M. & Dickinson R. E. 1989. A Physical Model for Predicting Bidirectional Reflectances over Bare Soil. Remote Sensing of Environment 27: 273–288.

Pinty, B., Verstraete, M. M. & Dickinson, R. E. 1990. A Physical Model of the Bidirectional Reflectance of Vegetation Canopies; Part 2: Inversion and Validation'. J. Geophys. Res. 95: 11 767–11 775.

Schlesinger, M. E. 1989. Model projections of the climatic changes induced by increased atmospheric CO_2. In: Climate and the Geo-Sciences: A Challenge for Science and Society in the 21st Century. Edited by A. Berger, S. Schneider and J. Cl. Duplessy. Kluwer Academic Publishers, 375–415.

Schlesinger, W. H., Reynolds, J. R., Cunningham, G. L.,

Huenneke, L. F., Jarrel, W. M., Virginia, R. A. Whitford, W. G. 1990. Biological feedbacks in global desertification. Science 247: 1043–1048.

Segal, M., Avissar, R., McCumber, M. & Pielke, R. A. 1988. Evaluation of vegetation effects on the generation and modification of mesoscale circulations. J. Atmos. Sci. 45: 2268–2292.

Sud, Y. C. & Smith, W. E. 1985. The influence of surface roughness of deserts on July circulation. A numerical study. Bound. Layer Meteorol. 33: 1–35.

Sud, Y. C., Shukla, J. & Mintz, Y. 1988. Influence of land surface roughness on atmospheric circulation and precipitation. A sensitivity study with a general circulation model. J. Appl. Meteorol. 27: 1036–1054.

Townshend, J. R. G. (Guest Editor) 1989. Comparison of passive microwave with near-infrared and visible data for terrestrial environmental monitoring. Special issue of Int. J. Rem. Sens. 10: 1575–1696.

UNCOD 1977. Desertification: Its Causes and Consequences. Prepared by the Secretariat of the United Nations Conference on Desertification. Pergamon Press, Oxford, 448 pp.

Verstraete, M. M. & Schwartz, S. A. 1990. Desertification and Global Change. In this volume.

Verstraete, M. M., Pinty, B. & Dickinson, R. E. 1990. A Physical Model of the Bidirectional Reflectance of Vegetation Canopies; Part 1: Theory. J. Geophys. Res. 95: 11 755–11 765.

Vegetatio **91**: 73–86, 1991.
A. Henderson-Sellers and A. J. Pitman (eds).
Vegetation and climate interactions in semi-arid regions.
© 1991 *Kluwer Academic Publishers. Printed in Belgium.*

73

Sources and sinks of greenhouse gases in the soil-plant environment

O. T. Denmead
CSIRO Centre for Environmental Mechanics, GPO Box 821, Canberra,
A.C.T., 2601, Australia

Accepted 24.8.1990

Abstract

The paper is concerned mainly with nitrous oxide, methane and carbon dioxide, which account for more than 70% of predicted greenhouse warming. All three have significant sources in the soil-plant environment and principal sinks in the atmosphere or the oceans. The emphasis is on methodological problems associated with measuring source and sink strengths, but the biogeochemistry of individual gases and problems of scaling to longer times and larger areas are addressed also.

Nitrous oxide accounts for some 6% of predicted greenhouse warming. Its atmospheric concentration is 315 ppbv and is increasing at 0.25% per year. The principal sink appears to be destruction through photochemical processes in the stratosphere. The main causes of the N_2O increase are thought to be biomass burning, fossil fuel combustion processes, and what now seem to be substantial emissions from soils associated with increased nitrogen inputs, irrigation and tropical land clearing. Uncertainty about the strengths of the soil sources is due largely to our reliance on enclosure techniques for flux measurement, and the lack of appropriate scaling procedures. Methane now accounts for 18% of anticipated greenhouse warming. Its atmospheric concentration is 1.7 ppmv and is increasing at 1% per year. Its greenhouse effect seems likely to increase over the next 50 years. The biggest sink appears to be oxidation in the atmosphere, but some oxidation occurs in soils as well. The main sources are rice fields, wetlands, biomass burning, ruminants, land fills, natural gas production, and coal mining. As for N_2O, there is much uncertainty about individual source strengths and there are urgent needs for better measurement and scaling techniques.

Increased CO_2 concentrations account for 49% of the greenhouse effect. The present atmospheric CO_2 concentration is 350 ppmv, increasing at 0.4% per year. Over 80% of the increase is due to fossil fuels, and the rest to deforestation and biomass burning. Atmospheric fluxes of CO_2 can be measured much more precisely than those of N_2O and CH_4, by micrometeorological techniques, but the scaling problem still remains. The largest known sink for CO_2 is the oceans, but recent calculations point to a large 'missing' sink for CO_2, which may be as yet unidentified sequestering processes in terrestrial ecosystems.

Introduction

There is little doubt that the greenhouse effect will have significant impacts on the natural vegetation, agriculture and land use, and the hydrology of Australia through such factors as increased CO_2, enhanced temperatures, changes in rainfall and evaporation, and changes in the length of the growing season (Walker *et al.* 1989). Some possible consequences (for Australia) foreseen by these authors are:

The summer-dominant rainfall area will expand to cover perhaps three-quarters of the continent with increases in rainfall of up to 50%, and most of the winter rainfall zone of Victoria and Western Australia will change to a more uniform rainfall distribution.

Water resources in the Murray and its tributaries could be reduced, but discharge into the Darling could double or treble.

Reduced winter-spring rainfall could reduce production in the wheat-sheep zones of Western Australia. Increased winter temperatures could lead to increased pasture growth and animal production in the southern wheat-sheep zone in eastern Australia. In the northern wheat-sheep zone, increases in summer cropping of maize, sorghum and oilseeds are possible. Soil water supplies in this zone are expected to improve since increases in rainfall will more than offset the increased evaporation due to higher temperatures. Evaporation may even be reduced by increased cloudiness.

In the pastoral zone, an increase in woody plants is forecast in the summer rainfall area and a probable decrease in both plant and animal production in the southern winter rainfall regions. It is suggested that a significant summer cropping industry could develop in north-central Queensland.

Changes in species' habitats are anticipated, associated with changes in fire regimes and land use. The ranges of different species will shift and their abundances change.

Mediterranean and alpine type biomes will be severely reduced in extent, and the species composition and boundaries of other biomes will change.

As Walker *et al.* (1989) note, all this is speculation. Our ability to predict such effects with any precision is hampered by the adequacy of existing climate and ecological models. As well, there is uncertainty about future trends in the atmospheric concentrations of greenhouse gases due to inadequate knowledge of their source strengths and almost no knowledge of how their biochemical cycles interact or how they will be affected by changing land-use practices. Some possibilities are:

The trend to much increased use of nitrogen fertilisers in agriculture will lead to increased emissions of nitrous oxide, N_2O, as well as NO_x (nitric oxide and nitrogen dioxide, the precursors of tropospheric ozone).

The same trend and expanding livestock production will lead to increased emissions of ammonia. Since ammonia is in a closed terrestrial cycle and returns to the land in rainfall, aerosols or by gaseous diffusion, there could be large effects on the nitrogen balance of native ecosystems. The addition of a few kgN per ha to the nitrogen economy of our native forests and shrub and grassland communities could have profound effects on growth rates and species composition. Further, small additions of nitrogen to the soil from atmospheric deposition and litter decomposition could increase N_2O emissions and decrease the destruction of methane, CH_4. This point is elaborated later in this paper, but it can be noted that at present, there are no measurements of N_2O emissions/destruction in any natural ecosystem in Australia. Extrapolation of overseas results indicates that anywhere between 1 and 10% of global N_2O emissions could come from Australian terrestrial sources.

Stubble burning, cane fires, or controlled or inadvertent burns in grassland and forests, common occurrences in Australia, could be contributing significant amounts of nitrous oxide and methane to the atmosphere.

Increases in atmospheric CO_2 will change the carbon/nitrogen ratio of plant litter and soil organic matter, thereby affecting soil emissions of nitrous oxide, methane and NO_x.

Some of the last problems are addressed in this contribution. It is concerned specifically with the gases nitrous oxide, methane and carbon dioxide, which, between them, account for more than 70% of the predicted global warming referred to loosely as the greenhouse effect. However, some of what I have to say carries over to other atmospheric trace gases with sources and sinks in the soil-plant environment such as NH_3, NO_x, SO_2, and CO, particularly in regard to flux measurement and extrapolation.

The best documented parts of the greenhouse story are the increasing atmospheric concentrations of the greenhouse gases. Careful measurement programs at baseline monitoring stations and analysis of ice-cores from the polar regions have shown that not only are the atmospheric concentrations of many trace gases well above pre-industrial levels, but also they are increasing steadily. The excellent Australian record, which derives from work in the CSIRO Division of Atmospheric Research and at the Cape Grim Observatory in Tasmania (Pearman 1988), will be referred to when discussing each gas.

By and large, the anthropogenic sources and sinks of the greenhouse gases are well known. The microbiological and plant physiological processes in soils and plants making for production and consumption are well known also, at least conceptually, but there is still a good deal to be learned about their behaviour in nature: their responses to the daily and annual cycles of solar radiation, plant growth and temperature, their dependencies on soil water content, the influences of distributed aerobic and anaerobic sites in soils, interactions with carbon and mineral-nutrient supplies, and so on (Robertson et al. 1989; Rosswall et al. 1989). Still larger problems are associated with quantifying individual source and sink strengths and scaling up point or paddock measurements to whole ecosystems, and then to seasons, geographic zones and the globe (Stewart et al. 1989; Vitousek et al. 1989; Wesely et al. 1989).

Most of the paper is concerned with methodological problems associated with identifying sources and sinks using chamber or micrometeorological systems, but the biogeochemistry of individual gases and the scaling problem are addressed briefly. We deal with N_2O, CH_4 and CO_2 in turn, that order reflecting not only the increasing contribution of each gas to the greenhouse effect, but also the levels of sophistication and inference that apply to their measurement.

Nitrous oxide

About 6% of the predicted warming is attributed to N_2O. The evidence from Cape Grim is that the atmospheric N_2O concentration is increasing at 0.2% to 0.3% per year (Fig. 1 and Pearman 1988). It appears to have increased from a pre-industrial level of 285 ppbv to about 315 ppbv in the southern hemisphere today (Pearman et al. 1986). Recent calculations suggest that the atmospheric sink for N_2O, viz., photochemical processes in the stratosphere, is about 12 Tg N_2O-N y^{-1}, but the observed rate of increase in atmospheric concentration represents a net annual addition to the global atmosphere of 3.5 Tg N_2O-N y^{-1}, which implies a current atmospheric loading almost 30% more than the probable steady-state destruction rate (Stewart et al. 1989).

The source of the additional N_2O is not well documented. Some attribute it almost wholly to biomass burning and fossil fuel combustion processes, but this seems to ignore what many believe to be substantial emissions from soils, e.g., in Bouwman (1990) where it is estimated that soils might contribute some 90% of total N_2O emissions. In soils, N_2O is formed by microbiological processes during nitrification, the oxidation of NH_3 and NH_4^+ to nitrate, and denitrification, the reduction of nitrate. The first is an aerobic process occurring in unsaturated soils; the second is an anaerobic process, usually occurring in saturated soils (Fig. 2 and Firestone & Davidson 1989). Microbial processes can also consume N_2O in soils, the net emission being a balance between production and consumption. Increases in atmospheric N_2O concentration are believed to be linked with increasing use of nitrogen fertilisers and irrigation in agriculture, and perhaps to increased atmospheric deposition of nitrogen on forest soils (Melillo et al. 1989). It has been pointed out that the inputs of nitrogen to

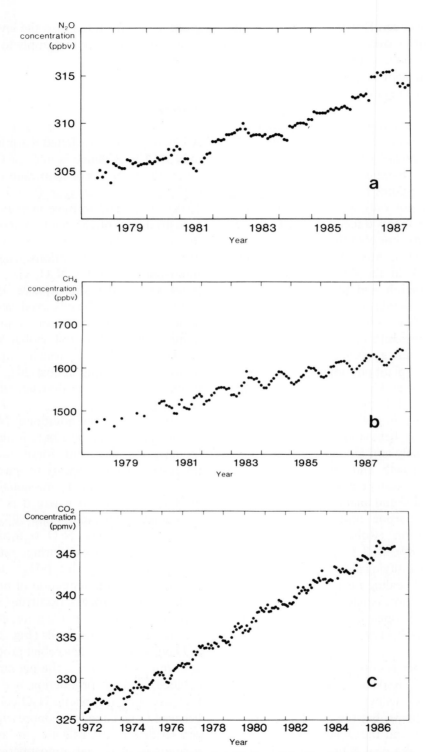

Fig. 1. (a) Atmospheric nitrous oxide concentrations measured at the Cape Grim Baseline Station, Tasmania; (b) Atmospheric methane concentrations at Cape Grim; (c) Atmospheric CO_2 concentrations measured in the mid-troposphere over southeastern Australia. From Pearman (1988).

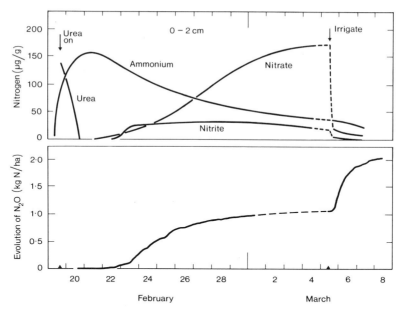

Fig. 2. Emissions of N_2O from soil during nitrification and denitrification. Urea fertilizer was applied to the soil, dissolved in irrigation water. The first emissions of N_2O were associated with the formation of nitrate following hydrolysis of the urea and transformation of the nitrogen through ammonium and nitrite; the later emissions occurred during denitrification after the addition of more irrigation water. From Freney *et al.* (1985).

agricultural systems through biological and industrial fixation are currently about 150 Tg N y^{-1} and the loss of 2 to 3% of this as N_2O would account for the entire observed increase in atmospheric N_2O of 3.5 Tg N y^{-1}. However, Stewart *et al.* (1989), in a recent exercise, estimated that the combined emissions of N_2O from combustion processes and fertilisers are not that large.

Biomass burning is certainly one candidate for the additional N_2O (Crutzen *et al.* 1979), but forest soils are now being advanced as another. Some recent budgets have recognised tropical forest soils as the single most important source, at 6 Tg N y^{-1}, e.g., Keller *et al.* (1986). Tropical land clearing is also thought to be a potentially important source (Robertson *et al.* 1989). In the last case, the N_2O production is believed to be associated with the rapid disappearance of carbon leading to high rates of nitrogen conversion from organic to inorganic forms after disturbance. As well, much mineral soil nitrogen is likely to be nitrified following land conversion. Finally, we note that unlike the baseline records for CH_4 and CO_2, gases for which there are significant biologi-

cal sources and sinks, the record for N_2O lacks any clear annual cycle, although it does show some interesting perturbations (Fig. 1). This might well indicate that a whole mixture of processes is contributing to the atmospheric increase.

As indicated by the above, there is much uncertainty about the strengths of the soil sources. Our present knowledge is based on a relatively few chamber measurements, and a fairly uncritical extrapolation procedure. The estimated contribution of 6 Tg N y^{-1} from tropical forests, for instance, is based on the multiplication of the entire area of tropical and subtropical forests and woodlands by an average flux derived from about 60 chamber measurements, each of 20 to 40 minutes duration. It is recognised that there is much room for improvement in sampling and statistical procedures, and considerable effort is now being spent on this problem, including the use of stratified sampling along soil fertility gradients and global data-bases (Stewart *et al.* 1989).

Due to sensor limitations, chambers have been the only feasible means for routine measurement of N_2O emissions in the soil-plant environment. (Australian scientists have been prominent in the design and evaluation of emission chambers for this purpose, e.g., Galbally & Roy 1978; Denmead 1979; and Galbally *et al.* 1985. The advantages of chambers include the abilities to detect very small emissions, to examine related soil properties and processes in some detail, and to experiment with a wide range of soil treatments. Their disadvantages include the facts that they necessarily interfere with the transfer processes that normally operate in the natural environment; they create artificial microclimates; they sample only a very small area of the surface, while the point to point variability of soil gas emissions is notoriously large, variations by a factor of two within distances of a few meters being commonplace and ten-to-one variations frequent; and the sampling periods are usually short, often less than 1 hour, whereas minimum sampling periods needed to characterise natural flux events are usually 1 to many days. Detailed discussions of these problems are given in the references above as well as by Mosier (1989), Vitousek *et al.* (1989) and Wesely *et al.* (1989).

Methane

Although present in the atmosphere at only 1/200 the concentration of CO_2, methane already accounts for some 18% of greenhouse warming, and it has been predicted that within 50 years, it could be the prime greenhouse gas (Pearce 1989). More conservative estimates give it a lesser, but still increased, greenhouse effect over that time span (Dickinson & Cicerone 1986). Molecule for molecule, CH_4 traps 25 times as much thermal radiation as CO_2 and its concentration is increasing more than twice as fast. The evidence from Cape Grim is that atmospheric CH_4 is increasing at about 1% per year (Fig. 1 and Pearman 1988). Ice-core data show that its concentration has increased from a pre-industrial level of 650 ppbv to a present level of 1650 ppbv in the southern hemisphere (Pearman & Fraser 1988).

The biggest sink for CH_4 appears to be photo-chemical oxidation by OH radicals in the troposphere and stratosphere, estimated by Pearman & Fraser (1988) to be 400–600 Tg y^{-1}, with a much smaller biological sink in microorganisms in aerobic soils, estimated at 32 Tg y^{-1} by Melillo *et al.* (1989). However, evidence is emerging to suggest both that OH levels in the atmosphere may be decreasing due to increasing concentrations of CH_4 itself and increasing carbon monoxide, and that CH_4 oxidation in soils is being suppressed in systems which receive high inputs of nitrogen, either through fertilisation or from deposition from the atmosphere (Pearman & Fraser 1988; Melillo *et al.* 1989; Steudler *et al.* 1989). In soils, it may be that the extra nitrogen inputs have a two-fold greenhouse effect: lower CH_4 destruction is accompanied by higher production of N_2O due to competition between carbon and nitrogen sources for oxidising microorganisms (Melillo *et al.* 1989).

Unlike N_2O, the candidate sources for CH_4 are known fairly well, although individual strengths are uncertain. A recent estimate gives a global emission of 407–672 Tg y^{-1} (Schütz & Seiler 1989), with 286–495 Tg coming from biogenic sources (mostly rice paddies, natural wetlands, the digestive tracts of animals and termites, and organic wastes), and 121–177 Tg from nonbiogenic sources (biomass burning, natural gas and coal mining). In soil systems, the CH_4 is produced microbially in anaerobic conditions, but as for N_2O, the net emission is the result of a complex chain of processes of production and consumption. It has been estimated that in rice paddies, for instance, only about 20% of the CH_4 produced is emitted to the atmosphere; the rest is oxidised (Holzapfel-Pschorn *et al.* 1986; Schütz *et al.* 1989).

Methane produced in wetland soils and sediments can follow a number of pathways to the overlying water and air. These include molecular diffusion or pressurised ventilation into the water and diffusion of the dissolved CH_4 across the air-water interface, ebullition of gas bubbles, or vascular transport through the aerenchyma tissues of plants. Ebullition seems the dominant mechanism in shallow water bodies and wetlands, but vascular transport appears to be the main pathway in rice paddies or wetland plant communities (Holzapfel-Pschorn *et al.* 1986; Schütz *et al.* 1989). Oxygen diffuses from the atmosphere through the vascular system into the roots and the rhizosphere, and CH_4 diffuses back along the same pathway. Emission from the plant does not seem to be under stomatal control, at least in rice plants, since emissions continue at night when they have appeared to be largest (Fig. 3 and Seiler *et al.* 1984). The emissions seemed to be in phase with the temperature of the surface soil, in keeping with the demonstrated temperature dependence of methanogenesis (Seiler *et al.* 1984). In other studies, however, that correlation has not been evident and two daily maxima have been observed in emission rates (Schütz *et al.* 1989).

Rice paddies are now considered to be the most important individual source of CH_4 at 22% (Schütz & Seiler 1989), but almost all our knowledge of emissions from rice paddies and wetlands, the latter estimated at 20%, comes from a handful of investigations employing chamber measurements backed up by some isotopic evidence involving ^{13}C and ^2H (Wahlen *et al.* 1989). As for N_2O, corroborative measurements which integrate over large areas and do not disturb the natural environment are much needed. Micrometeorological measurements represent one such alternative, but it is only now that sufficiently sensitive gas sensors are becoming available. There are great expectations for eddy-correlation techniques employing tunable diode lasers or gas lasers (He–Ne). The first measurements with these techniques are now appearing, but the technology is likely to be under development and to require specialist operators for some years yet. (Laser techniques are also under development for eddy-correlation measurement of N_2O fluxes, but the technology is not so advanced.)

An alternative micrometeorological technique for measuring CH_4 emissions, which employs existing technology, has been developed recently (Denmead & Freney 1990[1]). Based on a conventional aerodynamic approach, it uses a non-dispersive infra-red gas analyser to measure the vertical gradient of atmospheric CH_4 concen-

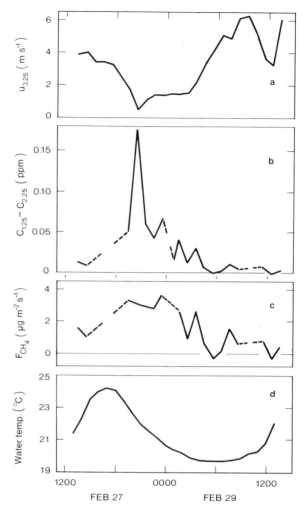

Fig. 3. Diurnal courses of: (a) wind speed at 3.25 m above water surface in rice field at Griffith, NSW; (b) corresponding differences in atmospheric methane concentration between heights of 1.25 m and 2.25 m; (c) calculated flux of methane from rice field; (d) temperature of flood water. From Denmead & Freney (1990); see footnote 1.

tration. Figure 3 shows an example of its use to measure CH_4 emissions from an Australian rice field. The success of the method hinges on how accurately the CH_4 gradient can be determined. The magnitude of the latter depends on the emis-

[1] O. T. Denmead & J. R. Freney, 1990. Micrometeorological measurement of methane emissions from flooded rice, Poster paper, SCOPE/IGBP Workshop on Trace Gas Exchange in a Global Perspective, Feb., 19–23, 1990, Sigtuna, Sweden.

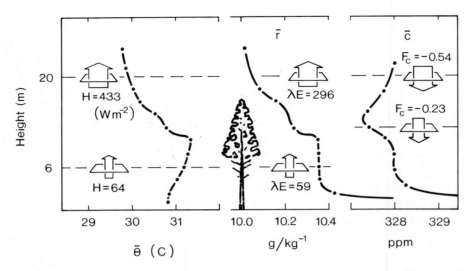

Fig. 4. Examples of apparent counter-gradient or zero-gradient transport in a pine forest 16 m high, from Denmead & Bradley (1985). The figure depicts mean profiles (over a period of one hour near noon) of temperature, $\bar{\theta}$, water vapour mixing ratio, $\bar{r}$, and CO_2 concentration, $\bar{c}$, and corresponding vertical flux densities of sensible heat, H (W m^{-2}), latent heat, λE (W m^{-2}), and CO_2, F_c (mg m^{-2} s^{-1}) at different heights. The flux densities were measured by eddy correlation. The flux of sensible heat in the bottom of the canopy was upwards, away from the forest floor, even though the gradient of temperature would suggest downward transport. At the same level, there was a large upward flux of water vapour, but no humidity gradient. And at mid-canopy, there was a substantial downward flux of CO_2, apparently against the concentration gradient. Above the canopy, the directions of the fluxes accord with the signs of the gradients. Gradient-diffusion concepts fail within the canopy space because most of the transport occurs at infrequent intervals when large eddies penetrate the canopy from above, rather than by steady diffusion along the mean concentration gradient.

sion rate and the wind speed. By day, with typical wind speeds of 4 m s^{-1}, gradients over the rice paddy were usually between 10^{-3} and 10^{-2} ppmv m^{-1}, and were sometimes difficult to distinguish from the system noise. By night, with lower wind speeds and apparently larger fluxes, gradients were almost an order of magnitude greater, but wind speeds were more difficult to determine accurately (Fig. 3a, 3b). Despite these difficulties, the approach is a feasible alternative to chambers and refinements in technique promise to make it a very useful tool for making nondisturbing, continuous measurements of CH_4 emissions on horizontal scales of 100 m or more. In this instance, the magnitudes of the emissions and their diurnal pattern (Fig. 3c) were remarkably similar to those measured elsewhere with chambers, including the occurrence of a night-time maximum lagging behind water temperature (Fig. 3d), as observed by Seiler *et al.* (1984).

Carbon dioxide

The baseline observations from Cape Grim show unequivocally that the atmospheric CO_2 concentration in Australia is increasing (at 0.4% or about 1.5 ppmv per year), and that the present concentration is close to 350 ppmv (Fig. 1 and Pearman 1988). Globally, increased CO_2 concentrations account for 49% of the anticipated greenhouse warming. Over 80% of the increase is due to fossil fuels and the rest to deforestation and biomass burning (Dickinson & Cicerone 1986). It is estimated that in 1989, man-made emissions of CO_2 were 5.7 Gt C due to fossil-fuel burning and 1.9 Gt C due to deforestation. Presently, about 40% of the emissions stay in the atmosphere; the remainder is taken up by oceans and possibly land ecosystems. However, the current estimates of global source and sink strengths do not balance: the atmospheric increase is less rapid than expected from carbon cycle models. The imbalance is substantial, about 1 Gt C y^{-1} or 13% of the estimated input. It appears that (i) the uptake of CO_2 by the oceans is underestimated, or (ii) there are important unidentified processes in terrestrial ecosystems that can sequester CO_2, or (iii) the release of CO_2 from tropical deforestation is being overestimated. The problem is in urgent need of research.

Carbon dioxide has special significance in the context of this paper because as yet, it is the only major greenhouse gas for which exchanges between the soil-plant environment and the atmosphere can be measured routinely by nondisturbing micrometeorological techniques, although there are still some substantial difficulties (Denmead 1984; Denmead & Bradley 1985, 1989). There are, for instance, few reliable measurements of CO_2 exchange at night when micrometeorological approaches, through either gradient-diffusion or eddy-correlation, are difficult. Further, the inapplicability of conventional gradient-diffusion approaches within the canopy air-space makes the identification of sources and sinks there, including identification of the separate contributions of soil and plant to the net CO_2 exchange, a difficult task also (Fig. 4 and Denmead & Bradley 1985, 1987). While eddy-correlation is still a feasible alternative procedure for measuring the CO_2 flux (or fluxes of other trace gases) within the canopy, eddy-correlation measurements made in situ must be corrected for density effects arising from the simultaneous transfer of heat and water vapour (Webb *et al.* 1980; Leuning *et al.* 1982). For CO_2 and the other trace gases considered here, these corrections may be substantial, sometimes as large as or larger than the true gas flux (Wesely *et al.* 1989). Figure 5, from Denmead & Bradley (1989), shows the magnitudes of these corrections for eddy-correlation measurements of the CO_2 flux over a forest.

A new micrometeorological approach based on Lagrangian analysis of scalar transport has been developed in the last few years, which has great potential for identifying source/sink distributions of heat, water vapour, CO_2, and other trace gases in plant canopies (Raupach 1989a, 1989b). The

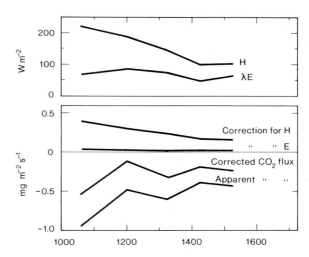

Fig. 5. Fluxes of sensible heat, H, and latent heat, λE, from a pine forest, and corrections to the apparent eddy-flux of CO_2 arising from associated density effects. From Denmead & Bradley (1989).

theory describes the space and time trajectories of 'marked' parcels of air originating at different points within the canopy. An inverse technique uses measurements of the turbulent wind field and the mean concentration profiles of the scalars to infer their source/sink distributions. Figure 6, from Raupach (1989b), illustrates the potential of the technique. In that instance, it was used successfully to infer the strength of an elevated heat source within a model canopy from measurements of the temperature profile. In the real world, it offers the possibility of inferring emissions of trace gases from leaves and soil within crops and forests from relatively simple measurements.

Some intriguing problems arise in assessing the impacts of increased atmospheric CO_2 concentrations on aspects of plant growth such as CO_2 assimilation, transpiration, and water-use efficiency (molecules of carbon assimilated per molecule of water lost through transpiration). Because of the rapid dilution of CO_2 released into the well-mixed natural atmosphere, the only feasible way to study these questions is in enclosures, but there are real difficulties in extrapolating results from that situation to the real world. First, enclosures change the microclimate and answers

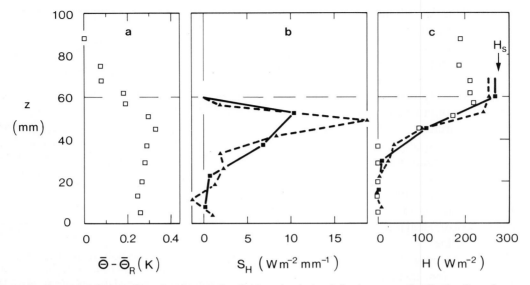

Fig. 6. Test of an inverse technique (based on Lagrangian fluid mechanics) to infer the source distribution from the concentration profile in an experiment on the dispersion of trace heat from an elevated plane source in a wind tunnel model canopy, from Raupach (1989b). The canopy was 75 mm high and electric heating wires at 60 mm gave a heat input there, H_s, of 250 W m^{-2}. (a) Measured temperature profile; (b) inferred profile of heat source density, S_H; (c) inferred cumulative heat flux profiles. Solid lines for calculations based on a 4-layer canopy; dashed lines for 8 layers. Squares in (c) are eddy-correlation measurements of the heat flux.

obtained in them must therefore be circumspect. This point is illustrated below by reference to recent work on eucalypt trees. Previously, Dunin & Greenwood (1986) had compared measurements of evaporation from eucalypt trees in a popular type of open-top, ventilated chamber (Greenwood et al. 1982) with micrometeorological measurements of evaporation from the surrounding eucalypt forest, and Leuning & Foster (1990) had made a detailed examination of the microclimate and transpiration of single leaves of eucalypt trees, inside and outside such a chamber. Both investigations concluded that evaporation rates were little affected by enclosure. However, Leuning & Foster (1990) showed clearly that the microclimate was changed very much: light in the chamber was diffuse rather than direct; net radiation was reduced by up to 50%; leaf boundary layer resistances were changed significantly. An important reason for the insensitivity of evaporation to enclosure seems to be that transpiration from eucalypt leaves is controlled more strongly by stomatal conductance and vapour pressure deficit than radiation level, and

those first two parameters were changed very little.

The light climate has a greater influence on CO_2 assimilation, however, and when Denmead et al. (1987)[2] examined CO_2 uptake by trees in the chamber and the forest, again using micrometeorological techniques for the latter, they found that the effects of enclosure were marked. Daily patterns of evaporation and CO_2 uptake in chamber and forest on two mostly clear and one partly cloudy day are shown in Fig. 7. On all days, evaporation rates from the chamber and the surrounding forest were quite comparable, as demonstrated by Dunin & Greenwood (1986). On the clear days, however, the assimilation rates were markedly different (Figs. 7a and 7b). Assimilation in the chamber proceeded at a high,

---

[2] O. T. Denmead, F. X. Dunin, S. C. Wong & E. A. N. Greenwood, 1987. Photosynthesis by trees in tents and forests. Poster paper. International Symposium on Flow and Transport in the Natural Environment: Advances and Applications, Aug. 31–Sep. 4 1987, Australian Academy of Science, Canberra, Australia.

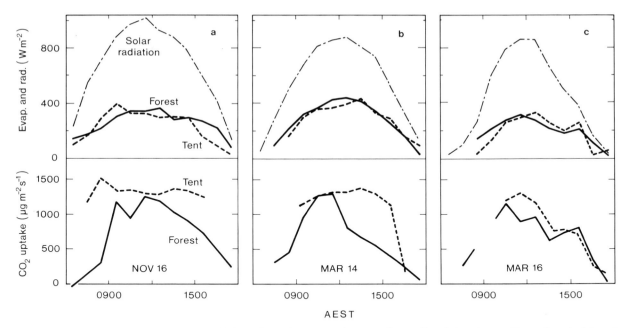

Fig. 7. Rates of evaporation and CO_2 assimilation of *Eucalyptus maculata* trees in ventilated chamber and in undisturbed forest. From Denmead *et al.* (1987); see footnote 2. AEST is Australian Eastern Standard Time.

constant rate for the main hours of sunlight, while in the unenclosed forest, it rose to a peak around mid-day and then decreased. Only on the partly cloudy day, Fig. 7c, were the assimilation patterns similar. This behaviour suggests that in direct sunlight, the light climate is more favourable for photosynthesis inside the chamber than outside. The difference appears to be that the plastic wall of the chamber acts as a giant diffuser, illuminating leaves throughout the canopy more uniformly than is the case outside, as noted by Leuning & Foster (1990).

Wong & Dunin (1987) found that the assimilation rates of individual leaves of *Eucalyptus maculata*, the dominant forest species in the CO_2 comparison, were about 90% of their maximum at about half full sunlight, and the same is true, near enough, for the trees in the chamber (Fig. 8). However, for the forest, illuminated mostly by direct sunlight, assimilation rates increase steadily with increasing solar radiation, attaining the same rate as the chamber only at the highest sun angles and deepest radiation penetration (Fig. 8). Although evaporation may not be much affected by enclosure, assimilation certainly can be. The

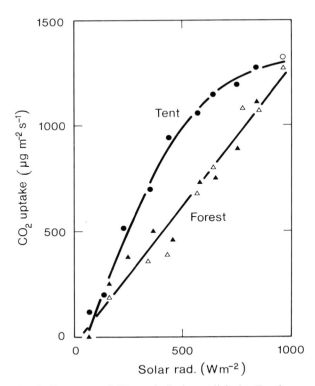

Fig. 8. Response of CO_2 assimilation to light by *Eucalyptus maculata* trees in ventilated chamber and in undisturbed forest. From Denmead *et al.* (1987); see footnote 2.

use of chambers like these for exploring questions of water use and productivity must be circumspect. As Fig. 7 shows, the act of enclosure, by itself, can change water use efficiency by as much as 100%.

A second problem associated with CO_2 enrichment experiments in enclosures is more complex. A number of physiological investigations indicate that CO_2 enrichment usually leads to decreased stomatal conductances in plant leaves. Doubling atmospheric CO_2 concentrations has often reduced conductance by about 60% (Rosenberg *et al.* 1989). In turn, we might expect this to reduce transpiration and increase the surface temperature and heat exchange of the foliage. In the real world, these new transpiration and heat exchange rates will eventually change the temperature and humidity of the air in contact with the leaves. Those changes then feed back on transpiration, and so on, until a stable microclimate is attained, in which temperature, humidity, heat and vapour exchange are all in equilibrium. Reaching that final state requires distances of many meters (Philip 1987; Cowan 1988).

In enclosures, the residence time of the air is short and equilibrium is not attained. Assuming no effects on the light climate, our enrichment experiment might well predict the right CO_2 exchange for tomorrow's environment, but the wrong transpiration rate because equilibrium has not yet been attained, and we will make false inferences about the probable water-use efficiency. This is a classic problem in advection and is amenable to biophysical and mathematical analysis, at least in a simplified way. Current work on the topic within the CSIRO Centre for Environmental Mechanics aims at predicting how water-use efficiency will change with distance as one moves from an area the size of a glasshouse, say, to a wheat belt hundreds of kilometres in extent.

Concluding remarks

The only confident assertion that we can make about the greenhouse gases is that their concentrations are increasing. For N_2O and CH_4, the highest and lowest estimates of most of the individual source strengths differ by a factor of 2. Even for CO_2, which has been studied longest and whose fluxes can be measured best, the difference between the estimated emissions and the calculated ocean sink exceeds the observed atmospheric increase by about 50%, which implies a large 'missing' sink.

Much of the uncertainty arises from our inability to measure fluxes over sufficiently large areas and long times, the heterogeneity of the source distributions, and the lack of statistically sound procedures for extrapolating local measurements to large space and time scales. There is much work to be done in improving flux measurement techniques, and devising appropriate sampling schemes and systems of inference. These latter should include modelling at the process level, stratified sampling, and synthesis with the help of geographic information systems.

As well, there is scope for experiments in which ecosystems are manipulated on various scales in order to 'preview' effects of global climate change. These will probably require use of enclosures, but as seen for CO_2, elucidating the effects of the enclosure itself on the microclimate and on relevant biological processes must be an important part of the exercise.

References

Bouwman, A. F. (ed.) 1990. Soils and the greenhouse effect. John Wiley & Sons Ltd., Chichester.

Cowan, I. R. 1988. Stomatal physiology and gas exchange in the field. In: Steffen, W. L. & Denmead, O.T. (eds.), Flow and transport in the natural environment: Advances and applications, pp. 160–172. Springer-Verlag, Heidelberg.

Crutzen, P. J., Heidt, L. E., Krasnec, J. P., Pollock, W. H. & Seiler, W. 1979. Biomass burning as a source of atmospheric gases CO_2, H_2, H_2O, NO, CH_3Cl and COS. Nature 282: 253–256.

Denmead, O. T. 1979. Chamber systems for measuring nitrous oxide emission from soils in the field. Soil Sci. Soc. Am. J. 43: 89–95.

Denmead, O. T. 1984. Plant physiological methods for studying evapotranspiration: problems of telling the forest from the trees. Agric. Water Manage. 8: 167–189.

Denmead, O. T. & Bradley, E. F. 1985. Flux-gradient relationships in a forest canopy. In: Hutchinson, B. A. &

Hicks, B. B. (eds.), The forest-atmosphere interaction, pp. 421–442. D. Reidel Publishing Co., Dordrecht.

Denmead, O. T. & Bradley, E. F. 1987. On scalar transport in plant canopies. Irrig. Sci. 8: 131–149.

Denmead, O. T. & Bradley, E. F. 1989. Eddy-correlation measurement of the CO_2 flux in plant canopies. Proc. Fourth Australasian Conf. on Heat and Mass Transfer, Christchurch, N.Z., pp. 183–192. Secretariat, Fourth Australasian Conf. on Heat and Mass Transfer, Christchurch.

Dickinson, R. E. & Cicerone, R. J. 1986. Future global warming from atmospheric trace gases. Nature 319: 109–115.

Dunin, F. X. & Greenwood, E. A. N. 1986. Evaluation of the ventilated chamber for measuring evaporation from a forest. Hydrol. Proc. 1: 47–62.

Firestone, M. K. & Davidson, E. A. 1989. Microbiological basis of NO and N_2O production and consumption in soil. In: Andreae, M. O. & Schimel, D. S. (eds.), Exchange of trace gases between terrestrial ecosystems and the atmosphere, pp. 7–21. John Wiley & Sons Ltd., Chichester.

Freney, J. R., Simpson, J. R., Denmead, O. T., Muirhead, W. A. & Leuning, R. 1985. Transformations and transfers of nitrogen after irrigating a cracking clay soil with a urea solution. Aust. J. Agric. Res. 36: 685–694.

Galbally, I. E. & Roy, C. R. 1978. Loss of fixed nitrogen from soils by nitric oxide exhalation. Nature 275: 734–735.

Galbally, I. E., Roy, C. R., Elsworth, C. M. & Rabich, H. A. H. 1985. The measurement of nitrogen oxide (NO, NO_2) exchange over plant/soil surfaces. CSIRO Australia Div. Atmos. Res. Tech. Paper No. 8, 23 pp.

Greenwood, E. A. N., Beresford, J. D., Klein, L. & Watson, G. D. 1982. Evaporation from vegetation in landscapes developing secondary salinity using the ventilated chamber technique. J. Hydrol. 58: 357–366.

Holzapfel-Pschorn, A., Conrad, R. & Seiler, W. 1986. Effects of vegetation on the emission of methane from submerged paddy soil. Plant Soil 92: 223–233.

Keller, M., Kaplan, W. A. & Wofsy, S. C. 1986. Emissions of N_2O, CH_4 and CO_2 from tropical forest soils. J. Geophys. Res. 91: 11,791–11,802.

Leuning, R., Denmead, O. T., Lang, A. R. G. & Ohtaki, E. 1982. Effects of heat and water vapour transport on eddy covariance measurement of CO_2 fluxes. Boundary-Layer Meteorol. 23: 255–258.

Leuning, R., & Foster, I. 1990. Estimation of transpiration by single trees: Comparison of a ventilated chamber, leaf energy budgets and a combination equation. Agric. For. Meteorol. (in press).

Melillo, J. M., Steudler, P. A., Aber, J. D. & Bowden, R. D. 1989. Atmospheric deposition and nutrient cycling. In: Andreae, M O. & Schimel, D. S. (eds.), Exchange of trace gases between terrestrial ecosystems and the atmosphere, pp. 263–280. John Wiley & Sons Ltd., Chichester.

Mosier, A. R. 1989. Chamber and isotope techniques. In: Andreae, M. O. & Schimel, D. S. (eds.), Exchange of trace gases between terrestrial ecosystems and the atmosphere,

pp. 175–187. John Wiley & Sons Ltd., Chichester.

Pearce, F. 1989. Methane: the hidden greenhouse gas. New Scientist, 6 May 1989, 19–23.

Pearman, G. I. 1988. Greenhouse gases: evidence for atmospheric changes and anthropogenic causes. In: Pearman, G. I. (ed.), Greenhouse planning for climate change, pp. 3–21, CSIRO, Australia.

Pearman, G. I., Etheridge, F., deSilva, F. & Fraser, P. J. 1986. Evidence of changing concentrations of atmospheric CO_2, N_2O and CH_4 from air bubbles in Antarctic ice. Nature 320: 248–250.

Pearman, G. I. & Fraser, P. J. 1988. Sources of increased methane. Nature 332: 482–490.

Philip, J. R. 1987. Advection, evaporation, and surface resistance. Irrig. Sci. 8: 101–114.

Raupach, M. R. 1989a. A practical Lagrangian method for relating scalar concentrations to source distributions in vegetation canopies. Q. J. R. Meteorol. Soc. 115: 609–632.

Raupach, M. R. 1989b. Applying Lagrangian fluid mechanics to infer scalar source distributions from concentration profiles in plant canopies. Agric. For. Meteorol. 47: 85–108.

Robertson, G. P., Andreae, M. O., Bingemer, H. G., Crutzen, P. J., Delmas, R. A., Duyzer, J. H., Fung, I., Harriss, R. C., Kanakidou, M., Keller, M., Melillo, J. M. & Zavaria, G. A. 1989. Group report – Trace gas exchange and the chemical and physical climate: critical interactions. In: Andreae, M. O. & Schimel, D. S. (eds.), Exchange of trace gases between terrestrial ecosystems and the atmosphere, pp. 303–320. John Wiley & Sons Ltd., Chichester.

Rosenberg, N. J., McKenney, M. S. & Martin, P. 1989. Evapotranspiration in a greenhouse-warmed world: a review and a simulation. Agric. For. Meteorol. 47: 303–320.

Rosswall, T., Bak, F., Baldocchi, D., Cicerone, R. J., Conrad, R., Ehhalt, D. H., Firestone, M. K., Galbally, I. E., Galchenko, V. F., Groffman, P. M., Papen, H., Reeburgh, W. S. & Sanhueza, E. 1989. Group report – What regulates production and consumption of trace gases in ecosystems: biology or physicochemistry? In: Andreae, M. O. & Schimel, D. S. (eds.), Exchange of trace gases between terrestrial ecosystems and the atmosphere, pp. 73–95. John Wiley & Sons Ltd., Chichester.

Schütz, H., Holzapfel-Pschorn, A., Conrad, R., Rennenberg, H. & Seiler, W. 1989. A 3-year continuous record on the influence of daytime, season, and fertilizer treatment on methane emission rates from an Italian rice paddy. J. Geophys. Res. 94: 16,405–16,416.

Schütz, H. & Seiler, H. 1989. Methane flux measurements: methods and results. In: Andreae, M. O. & Schimel, D. S. (eds.), Exchange of trace gases between terrestrial ecosystems and the atmosphere, pp. 209–228. John Wiley & Sons Ltd., Chichester.

Seiler, W., Holzapfel-Pschorn, R., Conrad, R. & Scharffe, D. 1984. Methane emissions from rice paddies. J. Atmos. Chem. 1: 241–268.

Steudler, P. A., Bowden, R. D., Melillo, J. M. & Aber, J. D. 1989. Influence of nitrogen fertilization on methane uptake in temperate forest soils. Nature 341: 314–316.

Stewart, J. W. B., Asselman, I., Bouwman, A. F., Desjardins, R. L., Hicks, B. B., Matson, P. A., Rodhe, H., Schimel, D. S., Svensson, B. H., Wassmann, R., Whiticar, M. J. & Yang, W.-X. 1989. Group report – Extrapolation of flux measurements to regional and global scales. In: Andreae, M. O. & Schimel, D. S. (eds.), Exchange of trace gases between terrestrial ecosystems and the atmosphere, pp. 155–174. John Wiley & Sons Ltd., Chichester.

Vitousek, P. M., Denmead, O. T., Fowler, D., Johansson, C., Kesselmeier, J., Klemedtsson, L., Meixner, F. X., Mosier, A. R., Schütz, H., Stal, L. J. & Wahlen, M. 1989. Group report – What are the relative roles of biological production, micrometeorology, and photochemistry in controlling the flux of trace gases between terrestrial ecosystems and the atmosphere? In: Andreae, M. O. & Schimel, D. S. (eds.), Exchange of trace gases between terrestrial ecosystems and the atmosphere, pp. 249–261. John Wiley & Sons Ltd., Chichester.

Wahlen, M., Tanaka, N., Henry, R., Deck, B., Seglen, J., Vogel, J. S., Southon, J., Shermesh, A., Fairbanks, R. & Broecker, W. 1989. Carbon-14 in methane sources and in atmospheric methane: the contribution from fossil carbon. Science 245: 286–290.

Walker, B. H., Young, M. D., Parslow, J. S., Cocks, K. D., Fleming, P. M., Margules, C. R. & Landsburg, J. J. 1989. Global climate change and Australia: effects on renewable natural resources. In: Global climatic change – issues for Australia, Papers presented at the first meeting of the Prime Minister's Science Council 6 October 1989, pp. 31–76, Australian Government Publishing Service Press, Canberra.

Webb, E. K., Pearman, G. I. & Leuning, R. 1980. Correction of flux measurements for density effects due to heat and water vapour transfer. Q. J. R. Meteorol. Soc. 106: 85–100.

Wesely, M. L., Lenschow, D. H. & Denmead, O. T. 1989. Flux measurement techniques. In: Lenschow, D. H. & Hicks, B. B. (eds.), Global tropospheric chemistry – chemical fluxes in the global atmosphere, pp. 31–46. National Center for Atmospheric Research, Boulder.

Wong, S. C. & Dunin, F. X. 1987. Photosynthesis and transpiration of trees in a Eucalypt forest stand. Aust. J. Plant Physiol. 14: 619–632.

Modelling

Introduction

The models and methods of modelling semi-arid regions reviewed here are primarily atmospheric though they span several spatial scales. First, Hunt describes the numerical simulation and prediction of drought using Atmospheric General Circulation Models (AGCMs). Raupach discusses the interaction between various types of surfaces and vegetation. This type of modelling methodology has considerable value in predicting the effects of surface change and surface energy balance at a variety of spatial scales. Cleugh uses a similar approach linked with field measurements to predict the evaporation from a semi-arid Australian catchment. These two papers provide a coherent means of attempting to predict the consequences of land cover change on evaporation, and thereby identifying areas likely to undergo excessive moisture loss which might lead to additional land degradation. The chapter by Pitman discusses a possible improvement to the hydrological regimes simulated by AGCMs based on the incorporation of the effects of sub-gridscale precipitation. Finally, Henderson-Sellers determines the potential for interactive biosphere models in AGCMs. In the context of the title of the book, these preliminary results suggest that the impact of land-clearing to date is very much greater than probable future greenhouse-induced desertification. It is shown that while hopes that interactive biosphere models may eventually prove successful, current AGCMs need considerable improvement before an interactive biosphere can be coupled to a climate model.

Vegetatio **91**: 89–103, 1991.
A. Henderson-Sellers and A. J. Pitman (eds).
Vegetation and climate interactions in semi-arid regions.
© 1991 *Kluwer Academic Publishers. Printed in Belgium.*

The simulation and prediction of drought

B. G. Hunt
CSIRO, Division of Atmospheric Research, PMB No. 1, Mordialloc, Vic. 3195, Australia

Accepted 24.8.1990

Abstract

Current capabilities for simulating and predicting drought and other rainfall anomalies are reviewed. Simulation of drought requires the insertion of suitable precursors into the model. The principal precursor is identified to be sea surface temperature anomalies, but other possibilities are discussed. Model global climatic models are able to simulate with commendable realism most features of observed drought, as illustrated by results from a number of such models. Importantly, the models also provide insight into the mechanisms responsible for the rainfall perturbations.

Both statistical and deterministic drought prediction methods are considered. The former can be used to make useful predictions for limited timeframes, but require separate relations to be derived for each region considered. A variety of statistical methods have now been developed, and some are being used operationally with mixed success. Deterministic drought prediction methods are still in their infancy, but have many attractions being physically based. In addition their ability to make predictions for the normal range of climatic variables usually required for practical utility provides a considerable advantage over statistical methods. The major requirement for accurate rainfall predictions is a good prediction of the spatial and temporal variability of the precursor. Examples of drought predictions are provided to illustrate the potential of the method. The numerous research problems which still have to be resolved are noted.

Introduction

Of all the natural environmental hazards on Earth, drought is perhaps the most insidious. By its nature it develops slowly, frequently occupying vast areas and persisting for lengthy periods. In addition to the loss of life and destruction of vegetation, serious damage can also occur to topsoil under drought conditions, thus resulting in enhanced desertification. Despite the widespread and repetitive nature of drought, research into its origins and precursor mechanisms has been surprisingly limited.

This low-level of interest in drought research appears to exist for two reasons. The first and principal one is that drought is not a major problem in western economies, and has therefore not excited a great deal of interest amongst atmospheric scientists. The second more deplorable reason is that drought prediction has generally been considered to be 'too difficult', and consequently does not justify extensive support. This lack of support has tended to produce a self-fulfilling prophecy, in that drought research has existed at a level such that the 'difficult' problems could not be tackled adequately. As a result, drought research has received only sporadic encouragement, normally at times of major drought episodes, and, in general, much research connected with drought has been incidental to other climatic problems. The possibility of deterministic drought predictions involving large-scale climatic models has been rather poorly regarded, until quite recently.

Fortunately, the current international concern with climatic problems, and the consequent emphasis on developing more adequate observational bases and climatic models, has greatly improved the environment for drought research. However, many major scientific problems exist as regards drought research in general, and drought prediction in particular. Hence this branch of climatic research remains one with major challenges to be resolved, and is undoubtedly one of the most exciting areas of science today. The status of drought research and prediction and the challenges ahead are reviewed in the subsequent sections. General articles on the problem of drought have been produced by Namias (1978), Landsberg (1982), Nicholls (1985), Rasmusson (1987) & Druyan (1989).

The simulation of drought

The simulation and deterministic prediction of drought require large-scale global climatic models. A complete model of the climatic system would incorporate atmospheric, oceanic, cryospheric and biospheric processes, each adequately represented and capable of mutual interactions in the spatial and temporal domains. Realistically, no such complete models exist at present, hence sub-sets have to be used for current experiments. In practice, both simulation and prediction of drought are undertaken using separate atmospheric and oceanic models as described below.

Model description

Typically an atmospheric model will be global in extent with between 2000 and 4000 gridpoints used to represent climatic variables, wind, temperature, specific humidity etc, for a given vertical level in the model, and with 2–15 such levels. This type of model would use the so-called primitive equations to predict the time variations of the fundamental variables, zonal and meridional winds, temperature, specific humidity and pressure, at each gridpoint. A typical timestep for these predictions is 10–30 minutes. From these variables subsidiary climatic properties are diagnosed such as cloud cover, convection, radiative heating and cooling, soil moisture, evaporation etc. Usually the sea surface temperature (SST) is specified at appropriate oceanic gridpoints from climatology held at monthly intervals, thus permitting daily values to be interpolated as required. The models can be integrated forward in the seasonal mode, or if desired for a fixed month of the year. Increasingly, models include the diurnal variation of the sun, and thus simulate the processes occurring during the daily cycle at the frequency of the chosen timestep. In order to commence a model experiment initial conditions are needed for the predicted variables, and these are obtained from observation or another model experiment.

In the case of oceanic models using the primitive equations, the procedures are similar except that the equation for specific humidity is replaced by an equation for salinity. Problems specific to the oceans occur because of land-sea boundaries, the need to have a finer horizontal resolution, and the difficulty of representing the physical processes because of our relative ignorance of the oceans. The much longer timescales associated with deep oceanic circulations also create computational problems.

Additional problems arise when the atmospheric and oceanic models are coupled together, as slow drifts away from observations occur. While procedures are available to correct for these situations, no satisfactory simulation of the climate with a coupled model has been made to date.

In practice, sub-sets of both the atmospheric and oceanic models may be used for particular studies as discussed below.

Drought precursors

One of the most challenging aspects of drought research is the identification and evaluation of the precursors responsible for the initiation, development and cessation of drought.

The most widespread and effective drought precursor is undoubtedly large-scale SST anomalies, Streten (1981), Nicholls (1985), Rasmusson (1987) and Owen & Ward (1989). By far the best known example of an SST anomaly is the El Niño phenomenon which involves a warming of 2–4 °C in the central and eastern tropical Pacific Ocean, Fig. 1. The resulting widespread rainfall anomalies, both droughts and enhanced rainfall, are shown in Fig. 2. This is a schematic figure based on the composite of a number of El Niño events by Ropelewski & Halpert (1987). These events are the single most important global climatic perturbation, and are a major cause of observed inter-annual variability. The opposite (cold) phase of El Niño events is also the precursor of widespread rainfall anomalies, involving enhanced rainfall over many regions, Ropelewski & Halpert (1989) and Kiladis & Diaz (1989). The study of this phase of the events is of equal importance as that of the more-intensely studied warm phase. In this regard, the suggestion by Trenberth *et al.* (1988) that the 1988 drought in the USA was attributable to the SST anomalies in the Pacific Ocean, associated with the change from the warm to the cold phase of the latest El Niño event, emphasises the growing appreciation of the role of SST anomalies as drought precursors. Since El Niño events are potentially predictable, Cane *et al.* (1986), this implies that the associated rainfall perturbations are also predictable, assuming that atmospheric models respond adequately to the SST anomalies.

Similarly, SST anomalies in the Atlantic Ocean have been identified as drought precursors. For example, Markham & McLain (1977) have related droughts in north-east Brazil with anomalies in the tropical Atlantic Ocean. Perhaps more importantly, Owen & Ward (1989) have conducted model studies which suggest that the long-term rainfall decline in the Sahel starting in 1968 was due to SST changes in the Atlantic Ocean. Finally, Nicholls (1989) has associated winter rainfall variations over central Australia to SST

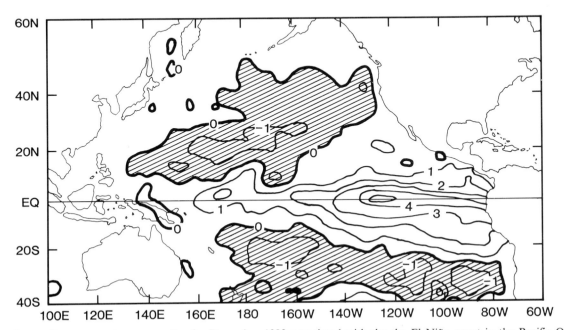

Fig. 1. Sea surface temperature anomalies for December 1982 associated with the the El Niño event in the Pacific Ocean. (Redrawn from Rasmusson & Hall, 1983). Units are in °C, with negative areas shaded. Contour interval is 1°.

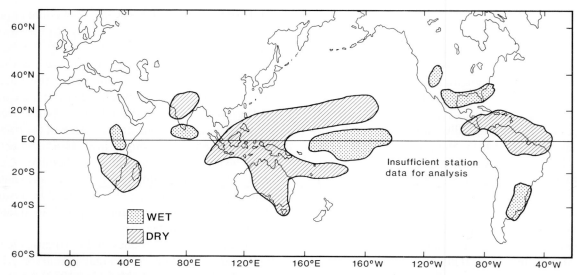

Fig. 2. Schematic diagram illustrating regions experiencing rainfall changes owing to El Niño events. Hatched areas indicate decrease, speckled areas increases. Redrawn from Ropelewski & Halpert (1987).

gradients between the Indian Ocean and Indonesia indicating the universality of SST anomalies as drought precursors. However, it needs to be stressed that these anomalies appear to be most effective when located at low altitudes. As a cautionary note it should be emphasized that it is uncertain currently whether SST anomalies other than those produced by El Niño events are predictable.

Other possible drought precursors are more difficult to quantify at this time. However, an important possibility is the interannual variability of Eurasian snowcover, which according to Barnett *et al.* (1989) can impact the timing of El Niño events. A more direct influence on drought can arise through soil moisture variations as drying conditions sometimes prevail for a year or more prior to a major drought. Such dryness preconditions and enhances the intensity of the subsequent drought and probably enlarges the area affected. It is desirable that the cause of this dryness should be identified. An interesting anthropogenetically-produced drought precursor was suggested by Charney (1975), related to the increase in surface albedo in semi-arid regions owing to over-grazing. This was subsequently evaluated in a general circulation model experiment by Charney *et al.* (1977).

A fascinating drought precursor mechanism may exist for the High Plains of the USA and other parts of the world according to Currie (1981). He has conducted careful statistical studies which identify an 18.6 year periodicity in drought related to the lunar nodal tide. The mechanistic links involved in such a climatic perturbation are not clear, but Currie (1984) has suggested in the case of the High Plains it may arise from an interaction between the resonant nodal wave and the Rocky Mountains.

Not all droughts necessarily have a distinct precursor mechanism. For example, Gordon & Hunt (1987) have discovered what they termed 'naturally-occurring drought' in a multi-annual simulation, see Fig. 3. Such drought is produced by the nonlinearities inherent in the atmospheric component of the climatic system. (Gordon & Hunt used seasonally varying climatological SST which were constant from year-to-year in their model). Such droughts are by definition unpredictable, but fortunately of local extent only. They can be readily identified, for example, in drought patterns over Australia.

Undoubtedly, other drought precursors will be identified as research proceeds, possibly associated with inter-annual variability of sea ice or sunspot cycles (?). Certainly, at the present time,

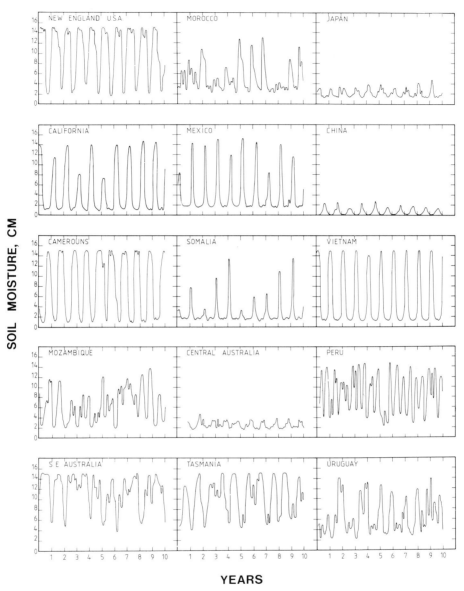

Fig. 3. The interannual variation of soil moisture for various geographical points over a 10 year period of an integration with a global climatic model. Soil moisture is a good approximation of rainfall, hence the figure effectively indicates rainfall variability. Note the very marked variability for the Somalian point in the figure which is indicative of 'naturally-occurring drought'. From Gordon & Hunt (1987).

SST anomalies would generally be considered to account for the majority of the observed variance in large-scale, interannual rainfall, and systematic modelling should be able to quantify how important other precursor mechanisms are. Such knowledge will greatly clarify the prospects for timely drought prediction.

Drought simulations

Numerous general circulation model experiments have been performed to investigate the impact of a great variety of SST anomalies (Mechoso & Lyons, 1988; Kitoh, 1988 & Pitcher *et al.* 1988). Many of these experiments have been concerned with identifying teleconnection patterns linking the SST anomaly to distant responses in climatic variables. Although changes in rainfall patterns have been presented in some of these studies, in general the objective of the experiments was not to investigate drought.

The first attempt to simulate specifically drought using a general circulation model was made by Moura & Shukla (1981). They inserted fixed SST anomalies with a maximum value of 2°C into the climatological SST distribution of their model, so as to obtain a warming in the northern tropical Atlantic Ocean and a cooling in the south tropical Atlantic Ocean. They initialised their model with observations for 1 January 1975 and ran out the experiment for 90 days. The smoothed rainfall differences they obtained for the last 60 days of their experiment are shown in Fig. 4. While the biggest rainfall changes occurred over the ocean, they obtained a decrease in rainfall over north-east Brazil and increased rainfall to the north. The latter result is consistent with the observations of Hastenrath & Heller (1977) for drought conditions in north-east Brazil. Importantly they carried out experiments to identify the mechanisms involved in the rainfall perturbations, and were able to conclude that diabatic circulations associated with the SST anomalies were the principal cause.

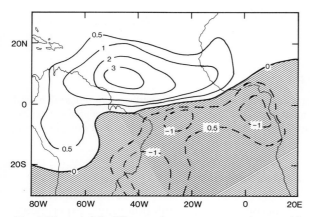

Fig. 4. The rainfall difference between an experiment with Atlantic SST anomalies and a control run averaged over sixty days. Redrawn from Moura & Shulka (1981). Units are mm d⁻¹, with negative areas shaded. Contour interval is 1 mm d⁻¹, with the 0.5 contour also shown.

The first general circulation experiment specifically designed to simulate ENSO-related drought was performed by Voice & Hunt (1984). They ran experiments for fixed January conditions and fixed SST anomalies, and obtained droughts over parts of Australia, southern Africa, South America and North America. They were able to relate the rainfall changes to perturbations induced in the low-level synoptic wind distributions. In particular they showed that while synoptic patterns were disturbed in the Australia region, where the largest forcing associated with the SST anomalies occurred, elsewhere only secondary changes were produced. In these situations the amplitudes of the synoptic patterns were altered, but not the patterns themselves. They also investigated the transient behaviour of the model's response and found very considerable variability between different 10-day periods inside the model.

Systematic studies of drought have also been carried out by the British Meteorological Office using various versions of their general circulation model, Owen & Folland (1988) and Folland *et al.* (1986). Their main area of interest has been the Sahel, and, again, the drought precursor used was SST anomalies. Folland *et al.* composited the SST for the global oceans for July to September for a number dry years in the Sahel minus the SST for a number of wet years. The resulting SST anomalies showed warmings, generally less than 1°C, for the South Atlantic and Indian Oceans and south-east Pacific Ocean, with the North Atlantic and North Pacific Oceans being colder. Folland *et al.* ran their general circulation model in perpetual July mode with these SST anomalies and obtained the rainfall changes shown in Fig. 5. Substantial rainfall reductions occurred throughout the Sahel, indicating the critical impact of the SST anomalies for drought in this region. The principal cause of the rainfall reduction was shown to be a decrease in the low-level flux of moisture from the south Atlantic into the Sahel.

The utility of using general circulation models for drought simulation and prediction is illustrated in Fig. 6, (Hunt & Gordon, 1990a). In this figure, 7-day time sequences are compared for

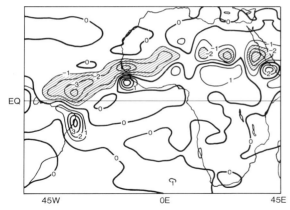

Fig. 5. The rainfall difference between an experiment with global SST anomalies and a control run based on a 180 day experiment. Redrawn from Follard *et al.* (1986). Units are mm d^{-1}, with negative areas shaded. Contour interval 1 mm d^{-1}.

control and drought simulations for an individual, inland point in southern central Australia, from a global simulation set up for fixed January conditions. In the control run very heavy rain occurred during this sequence, with related impacts on the surface temperature and insolation at the surface owing to variations in the model cloud cover. The corresponding drought run had an El Niño-type SST anomaly in the Pacific Ocean, which produced extensive drought over eastern Australia and rainfall perturbations in agreement with observation at other low latitude sites. No rain occurred in the time sequence shown in Fig. 6(b) for the drought run, as a consequence much higher surface temperatures were obtained and cloud cover was zero, as indicated by the uniformity of the insolation curve. The soil moisture and evapo-

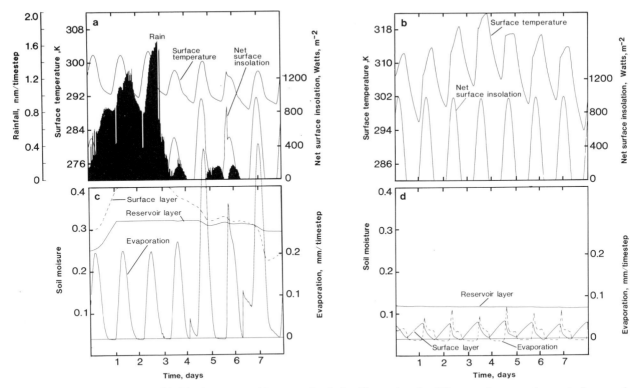

Fig. 6. Diurnal plots for days 92–99 of a perpetual January simulation illustrating the differences between the control run and an experiment with an El Niño type warm anomaly in the Pacific Ocean. The control run is shown in the left hand panels, the drought conditions of the experiment in the right hand panels. Results are for a model grid point in southern central Australia. From Hunt & Gordon (1990a).

rative flux were quite different in the two runs, Fig. 6(c) and 6(d), again indicating the information content of the model output. Note particularly the sudden enhancement of the evaporation in the control run on the first sunny day, which resulted from the higher surface insolation being able to provide the requisite energy. Clearly the type of information shown in Fig. 6 would greatly enhance the value of drought predictions to the general community, as requirements are not limited just to expected changes in rainfall.

The final example of drought simulation is concerned with the 1988 drought in the USA. Trenberth *et al.* (1988), on the basis of a simplified atmospheric model, have suggested that the precursor of this drought was the SST anomaly patterns in the Pacific Ocean during 1988. These reflected the changeover from an El Niño to an anti-El Niño situation. They showed, for a fixed anomaly appropriate to May–June 1988, that a wavetrain could be set up across the North Pacific and North America similar to that observed, and thus consistent with the occurrence of the drought.

Recently, Hunt & Gordon (1990b) have conducted global model experiments designed to simulate the US 1988 drought. They inserted the observed monthly SST anomalies for the Pacific Ocean in 1988 into their model and obtained rainfall perturbations in broad agreement with observations over North America. Figure 7 shows rainfall changes over the American continent between the equator and 60°N produced in the model for May, June, July and August 1988 and observed *extreme* changes. For each of these months rainfall reductions of up to 2 mm d^{-1} occurred over the broad area of the Great Plains, hence the *general* characteristics of the drought were quite well simulated. Detailed comparisons revealed flaws in the simulation, such as the enhanced rainfall from Florida to Mississippi in May, but the increased rain over northwest Canada was well reproduced. Similarly, in June too much rainfall occurred over the eastern USA, but the reduced rainfall in the central USA and Mexico was realistic. The July simulation was quite accurate overall, while that for August re-

produced the enhanced rainfall observed from Mexico to Venezuela.

Only qualitative comparisons have been made here as further experiments are planned, but the general impression obtained is that a fairly acceptable simulation of the US 1988 drought was obtained. A second simulation starting from a different control year produced somewhat less intense rainfall perturbations, but tended to have more realistic results for regions with increased rainfall. This may indicate that a Monte Carlo approach may be needed to cope with the sensitivity of drought simulations to initial conditions. However, initialising the model with observed data for January 1988, including soil moisture amounts, would be expected to improve further the quality of the simulations obtained.

In summary, it is clear that quite realistic simulations of drought conditions can now be obtained with current general circulation models of the atmosphere, by inserting observed SST anomalies into their climatologically defined SST distributions. These simulations are basically capable of reproducing the whole range of climatic effects associated with drought, i.e. soil moisture, surface temperature, evaporation changes etc., as well as simply the rainfall perturbations. This capability is very encouraging, not only as regards the potential for drought prediction, but also for understanding the mechanisms involved in the life cycle of typical droughts.

Drought prediction

Other than the interpretation of Pharoah's dreams by Joseph, the only accepted forms of drought prediction currently in use are based on either statistical or deterministic (mathematically formulated models) approaches.

Statistical methods

The basis of these methods is to derive from observations correlations between some index of climatic variability (SST anomalies, Southern

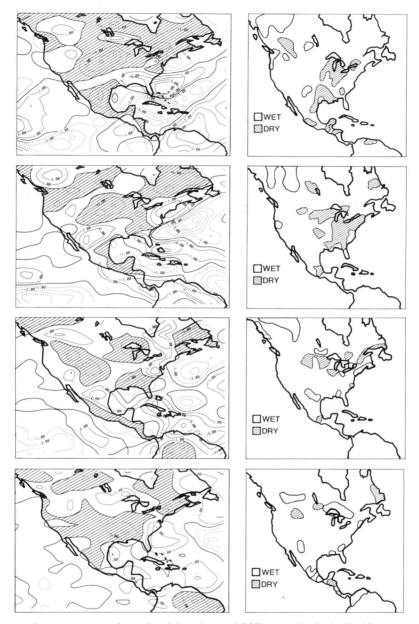

Fig. 7. Rainfall differences between an experiment involving observed SST anomalies in the Pacific Ocean in 1988 and a control run are shown for May, June, July and August in the left hand panels. Observed rainfall anomalies are shown in the right hand panels. The shaded regions for the observations define areas where estimated rainfall anomalies were within the wettest or driest 10% of climatological occurrences. Hence only extreme areas of anomalies are depicted. Units are mm d^{-1} for the model results, with negative areas shown dashed. The contour interval is 1.0 mm d^{-1}.

Oscillation Index (SOI)) and, usually, rainfall. On the basis of such correlations the current observed value of the predictor is then used to make predictions, generally for a few months ahead. As might be expected the success rate of such predictions is rather variable, with the better performing schemes being those which concentrate on selected areas, rather than large regions.

For example, Chu (1989) has shown that winter rainfall is Hawaii is highly correlated with the SOI, and can be used to predict drought conditions up to two seasons in advance with a simple regression model. This regression method does not work for rainfall predictions for non-El Niño SOI situations. Chu's results are a good example of the capability and specificity of statistical methods.

Barnett (1984) claims that the 1982–83 El Niño was inherently predictable 4–5 months in advance, using a prediction model based on regressions between wind fields in selected areas of the Pacific Ocean and SST. Other authors have developed various statistical methods for specific purposes, but rather than discuss these attention will be focussed on two methods which are in operational use.

The rainfall variability over eastern Australia is quite highly correlated with the SOI, Nicholls (1989), thus providing a framework for predictions in this region, particularly of drought, Nicholls (1985). The derived statistical relations are such that maximum correlations occur in different regions of eastern Australia in different seasons. On the basis of such correlations the Australian Bureau of Meteorology in 1989 commenced issuing rainfall predictions at monthly intervals for 3 months ahead, averaged over the predicted period. The current SOI is also used to identify earlier years, with similar values of the SOI, to add further weight to the predictions as analogues. The predictions are presented in three categories of below average, average or above average rainfall. Preliminary assessment of the forecasts for the winter and spring of 1989 indicate mixed success with the skill varying from 9 to 72% (Dr. I. Smith, Private Communication). For the five forecasts issued for the period June to October, the overall skill was 40%, against a persistence value of 44%. These predictions were for anti-El Niño conditions and it will be interesting to see how they perform under an El Niño situation.

The British Meteorological Office has now been issuing statistically-based rainfall predictions for the Sahel and north east Brazil for the last two to three years. An indication of the procedures used for the Sahel has been given by Owen & Ward (1989). One technique involves the identification of two characteristic SST anolomaly patterns (empirical orthogonal functions) relevant to Sahelian rainfall. Covariances C1 and C2 are calculated between a given SST anomaly pattern for any selected year and the characteristic patterns. A 'training period' involving a number of years is used to assign the C values to each of five equi-probable rainfall ranges, or quints, spanning between extremely dry and extremely wet. A discriminant analysis technique produces forecast equations from this information. Observed spring SST anomalies are then used to calculate current C1 and C2 values and the forecast equations determine the probability of the total rainfall for that year lying within each quint.

An alternative method developed by the British Meteorological Office for predictions of Sahelian rain (Owen & Ward 1989), involves multiple regression between standardised rainfall and C1 and C2. Further discussion of these methods and an assessment of their viability has been presented by Parker et al (1988). More complicated prediction methods, being developed for north east Brazil, are given by Ward & Folland (1990). Operational forecasts for Sahelian rainfall were made by the British Meteorological Office using their statistical techniques commencing in 1986. The forecasts were for dry conditions in both years and were substantially correct (Parker *et al.* 1988). Other statistical forecasts for north east Brazil for 1987–1989 were also substantially correct (Ward & Folland, 1990).

Clearly statistical forecasts offer considerable hope for rainfall predictions. The limitations of the techniques are that the time range for most predictions is only a few months, which restricts their utility. Also forecasts tend to be for seasons, whereas monthly values are preferable, because an overall dry season can commence with a very wet first month, as happened in north east Brazil in 1987. Currently most drought predictions are limited to rainfall estimates, whereas surface temperatures, evaporation rates, soil moisture amounts are also necessary for adequate decision making. The methods also seem to work best for rather limited areas, which need not be a disadvantage if individual schemes can be developed. Of course, there will be occasions when such schemes fail miserably but this is likely to happen with any prediction method. The great attraction of statistical schemes is that when developed they are usually very cheap and fast to use. They are therefore particularly valuable for under-developed countries.

Deterministic methods

Some quantification of the term deterministic is needed here as it should not be taken to mean unique. Due to the nonlinear nature of the climate system an element of chaos exists, resulting in somewhat different answers in a numerical model for a given experiment starting from slightly different initial conditions.

Deterministic predictions require the identification and prediction of drought precursors. For brevity, only SST anomalies will be considered as drought precursors here, in order to illustrate the potential of the deterministic approach.

This potential is demonstrated in Fig. 8 by the observed and predicted rainfall anomalies for the Pacific region simulated in a general circulation model by Shukla & Fennessy (1988). They generated a control run by compositing results from three 60 day experiments initialised from conditions for 15th, 16th and 17th December 1982. Three perturbation runs for the same initial conditions were also produced, by inserting observed monthly SST anomalies for December 1982, January 1983 and February 1983 during the course of the model integration. The rainfall anomaly between the perturbation and control runs is in quite good agreement with that deduced from observation. Although this is a simulation rather than a prediction, as the observed SST anomalies were used, it clearly indicates the capabilities of current models to *predict* rainfall anomalies if the corresponding SST precursor anomalies can be *predicted*.

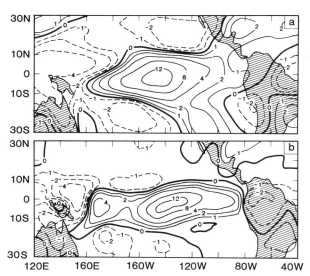

Fig. 8. In the upper panel of the figure the observed rainfall anomaly derived from outward longwave radiation anomalies is shown, averaged over the period December 1982 to February 1983. The lower panel illustrates the averaged rainfall difference between an experiment with SST anomalies and a control run, meaned over three runs starting from different initial conditions. Redrawn from Shulka & Fennessy (1988). Units are mm d^{-1}, with negative areas shaded.

use of three sets of initial conditions, and the subsequent compositing of results. Monte Carlo experiments of this type may be a necessity for drought experiments as personal experience has shown that predictions can be very sensitive to initial conditions, see also Palmer & Brankovic (1989).

Deterministic predictions of Sahelian rainfall have been attempted by Owen & Ward (1989). These predictions involved global SST anomalies. As a test of their procedures hindcasts were made by taking observed SST anomalies at the beginning of the test period and holding them constant for the duration of the test. In practice the anomalies were held fixed at their mid-July values, thus permitting forecasts of the August to October rainfall, which is when most of the annual rainfall occurs in the Sahel. They experimented on wet and dry years, and, as shown in Fig. 9, their model was able to differentiate between these conditions. It produced quite good rainfall predictions except for 1984. However, Owen & Ward

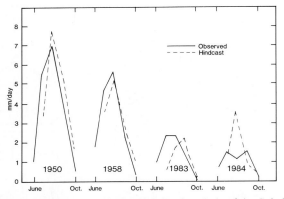

Fig. 9. Monthly averaged rainfall for the whole of the Sahel for two dry years and two wet years. Observed and hindcast rainfalls are compared in the figure. The hindcast results were obtained by running a model with global SST anomalies appropriate to each individual year. Redrawn from Owen & Ward (1989).

(1989) state that in practice such predictions would need to be made in May or earlier to be of use, and therefore predictions of SST anomalies would be required.

An operational prediction was attempted in 1988, which considerably underestimated observed rainfall, a fault they attributed to worldwide changes in SST anomalies at that time. Annual operational predictions are now being made, but these are the only ones attempted by any group world-wide.

A more elaborate experiment in drought prediction has recently been conducted by the author and his colleagues – Dr. H.B. Gordon, Dr. R.L. Hughes & Dr. R. Kleeman. A global climatic model was used to hindcast the great Australian drought of 1982 caused by the El Niño event in the Pacific Ocean. A simple model of the Pacific Ocean, extending from 30°S to 30°N was developed to predict the El Niño event. This model was driven by observed winds for a 20 year spinup period prior to January 1st 1982. It was then assumed that no further observational information was available and a simple Gill (1980) type atmospheric model was coupled to the oceanic model. The coupled model was integrated for eighteen months to predict the development of the El Niño event for the period January 1982 to June 1983. Monthly SST anomalies were derived

from this prediction. A typical example is shown in Fig. 10.

These predicted anomalies were inserted into the global atmospheric model starting from January conditions of a control run. This permitted rainfall anomalies created by the El Niño event to be predicted as differences from the control run. Although rainfall anomalies were produced extensively over the Pacific basin only those for the Australian region will be presented here. Results for July and October are shown in Fig. 11, with an indication of the observed rainfall variability. Only qualitative comparisons can be made from Fig. 11. Quantitative comparisons will be made when more refined experiments have been performed. The selected months shown were particularly good examples, other months such as June and August were much poorer. It should be noted that, in general, responses only over the eastern half of Australia can be expected from El Niño events, Nicholls (1989), hence comparisons in Fig. 11 should be restricted accordingly. Despite the continued existence of the predicted and observed SST anomalies into 1983, the model predicted a breaking of the drought early in 1983, as was observed.

A repeat of the above experiment commencing from a different year of the control run produced poorer results overall, indicating the sensitivity of the experiments to initial conditions, as discussed above.

Since no attempt was made to initialise the model with actual conditions for January 1982, the current results represent a lower limit of potential drought predictions. In particular, the effect of initialisation with more appropriate soil

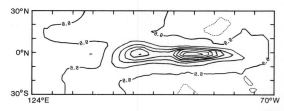

Fig. 10. Sea surface temperature anomalies predicted for September 1982 using a simple coupled atmospheric-oceanic model. (Figure supplied by Dr. R. Hughes & Dr. R. Kleeman). Units are °C, contour interval 0.5.

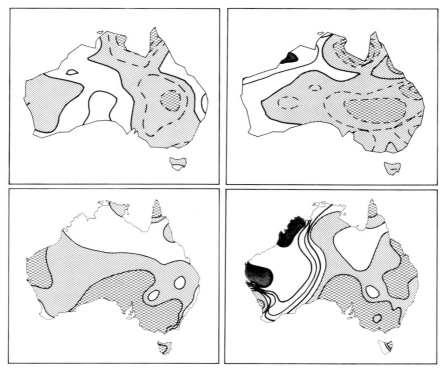

Fig. 11. Hindcasts of the rainfall anomalies over Australia for July and October 1982 are shown in the left hand panels. These anomalies are the differences between an experiment using predicted SST anomalies and a control run. Observations are shown for the same months in the right hand panel. These are based on decile ranges and can only be compared qualitatively with the model. Units are mm d^{-1} for the model results, with negative areas over Australia shaded. The contour interval is 1 mm d^{-1}. (unpublished research of Hunt & Gordon).

moisture amounts will be explored in the future. Also considerable scope exists regarding improvement of the El Niño SST anomaly prediction. Research is continuing in this area. In summary, the possibility for deterministic prediction of droughts and other rainfall anomalies is encouraging, but much more research will be needed before routine operational predictions of practical utility are possible. The highest potential appears to exist in the Pacific area, because of the inherent predictability of El Niño SST anomalies. If suitable schemes can also be developed for other oceans, then the possibility of global-scale drought predictions would be greatly enhanced. Many questions needed to be answered concerning the importance of initial conditions, spatial and temporal variability, and accuracy of prediction, but the essential question concerning

the viability of the deterministic method appears to have been resolved in the positive.

Concluding remarks

The prospects for drought research are improving. Current knowledge has identified SST anomalies as the primary cause of drought world-wide, but much remains to be done to clarify the role of other possible drought precursors. Global climatic models provide an invaluable tool for the assessment of precursors and the evaluation of the physical mechanisms contributing to rainfall fluctuations. These models have clearly demonstrated their ability to simulate droughts, and their associated characteristics, with considerable accuracy. The precise role of SST anomalies in a given ocean in causing regional droughts has still to be quantified.

The prediction of droughts remains a major scientific problem. Timely drought predictions require the prediction of the temporal evolution of the precursors, and this is undoubtedly the heart of the problem. Only in the case of SST anomalies attributable to El Niño events in the Pacific are the prospects for such predictions reasonably secure. Other oceans lack such a clear physical mechanism, and it is unclear currently how predictable their SST anomalies are.

Statistical prediction of drought is a viable method for certain regions of the globe. The most effective methods concentrate on limited areas where substantial skill can sometimes be obtained. Limitations of the current methods include rather restricted timeframes, lack of adequate temporal resolution, and predictions currently for rainfall only. While other problems exist with these methods, they do provide a cheap and fast source of predictions which are adequate for a range of purposes.

Deterministic prediction methods are only just being developed, but offer considerable scope because of their ability to predict most climatic variables of interest. Many questions have still to be answered concerning the potential of these methods. For example, will the accuracy of the methods justify predictions on a monthly timescale or only seasonal? How important are initial conditions, especially such variables as soil moisture? Will Monte Carlo techniques be necessary to counteract sensitivity to initial conditions? Will it be possible to predict for at least a year ahead, and thus follow droughts from initiation to cessation? What spatial accuracy can be expected? Obviously, considerable research will be needed to resolve these and other drought-related questions, but current prospects can be viewed as quite encouraging. Possibly the combination of deterministic and statistical predictions for a given region may provide the optimal outcome.

References

Barnett, T. P. 1984. Prediction of the El Niño of 1982–83, Mon. Weath. Rev. 112: 1403–1407.

Barnett, T. P., Dumenil, L., Schlese, U., Roeckner E. & Latif, M. 1989. The effect of Eurasian snow cover on regional and global climate variations. J. Atmos. Sci. 46: 661–685.

Cane, M. Z., Zebiak, S.E. & Dolan, S.C. 1986. Experimental forecasts of El Niño. Nature 321: 827–832.

Charney, J. G. 1975. Dynamics of deserts and droughts in the Sahel. Quart. J. Roy. Meteor. Soc. 101: 193–202.

Charney, J. G., Quirk, W. J., Chow, S-H. & Kornfield, J. 1977. A comparative study of the effects of albedo change on drought in semi-arid regions. J. Atmos. Sci. 34: 1366–1385.

Chu, P-S. 1989. Hawaiian drought and the Southern Oscillation. Intl. J. Clim. 9: 619–632.

Currie, R. G. 1981. Evidence for 18.6 signal in temperature and drought conditions in North America since A.D. 1800. J. Geophys. Res. 86: 11055–11064.

Currie, R. G. 1984. Evidence for 18.6 year lunar nodal drought in western North America during the past millenium. J. Geophys. Res. 89: 1295–1308.

Druyan, L. M. 1989. Advances in the study of sub-Saharan drought. Intl. J. Clim. 9: 77–90.

Folland, C. K., Palmer T. N. & Parker, D. E. 1986. Sahel rainfall and worldwide sea temperatures 1901–85. Nature 320: 602–607.

Gill, A. E. 1980. Some simple solutions for heat-induced tropical circulation. Quart. J. Roy. Meteor. Soc. 106: 447–462.

Gordon, H. B. & Hunt, B. G. 1987. Interannual variability of the simulated hydrology in a climatic model – implications for drought. Clim. Dyn. 1: 113–130.

Hastenrath, S. & Heller, L. 1977. Dynamics of climate hazards in north east Brazil. Quart. J. Roy. Meteor. Soc. 103: 77–92.

Hunt, B. G. & Gordon, H. B. 1990a. Droughts, floods and sea surface temperature anomalies. Accepted by Intl. J. Clim.

Hunt, B. G. & Gordon, H. B. 1990b. Simulations of the U.S. drought of 1988. Accepted by Intl. J. Clim.

Kiladis, G. N. & Diaz, H. F. 1989. Global climatic anomalies associated with extremes in the Southern Oscillation J. Clim. 2: 1069–1090.

Kitoh, A. 1988. A numerical experiment on sea surface temperature anomalies and warm winter in Japan. J. Meteor. Soc. Japan 66: 515–533.

Landsberg, H. E. 1982. Climatic aspects of droughts, Bull. Amer. Meteor. Soc. 63: 593–596.

Markham, C. G. & McLain, D. R. 1977 Sea surface temperature related to rain in Ceara, north-eastern Brazil. Nature 265: 320–323.

Mechoso, C. R. & Lyons, S. W. 1988. On the atmospheric response to SST anomalies with the Atlantic warm event during 1984. J. Clim. 1: 422–428.

Moura, A. D. & Shulka, J. 1981. On the dynamics of

droughts in north east Brazil, observations, theory and numerical experiments with a general circularion model, J. Atmos. Sci. 38: 2653–2675.

Namias, J. 1978. Recent drought in California and Western Europe. Rev. Geophys. Sp. Phys. 16: 435–458.

Nicholls, N. 1985 Towards the prediction of major Australian droughts Aust. Meteor. Mag. 33: 161–166.

Nicholls, N. 1989. Sea surface temperatures and Australian winter rainfall. J. Clim. 2: 965–973.

Owen, J. A. & Folland, C. K. 1988. Modelling the influence of sea-surface temperatures on tropical rainfall, Recent climatic change – a regional approach, ed. S. Gregory Bellhaven Press, London, p 141–153.

Owen, J. A. & Ward, M. A. 1989. Forecasting Sahel rainfall, Weather 44: 57–64.

Parker, D. E., Folland C. K. & Ward, M. N. 1988. Sea surface temperature anomaly patterns and predictions of seasonal rainfall in the Sahel region of Africa, Recent climatic change – a regional approach, ed. S. Gregory. Bellhaven Press, London, p. 166–178.

Palmer, T. N. & Brankovic, C. 1989. The 1988 US drought linked to anomalous sea surface temperature. Nature 338: 54–57.

Pitcher, E. J., Blackmon, M. L., Bates, G. T. & Munoz, S. 1988. The effect of north Pacific sea surface temperature anomalies on the January climate of a general circulation model. J. Atmos. Sci. 45: 173–188.

Rasmusson, E. M. & Hall, J. M. 1983. El Niño – the great equatorial Pacific Ocean warming event of 1982–83. Weatherwise 36: 166–175.

Rasmusson, E. M. 1987. The prediction of drought, a meteorological perspective. Endeavour 11: 175–182.

Ropelewski, C. F. & Halpert, M. S. 1987. Global and regional scale precipitation patterns associated with the El Niño/Southern Oscillation. Mon. Wea. Rev. 115: 1606–1626.

Ropelewski, C. F. & Halpert M. S. 1989. Precipitation patterns associated with the high index phase of the Southern Oscillation. J. Clim. 2: 268–284.

Streten, N. A. 1981. Southern Hemisphere sea surface temperature variability and apparent associations with Australian rainfall. J. Geophys. Res. 86: 485–497.

Shukla, J. & Fennessy, M. J. 1988. Prediction of time-mean atmospheric circulation and rainfall: influence of Pacific sea surface temperature anomaly. J. Atmos. Sci. 45: 9–28.

Trenberth, K. E., Branstator G. W. & Arkin, P. A. 1988. Origins of the 1988 North American drought, Science, 242: 1640–1645.

Voice, M. E. & Hunt, B. G. 1984. A study of the dynamics of drought initiation using a global general circulation model. J. Geophys. Res. 89: 9504–9520.

Ward, M. N. & Folland, C. K. 1990. Statistical prediction of north east Brazil seasonal rainfall from sea surface temperatures. Submitted to Intl. J. Clim.

Vegetatio **91**: 105–120, 1991.
A. Henderson-Sellers and A. J. Pitman (eds).
Vegetation and climate interactions in semi-arid regions.
© 1991 *Kluwer Academic Publishers. Printed in Belgium.*

Vegetation-atmosphere interaction in homogeneous and heterogeneous terrain: some implications of mixed-layer dynamics

M. R. Raupach
CSIRO Centre for Environmental Mechanics, GPO Box 821, Canberra ACT 2601, Australia

Accepted 24.8.1990

Abstract

This paper considers the modelling of available energy partition into sensible and latent heat at land surfaces, *inter alia* for GCM applications. In a preliminary discussion, processes at canopy scale are reviewed briefly by outlining two classes of model: comprehensive, multilayer Canopy-Atmosphere Models (CAMs) which attempt to include all physical and biological processes influencing the canopy microclimate and atmospheric exchanges, and single-layer Simplified Canopy-Atmosphere Models (SCAMs) which attempt a physically acceptable description of energy partition with the fewest possible parameters. Details of a four-parameter SCAM are outlined. It is suggested that CAMs are necessary for understanding the 'downward' influence of climate on a canopy, but that SCAMs may be useful in modelling the 'upward' influence of vegetation on climate with GCMs.

The main part of the paper considers the generalisation of land-surface models from homogeneous to flat, heterogeneous terrain in which local advection is prominent. The approach is to model the planetary boundary layer as well as the surface layer. A simple mixed-layer model is outlined for the daytime convective boundary layer (CBL). Boundary conditions for sensible and latent heat transfer at the ground are made separable by defining two new conserved scalar variables, a total energy content and a linearized saturation deficit. A new analytic solution of the energy partition problem is developed for the case of an encroaching CBL with the depth h proportional to the square root of time. The general model, with this analytic solution as a particular case, is then extended from homogeneous to heterogeneous surfaces by defining a CBL horizontal length scale $X = hU/w_*$ (where U is the horizontal velocity and w_* the convective velocity scale). When individual terrain patches have scales much less than X, the inhomogeneity is *microscale* and is averaged out by the CBL itself; *mesoscale* inhomogeneity occurs when patches have scales much greater than X, leading to separate CBL development over each patch. The CBL equations yield a statement of the partition of surface sensible and latent heat fluxes among the surface patches in a heterogeneous landscape, and averaging operators for surface properties.

Introduction

In the study of short-term climatic perturbations such as droughts, and longer-term changes such as possible greenhouse warming, vegetation-atmosphere interactions encompass both 'up-

ward' and 'downward' influences. The former is the influence of given types and distributions of vegetation upon the large-scale climate, while the latter is the way that a given large-scale climate is manifested at the local scale of a vegetation community. At present, the 'upward' influence is investigated almost solely with global climate models (GCMs), which dynamically model atmospheric (and sometimes also oceanic) large-scale circulations. However, GCMs operate at far too large a scale to address the equally important problem of the 'downward' influence, which requires physical understanding at a variety of smaller scales.

In fact, this understanding is indispensable in either case: the same physical processes are at work in both directions, so a good physical model of vegetation-atmosphere interaction should be 'reversible' enough to describe both – though perhaps with different emphases. Such a model must incorporate, either explicitly or implicitly, processes at a range of length scales extending from the mesoscale to the scale of a vegetation canopy. This is the scale range required to bridge between processes in a vegetation canopy and the smallest-scale processes explicitly resolved in GCMs, which presently have grid cells of order 500 by 500 km horizontally and 4 to 15 layers in the vertical, of which only the lowest two or three (at most) lie below a height of about 3 km.

It is now recognised from a number of GCM sensitivity studies (Dickinson & Henderson-Sellers 1988; Garratt 1990) that a realistic 'upward' parameterization of vegetation-atmosphere interaction is among the essential requirements for a good GCM. Several land-surface models for GCMs have been developed over the last decade, beginning with the work of Dickinson (1983, 1984) which led directly to BATS, the Biosphere-Atmosphere Transfer Model (Dickinson *et al.* 1986, Wilson *et al.* 1987, Dickinson & Henderson-Sellers 1988). A similar but more complicated model is SiB, the Simple Biosphere model of Sellers *et al.* (1986) & Sellers (1987). Other models along similar lines have been developed by Choudhury & Monteith (1988) and Noilhan & Planton (1989). Land-surface models of this type attempt a detailed resolution of the *vertical* structure of the soil, vegetation canopy and atmospheric surface layer, and therefore require a large number of parameters to specify the physical and biological properties of both the vegetation and the soil: BATS requires about 20

parameters and SiB about 50, all of which must be specified for every grid point in the GCM.

In the commentary on one of the SiB papers, McNaughton (1987) argued that this type of land-surface model faces three difficulties: (1) critical biological and aerodynamic processes are handled crudely (mainly by bulk resistances), making irrelevant a detailed physical model of other parts of the system such as the lower soil layers; (2) it is impossible in practice to measure such a large number of parameters for every land-surface grid point in a GCM, so most of the parameters will necessarily be guessed; and (3) even if the model did work at one point in space with properly measured parameters, there is no logical way of spatially averaging the soil and vegetation parameters over the many vegetation and other surface types within a single grid cell in a GCM, typically 500 by 500 km.

These criticisms identify several weaknesses with current land-surface models, and can therefore be interpreted as indicators of useful research directions. The main problems to be solved fall into two classes: those concerned with aspects of vegetation-atmosphere interaction in spatially *homogeneous* conditions which are inadequately dealt with in current models, and those concerned with generalising current and future homogeneous-surface models to *heterogeneous* land surfaces. Of the 'homogeneous' problems, two of the most important are the specification of stomatal resistances and the description of turbulent transfer; a third is the investigation of the question of model complexity by developing and testing a range of models from simple to complex, so that in future the complexity of the land-surface model can be matched rationally to the application. The 'heterogeneous' problems are more urgent in practice because far less work has been done on them and understanding is correspondingly weaker. They include the effects of variability in surface type on flat land, leading to local advection, and the influence of hilly terrain. This last area encompasses several problems for which the necessary physics is still poorly understood: the aerodynamics of canopies on hills, separation phenomena, drag partition between vegetation and topography, and the effects of gravity waves in stable conditions.

The present paper focuses on just one of this plethora of problems, that of generalising land-surface models from homogeneous surfaces to heterogeneous (but still flat) terrain in which local advection is prominent. The basic theme is that the most appropriate way of treating land-surface heterogeneity is at the scale of the planetary boundary layer (PBL), rather than at surface-layer scales. This is so because local advection phenomena extend through much or all of the surface layer but are essentially confined to the PBL, a well defined, strongly mixed region of the atmosphere (at least in daytime) which naturally smooths out small-scale surface heterogeneity and responds *in toto* to large-scale heterogeneity. From the climate modelling viewpoint, consideration of surface heterogeneity therefore implies that the PBL is a more suitable atmospheric entity to model as a discrete component of a GCM than the surface layer, which is the focus of current land-surface models such as SiB and BATS. Because a PBL model necessarily includes a description of vegetation-atmosphere interaction as its lower boundary condition, canopy-scale processes are fully included in a PBL-based model, and can be represented at any appropriate level of complexity.

Explicit incorporation of the PBL into a GCM is not a new idea; it dates back to the notable paper by Deardorff (1972). However, the possibilities for treating surface heterogeneity in this way have not been explored before. It is also noteworthy that there is a second, independent argument for modelling the PBL properly in GCMs: to improve the description of boundary-layer stratus, stratocumulus and cumulus clouds, both terrestrial and marine. A minority of GCMs include PBL physics for this reason (Suarez *et al.* 1983; Garratt 1990), but most present GCMs do not model the PBL depth.

The plan for the rest of the paper is as follows. The next (second) section discusses models of vegetation-atmosphere interaction at canopy scale, canvassing a spectrum of possible approaches ranging from very simple to very detailed and complex. Attention then shifts to PBL-scale processes: the third section considers the most common form of daytime PBL, the convective boundary layer (CBL), by developing a CBL model for homogeneous terrain using well-known mixed-layer growth concepts. The fourth section develops boundary conditions for sensible and latent heat transfer at the ground, using the simplest possible 'big-leaf' description of vegetation-atmosphere interaction, while the fifth section derives two analytic solutions (one old, one new) to the CBL energy partition problem, under some idealisations including horizontal homogeneity. Finally, the sixth section uses the CBL framework to analyse the effect of surface heterogeneity at both small and large scales compared with a horizontal length scale defined by the CBL depth. In the case of small-scale heterogeneity, this leads to (1) a statement of the partition of surface sensible and latent heat fluxes among the surface types making up a heterogeneous landscape, (2) an averaging scheme for surface properties, and (3) some insight into how local advection effects are smoothed over by the CBL.

Modelling the canopy

Before embarking on a discussion of PBL processes, it is useful to review briefly the options for modelling canopy-scale processes; specifically, the exchange of energy, mass and momentum between the atmosphere and a uniform canopy on level ground. Two approaches are current: multilayer models resolve explicitly the vertical structure of the canopy and the underlying soil, while single-layer models do not. The multilayer approach is necessary for a complete resolution of aspects such as the microclimate within the canopy, the partition of sensible and latent heat exchange between soil, understorey and overstorey, and the vertical distribution of soil water. However, single-layer models are a great deal simpler and can often do a good job of representing the overall energy, momentum and mass exchanges between the atmosphere and the composite surface, provided that simple expressions for the semi-empirical 'bulk resistances' or exchange coefficients are available from observations or from calibration with a more detailed (multilayer) model.

In a detailed multilayer Canopy-Atmosphere Model or CAM (the acronym is intended to cover a class of models rather than any particular realisation), six separate physical or biological components are needed:

(1) *Radiation physics*: to specify, for each leaf (or layer) in the canopy and for the underlying soil, the radiant energy at the surface. This is partitioned into sensible and latent heat fluxes and into stored heat.

(2) Soil physics: to specify the heat and water transport through and out of the soil, and the soil moisture status.

(3) *Interception*: to represent the storage and movement of water intercepted on the canopy from rainfall.

(4) *Plant physiology*: to specify the water transport through roots and stems, and the stomatal resistance governing gaseous exchange through leaf surfaces.

(5) *Aerodynamics*: to specify the turbulent wind field and the leaf boundary-layer resistances for momentum, heat and mass exchange.

(6) *Turbulent transfer of scalars*: to specify the scalar concentration (temperature, humidity and other) fields in the canopy, which are maintained by the turbulent mixing of scalar entities (heat, water vapour and others) released from or absorbed by distributed sources or sinks in the canopy.

For at least the last three of these components, the necessary basic physical or biological knowledge is still barely adequate to build a comprehensive CAM, and efforts to improve this situation are continuing. For example, recent work on (6) is described by Raupach (1989a, b, c).

Problems of physical and biological understanding aside, a CAM is necessarily complicated. Each of its six components requires several parameters to specify the necessary canopy, soil or plant properties. Hence, the 20-odd parameters of BATS and the 50-odd of SiB are not unreasonable totals for this class of model. Neither BATS nor SiB has more than two canopy layers, so the parameter count for a truly multilayer CAM may well be higher still. Nevertheless, if a detailed prediction or understanding of the dynamics of a well-specified uniform canopy is required, then this is the magnitude of the modelling task.

At the opposite pole in complexity to a CAM lies a SCAM (Simplified Canopy-Atmosphere Model): a single-layer model of surface energy exchange based on the bulk combination (Penman-Monteith) equation, with semi-empirical rules for finding the necessary bulk resistances from basic canopy and soil properties. The bulk combination equation itself, far from being a scam in the literal sense, is completely sound within its idealisations; however, scam-like elements in such a model inevitably arise from simple specifications of the bulk resistances (Raupach & Finnigan 1988). The details of a possible SCAM with just four surface parameters, complete with rules for the bulk resistances, are outlined in the Appendix.

It is proper to examine the appropriateness of a CAM and a SCAM in the 'upward' and 'downward' context mentioned at the outset. For the 'downward' task of inferring the detailed microclimatic response of a given plant community to an imposed (modelled or measured) external large-scale climate, a CAM is indeed appropriate, because of its detailed resolution of the canopy microclimate. Correspondingly, a SCAM is much too oversimplified for this task. However, for the 'upward' problem of specifying climate response to vegetation type by parameterizing vegetation-atmosphere interactions in a GCM, there are practical reasons for using a SCAM – at least for the time being. Three such reasons, besides the obvious one of economy, were implicit in McNaughton's (1987) criticisms of current land-surface models for GCMs (see the Introduction): many CAM parameters are (1) irrelevant in practice, (2) cannot be measured in a GCM context, and (3) cannot be spatially averaged.

Of these, the most fundamental in a conceptual

sense is the last. Models of neither the CAM nor the SCAM type attend to the critical problem of averaging the variety of land-surface types found in most land grid cells. To incorporate this averaging requires an extra component of physical understanding at a scale larger than the canopy, that of the PBL. In order to explore vegetation-atmosphere interactions in a PBL context, it is *necessary* for conceptual reasons to use a simple description of vegetation-atmosphere interaction as a lower boundary condition; in the following, the simplest possible SCAM (the bulk combination equation with prescribed bulk resistances) is used. It is also possible that a SCAM rather than a CAM may turn out to be *sufficient* in practice, because feedbacks at the PBL scale turn out to make sensible and latent heat transfer less sensitive to surface properties than appears to be the case from the surface-layer viewpoint.

Modelling the mixed layer

Attention now shifts to the PBL, specifically the daytime convective boundary layer (CBL) over flat land. The nocturnal PBL and the problems of hilly terrain are plainly important but are outside the present scope, though it is hoped that some aspects from later sections will be relevant.

Mixed-layer or 'slab' models for the daytime growth of the CBL have antecedents in the work of Ball (1960), and were developed in essentially their present form by Deardorff (1972), Tennekes (1973), Tennekes & Driedonks (1981) and others cited by them. The mixed-layer model used here has three ingredients, the first being the assumption of a growing, horizontally homogeneous CBL capped by a thin inversion at height $h(t)$, and well mixed by convective turbulence except possibly in a thin surface layer. The second is the conservation equation for an arbitrary (conserved) scalar entity with mean absolute concentration $C(t)$ and vertical flux density $\overline{wc}(z, t)$, where w is the fluctuating vertical wind component, c the fluctuating scalar concentration and the overbar denotes a time average long enough to smooth over time scales for convective turbulence but short com-

pared with the time scale for CBL growth. The well-mixed assumption means that C is a function of time t only, independent of z. The scalar conservation equation, integrated over the CBL depth, is

$$\frac{dC}{dt} = \frac{F_c}{h} + \left(\frac{C_+(h) - C}{h}\right)\frac{dh}{dt}, \tag{1}$$

where $F_c(t) = \overline{wc}(0, t)$ is the scalar flux density at the surface ($z = 0$), and $C_+(z)$ is the scalar concentration profile in the undisturbed atmosphere above the CBL. Algebraic rearrangement gives another useful form of Equation (1) (Tennekes 1973)

$$\frac{d(h\Delta c)}{dt} = \gamma_c h \frac{dh}{dt} - F_c, \tag{2a}$$

where $\Delta_c(t) = C_+(h(t)) - C(t)$ is the concentration jump across the inversion and $\gamma_c = \lim(z \downarrow h) dC_+/dz$ is the slope of the concentration profile just above the inversion. When γ_c is constant, Equation (2a) has the integral form

$$h(t)\Delta_c(t) - h(0)\Delta_c(0) =$$

$$\frac{\gamma_c[h^2(t) - h^2(0)]}{2} - \int_0^t F_c \, dt, \tag{2b}$$

where $t = 0$ is at sunrise.

With $F_c(t)$ and $C_+(z)$ regarded as given, either of Equations (1) or (2) contains two unknowns, $C(t)$ and $h(t)$, so a solution requires another equation or closure assumption, which is the third ingredient of the model. One of the simplest of many possibilities is derived from a parameterization for the entrained buoyancy flux density at the inversion (Tennekes 1973):

$$\overline{w\theta_v}(i, t) = -\alpha\overline{w\theta_v}(0, t), \tag{3}$$

where α is an empirical dimensionless parameter (typically about 0.2), i denotes a level just below the inversion at $z = h(t)$, and θ_v is the fluctuating virtual potential temperature (defined by $\theta_v = \theta + 0.61qT$, θ being fluctuating potential temperature, q fluctuating specific humidity and T mean absolute temperature). The quantity $g\overline{w\theta_v}/T$ (g being gravitational acceleration) is the buoyancy flux density. This 'entrainment hypothesis' applies in the usual situation that the atmosphere is stable above the inversion and the upward buoyancy flux at the surface is strong enough to induce penetrative convection which entrains overlying, less dense air into the boundary layer at $z = h$, maintaining a downward buoyancy flux at the inversion and deepening the CBL. Equation (3) was advanced by Tennekes (1973) with θ in place of θ_v, thus neglecting the buoyancy contribution of moisture; it was written in the form above by McNaughton & Spriggs (1986).

Equation (3) represents only one of many possible entrainment hypotheses which have been suggested to close Equations (1) or (2); see Tennekes & Driedonks (1981) for further discussion. However, the choice is not critical because the model predictions for $C(t)$ and $h(t)$ are rather insensitive to the entrainment hypothesis.

To obtain an equation for $h(t)$ from Equation (3), one writes Equation (1) with $C = \theta_v$, the mean virtual potential temperature (defined by $\Theta_v = \Theta + 0.61QT$, with mean potential temperature Θ and mean specific humidity Q). The second term in Equation (1) can be replaced by Equation (3), since

$$-\overline{w\theta_v}(i, t) = (\theta_{v+} - \theta_v)\frac{dh}{dt}. \tag{4}$$

Thus, $\Theta_v(t)$ and thence $h(t)$ can be found from a given virtual potential temperature (or buoyancy) flux density at the ground, $F_{\theta v}(t) = \overline{w\theta_v}(0, t)$. The solution is particularly simple when $F_{\theta v}(t)$ is constant and $\alpha = 0$, so that $\overline{w\theta_v}(i)$ and the virtual potential jump $\Theta_{v+} - \Theta$ at the inversion are both zero. Then Equation (2b) becomes (with $\gamma_{\theta v} = d\Theta_{v+}/dz$):

$$h^2(t) - h^2(0) = \frac{2F_{\theta v}t}{\gamma_{\theta v}}, \tag{5}$$

so that $h(t)$ grows as $t^{1/2}$ when the boundary layer has grown sufficiently that $h(t) \gg h(0)$, $h(0)$ being the initial (sunrise) depth. This is the 'encroachment model' for CBL depth.

More realistically, Tennekes (1973) identified three phases during a typical diurnal cycle in which $F_{\theta v}(t)$ varies roughly sinusoidally ($F_{\theta v}(t) = F_{\theta v max}\sin(\pi t/T_{day})$, T_{day} being the time from sunrise to sunset). First is the 'morning transient', during which the shallow but strong nocturnal inversion ($\Delta_{\theta v}(0)$ several degrees, $h(0)$ a few tens of metres) is 'filled in', without growth of h, to the point where penetrative convection can occur. Next is a phase during the mid-morning hours in which $F_{\theta v}(t)$ grows approximately linearly, leading to linear growth of $h(t)$ with t. Thirdly, around noon, $F_{\theta v}(t)$ is roughly constant and $h(t)$ grows approximately with $t^{1/2}$, as in Equation (5). The growth rate decreases below this rate as $F_{\theta v}(t)$ decreases during the afternoon. Eventually, convective activity ceases with the collapse of the CBL at evening transition, leading to a stable layer with very little turbulent mixing, in which the temperature structure is a 'frozen' version of the structure just prior to evening transition.

Boundary conditions of sensible and latent heat

To apply this model of CBL dynamics to sensible and latent heat exchange, it is necessary to establish suitable boundary conditions at the ground. For flat, homogeneous terrain in which the surface is supplied with an available energy flux density F_A, the energy balance at the surface is

$$F_A = R_n - G = H + \lambda E, \tag{6}$$

where R_n is the downward net irradiance, G the heat flux density into the ground, H the upward sensible heat flux density and λE the upward latent heat flux density at the surface. The partition of energy between H and λE is given by the bulk combination equation

$$\lambda E = \frac{\varepsilon F_A + \rho\lambda D/r_a}{\varepsilon + 1 + r_c/r_a}, \tag{7}$$

where ρ is the air density, $\varepsilon = (\lambda/c_p)\, dQ_{sat}/dT$ the dimensionless slope of the saturation specific humidity $Q_{sat}(T)$ as a function of temperature T, λ the latent heat of vaporisation of water, c_p the specific heat of air at constant pressure, D the potential saturation deficit in the mixed layer (independent of height because the layer is fully mixed), r_a the bulk aerodynamic resistance of heat or water vapour transfer from the surface to the mixed layer (through the surface layer) and r_c the bulk surface or canopy resistance. The potential saturation deficit is the saturation deficit of an air parcel brought adiabatically to a reference level z_R, conveniently the level by which potential temperature is defined. In this case, $\Theta = T + \Gamma(z - z_R)$ (where $\Gamma = g/c_p = 0.01$ K m^{-1} is the dry adiabatic lapse rate), and

$$D = Q_{sat}(\Theta) - Q$$

$$= Q_{sat}(T + \Gamma(z - z_R)) - Q$$

$$= Q_{sat}(T) - Q + \frac{c_p \varepsilon}{\lambda} \Gamma(z - z_R) \qquad (8)$$

(linearizing $Q_{sat}(T)$ in the last line) which defines the relationship between the potential saturation deficit D and the actual saturation deficit $Q_{sat}(T) - Q$.

Equation (7) is the boundary condition for solving a mixed-layer model to determine the evolution of the surface evaporation and mixed-layer humidity ($E = \rho \overline{wq}$ and Q), and also the surface heat flux and mixed-layer temperature ($H = \rho c_p \overline{w\theta}$ and Θ). However, there is an analytical difficulty: the boundary condition for the water vapour flux (evaporation) involves not only the water vapour concentration but also the heat concentration (temperature) through the saturation deficit D, while the boundary condition for the heat flux is likewise coupled to the water vapour field.

McNaughton (1976) showed how this problem can be overcome by defining two new conserved scalar entities, both linear combinations of Θ and Q, with separated boundary conditions; the mean concentrations of these scalars will be denoted A and B, and the fluctuating concentrations a and

b. The necessary linear combinations of Θ and Q are[1]

$$A = \rho c_p \Theta + \rho \lambda Q \qquad (9a)$$

$$B = \varepsilon \rho C_p(\Theta - \Theta_R) - \rho \lambda (Q - Q_{sat}(T_R)). \quad (9b)$$

Both of these scalars have the dimensions of energy per unit volume, which simplifies later algebra. Scalar A is the total (sensible plus latent) heat energy per unit volume. Scalar B is the potential saturation deficit in energy units, since

$$B = \rho \lambda D, \qquad (10)$$

which is obtained by combining Equations (8) and (9) and linearizing $Q_{sat}(T)$ about T_R.

The flux densities of A and B at the surface are found by forming the covariance of the fluctuating counterpart of Equation (9) with the vertical velocity w. They are:

$$F_A = \overline{wa} = \rho C_p \overline{w\theta} + \rho \lambda \overline{wq} = H + \lambda E \qquad (11a)$$

$$F_B = \overline{wb} = \varepsilon \rho c_p \overline{w\theta} - \rho \lambda \overline{wq} = \varepsilon H - \lambda E. \qquad (11b)$$

Combining these with Equations (6) and (7), the boundary conditions for A and B are obtained:

$$F_A = R_n - G \qquad (12a)$$

$$F_B = \frac{\varepsilon F_A r_c/r_a - (\varepsilon + 1)B/r_a}{\varepsilon + 1 + r_c/r_a}. \qquad (12b)$$

These boundary conditions for A and B are now separated. The boundary condition for A is a simple flux boundary condition, since for the present problem, the available energy flux density $R_n - G = F_A$ is externally given in both Equations (12a) and (12b). The condition for B is a mixed boundary condition involving both the concen-

---

[1] (although note that in fact, A and B are non-unique to within height-independent multiplicative and additive constants; in Equation (9), these constants have been chosen so that A and B have the simple physical meanings stated).

tration B (in essence, the saturation deficit) and the surface flux density F_B (a flux of saturation deficit, or 'dryness', for the surface), together with F_A, which is given.

In making use of the boundary conditions for A and B, it is very helpful, both for notational convenience and for understanding, to introduce the *equilibrium* latent heat flux density defined by $\lambda E_{eq}/F_A = \varepsilon/(\varepsilon + 1)$. The evaporation rate E_{eq} occurs for a wet surface in saturated air, so that $r_c = 0$ and $D = 0$ (Slatyer & McIlroy 1961). It is also the limit of Equation (7) as $r_a \to \infty$ (Thom 1975; Finnigan & Raupach 1987). More fundamentally, it is the limiting evaporation rate from an evaporating surface supplied with a constant available energy in an enclosed region (McNaughton 1976; Perrier 1980; McNaughton & Jarvis 1983). In the last case, the corresponding equilibrium saturation deficit is given by:

$$B_{eq} = \frac{\varepsilon F_A}{\varepsilon + 1} \, r_c = \lambda E_{eq} r_c . \tag{13}$$

The quantity B_{eq} has a more general role, since the general boundary condition for B, Equation (12b), can be written in the compact form

$$F_B = \frac{B_{eq} - B}{r_{eq}} , \tag{14}$$

where r_{eq} is a weighted sum of the bulk resistances:

$$r_{eq} = r_a + r_c/(\varepsilon + 1) . \tag{15}$$

With this notation, the combination equation itself also becomes very compact. In terms of the scalars A and B, it is

$$\lambda E = \frac{\varepsilon F_A + B/r_a}{\varepsilon + 1 + r_c/r_a} , \tag{16a}$$

which can also be expressed in the illuminating form

$$\lambda E - \lambda E_{eq} = \frac{B - B_{eq}}{(\varepsilon + 1)r_{eq}} \tag{16b}$$

as discussed by McNaughton & Jarvis (1983).

With this formalism it is now possible to write the primary equation to be analysed in the following sections, the conservation equation for $B(t)$ in the mixed layer. From Equation (1) (with $C = B$) and Equation (14), $B(t)$ is governed by:

$$\frac{dB}{dt} = \frac{B_{eq}}{r_{eq}h} - \frac{B}{r_{eq}h} + \left(\frac{B_+ - B}{h}\right) \frac{dh}{dt} . \tag{17}$$

Here, B_{eq} and r_{eq} should be regarded in general as surface properties defined by Equations (13) and (15), respectively, and therefore dependent on time through the time dependence of F_A, r_a and r_c. It is worth noting that the most important aspects of Equation (17), in contrast to Equation (1), is the appearance of a negative term linear in $B(t)$ on the right hand side; this 'damping term' gives solutions of Equation (17) a tendency to evolve toward a state of thermodynamic equilibrium.

Energy partition in a mixed layer: homogeneous terrain

With the equations already developed, it is possible to obtain a complete numerical solution for $H(t)$, $\lambda E(t)$ and the mixed layer depth $h(t)$, from a specified available energy history $F_A(t)$ and profiles of potential temperature and humidity in the undisturbed atmosphere above the boundary layer. However, to understand thermodynamic constraints on the energy partition at the surface, it is helpful to regard the mixed-layer depth $h(t)$ as *externally* given, rather than being determined by a CBL growth model, and then to examine the response of the energy partition, surface fluxes and mixed-layer saturation deficit to energy supply and boundary-layer growth. The problem now comes down to solving only the conservation equation for $B(t)$ in the mixed layer, Equation (17), with $h(t)$ specified. From $B(t)$, the surface fluxes λE and $H = F_A - \lambda E$ can be found. This is the approach taken here. Two scenarios for the evolution of the (now given) $h(t)$ will be examined, in order to establish some of the properties of solutions to Equation (17) in homogeneous terrain (heterogeneous terrain will be discussed in the next section). In both cases the available energy and the bulk resistances (and hence B_{eq} and r_{eq}) will be taken as time-independent for simplicity.

Case 1 – h(t) is constant in time

When $h(t)$ is constant there can be no entrainment, from Equation (4), so this case is the well-known closed-box CBL considered by McNaughton (1976), Perrier (1980) and McNaughton & Jarvis (1983). The CBL has a fixed, impermeable lid at height h and is well-mixed apart from a thin surface layer with bulk aerodynamic resistance r_a. The ground is covered with a uniform evaporating surface with bulk surface resistance r_c. The second term in equation (17) vanishes and the solution is

$$B(t) = B_{eq}(1 - \exp(-t/T_{eq})) + \\ + B(0)\exp(-t/T_{eq}) \qquad (18a)$$

$$\lambda E(t) = \lambda E_{eq}(1 - \exp(-t/T_{eq})) + \\ + \lambda E(0)\exp(-t/T_{eq}), \qquad (18b)$$

where $T_{eq} = r_{eq}h$ is a time scale for approach to the final thermodynamic equilibrium state. In this state the CBL is continually warming and humidifying because of the energy supply to the surface, but with a constant evaporation rate $\lambda E = \lambda E_{eq}$ independent of both r_a and r_c, and a constant saturation deficit $B = B_{eq} = \lambda E_{eq} r_c$ dependent linearly on r_c.

Values of the time scale T_{eq} are shown in Table 1 for several surface types for which typical values of r_a and r_c have been chosen. In all cases the time scale is much greater than the daytime span of about 12 hours; hence, even if the closed-box CBL model were dynamically realistic (which it is not), thermodynamic equilibration of the CBL could not be achieved over natural surfaces.

Case 2 – h(t) grows as square root of time

This case is somewhat more realistic than the closed-box CBL, as it incorporates the CBL growth obtained from an encroachment model, Equation (5), and includes entrainment of (usually dry) air from above the CBL. Let $h(t) = (Kt)^{1/2}$, where K is a growth-rate parameter with the dimensions of diffusivity. To account for the entrainment of overlying air, it is necessary to specify the profile of B in the undisturbed atmosphere above the CBL; the simple linear profile $B_+(z) = \gamma_B z$ is assumed. The initial conditions at $t = 0$ (sunrise) are $h(0) = 0$ and $B(0) = 0$ (dew at dawn), which are consistent with the assumed profile of $B_+(z)$.

Putting these assumptions into Equation (17) for $B(t)$, the following equation is obtained:

$$\frac{dB}{dt} = \frac{1}{(Kt)^{1/2}}\left(\frac{K\gamma_B}{2} + \frac{B_{eq}}{r_{eq}}\right) -$$

$$- B\left(\frac{1}{(Kt)^{1/2}r_{eq}} + \frac{1}{2t}\right). \qquad (19)$$

This is a linear, first-order ordinary differential equation in $B(t)$ with non-constant coefficients (unlike the equivalent equation in Case 1, which had constant coefficients). The solution (with the initial condition $B(0) = 0$) is not as straightforward as for Case 1, but produces a fairly simple result:

$$B(t) = \left(B_{eq} + \frac{K\gamma_B r_{eq}}{2}\right)\left(1 - \frac{1 - e^{-\tau}}{\tau}\right), \qquad (20)$$

where $\tau = 2t^{1/2}/(K^{1/2}r_{eq})$ is a dimensionless temporal variable proportional to the square root of time.

In Equation (20), the saturation deficit at large time approaches the steady limiting value $B = B_{eq} + K\gamma_B r_{eq}/2$. The corresponding steady evaporation rate at large time is $\lambda E = \lambda E_{eq} + K\gamma_B/2(\varepsilon + 1)$. For both quantities, the long-time limits are the equilibrium values with added terms accounting for the entrainment of dry air from above the boundary layer. The appearance of the equilibrium values emphasises their generality beyond the closed-box analysis.

The rate of approach to the limiting value is determined by a function of time,

$f(\tau) = (1 - (1 - e^{-\tau})/\tau)$, which is 0 when $\tau = 0$, 0.5 when $\tau = 1.5935$, and 1 when $\tau \to \infty$. Its asymptotic limits are $f(\tau) \sim \tau/2$ at small τ and $f(\tau) \sim 1 - \tau^{-1}$ at large τ. The slow (τ^{-1}) approach to 1 at large τ means that thermodynamic equilibration in a growing CBL is slower than in a closed-box CBL, where the approach is exponential in time. A measure of the rate of approach to equilibrium is the time T_g at which $f(\tau) = 0.5$ (so that equilibration is at the half-way mark). From the definition of τ, $T_g = K(1.5935r_{eq}/2)^2$.

The values in Table 1 show that T_g is indeed substantially larger than the corresponding closed-box time scale T_{eq}. Hence, thermodynamic equilibration within a daily cycle is out of the question with a growing CBL.

Energy partition in a mixed layer: heterogeneous terrain

The aim of this final section is to examine to what extent the above concepts are applicable in heterogeneous as well as homogeneous terrain. The starting point is a scale analysis of the effect of the horizontal length scale of the surface heterogeneity.

Suppose that a CBL, in which the mean wind speed is U, encounters a surface discontinuity at $x = 0$ (x being the horizontal coordinate in the wind direction). There are two primary types of discontinuity which cause a change in the dynamical structure of the CBL, corresponding to an increase and a decrease in the surface buoyancy flux density $g\overline{w\theta_v}/T$, respectively. In order to retain the CBL analysis, $g\overline{w\theta_v}/T$ must always be positive, so that over all surface types the atmosphere is convectively unstable. Therefore, this discussion excludes the case where the buoyancy flux density becomes negative over a large area, inducing a horizontally extensive stable internal boundary layer near the ground.

An order the magnitude is required for a length scale over which the CBL adjusts to the new surface; a rough estimate can be obtained as follows. For the case of increasing buoyancy flux at the interface $x = 0$, convective plumes in the more strongly unstable, more turbulent downwind region have a velocity scale $w_* = (hg\overline{w\theta_v}/T)^{1/3}$ and intrude rapidly (with vertical velocity of order w_*) into the less turbulent upwind CBL. Therefore, the adjustment length scale X is estimated by

$$\frac{w_*}{U} \sim \frac{h}{X}, \qquad X \sim \frac{hU}{w_*}. \tag{21}$$

This estimate is motivated partly by the success of a similar simple estimate of the growth of the inner layer in thermally neutral, shear-driven flow over a hill, where the inner-layer depth grows as xu_*/U, u_* being the friction velocity (Finnigan et al. 1990).

In the opposite case of decreasing buoyancy flux at the interface, the strong convection in the upwind CBL has its energy supply attenuated at the ground as the flow crosses the interface. A rough estimate for the decay rate of the strongly convective turbulence can be obtained by approximating the turbulent energy budget in the upwind, strongly convective region with

$$\text{production rate} \approx w_*^3/h \approx$$
$$\text{dissipation rate} \approx w_*^2/T_{\text{diss}}, \tag{22}$$

where T_{diss} is a dissipation time scale. The assumption that energy production and dissipation are locally in balance is incorrect from point to point because of the prominent role of both turbulent and pressure transport in the CBL, but as an order of magnitude for the energy budget integrated across the CBL, Equation (22) is reasonable (Tennekes & Driedonks 1981). It follows that $T_{\text{diss}} \sim h/w_*$, and if the adjustment length scale X is assumed to be of order UT_{diss}, then the estimate $X \sim hU/w_*$ is obtained. This is

---

[1] It is significant that the *dynamic* time scale h/w_*, of order half an hour, is at least an order of magnitude less than the *thermodynamic* time scales T_{eq} and T_g discussed in the last section. This means that, although a CBL does not reach thermodynamic equilibrium within a daily cycle, dynamic equilibrium is easily achieved.

the same as the estimate in Equation (21) for the case of increasing buoyancy flux at the interface. In both cases, $X = hU/w_*$ is in the range of 1 to 10 km for typical values of the relevant quantities (around 1 km for h, 5 m s^{-1} for U and 2 m s^{-1} for w_*).[1]

Disclaimers are necessary here; I do not pretend that this simple length scale is rigorously based or that it is any more than a rough estimate in either the case of increasing or decreasing buoyancy flux at the transition. However, an order of magnitude for the CBL adjustment length scale X is all that is necessary in the following.

Having defined the length scale X, two classes of heterogeneous surface can be considered. Let L_x be the length scale of the surface heterogeneity or patchiness; say the typical size of a field or forested area in a mixed landscape. If $L_x \ll X$, so that the surface patches are small relative to the length scale for CBL adjustment, then the heterogeneity is *microscale*; the CBL cannot adjust to successive surface types and develops as it would over homogeneous terrain with surface properties defined as averages of those of the individual surfaces. The nature of the averaging is indicated below. If $L_x \gg X$, then the CBL adjusts completely to successive surface types and the heterogeneity can be labelled *mesoscale*. Shuttleworth (1988) has made a similar distinction, calling these two classes of heterogeneity 'disordered' and 'ordered', respectively. Some attributes of each class are now discussed, concentrating mainly on the microscale case.

Microscale heterogeneity: A landscape which is heterogeneous on the microscale can be envisaged as a collection of n surface patches, of which patch i occupies an area fraction f_i of the total area, has available energy $F_{Ai} = R_{ni} - G_i$ and has bulk aerodynamic and surface resistances r_{ai} and r_{ci}, respectively. Because the CBL is well-mixed and does not respond to individual surface patches, there is a common saturation deficit B and a common CBL depth h for all patches.

The definition of r_{ai} requires some care, as it represents the resistance to scalar transfer from the surface patch i to the well-mixed layer, integrated over the whole patch. Because of micro-

scale advection, scalar fluxes and surface concentrations will both vary from point to point within the patch. Formally, if C_0 is the absolute surface concentration and F_c the surface flux density of a scalar (both functions of local position) and C the mixed-layer concentration (independent of position), then

$$r_{ai} = \frac{\langle\!\langle C_0 \rangle\!\rangle - C}{\langle\!\langle F_c \rangle\!\rangle}, \qquad (23)$$

where the double angle brackets denote a horizontal average over the patch. This is not a problem in principle. Indeed, on a different scale, the boundary-layer resistance to scalar transfer from individual leaves is almost always implicitly averaged in this way over the leaf surface (Cowan 1972; Finnigan & Raupach 1987), even though local advection at leaf scale causes the surface concentration and flux to vary with position across the leaf (for example, surface temperature at the leading edge of a leaf is lower than at the centre). The consequence is that both the leaf boundary-layer resistance in the canopy, and r_{ai} as defined by Equation (23) in the CBL, parameterize the effects of local advection at scales much smaller than the system under consideration. The determination of r_{ai} is a task for a theory of local advection, such as that of Philip (1959, 1987).

Defining r_{ai} in this way, it is simple to write boundary conditions for the transfer of sensible and latent heat from each element of the patchy surface into the CBL, or more usefully, for the transfer of the scalars A and B. The condition for A is that the surface flux density F_{Ai} is prescribed, while that for B is a mixed boundary condition from Equations (12b) and (14):

$$F_{Bi} = (B_{eqi} - B)/r_{eqi}, \qquad (24)$$

where $B_{eqi} = \varepsilon F_{Ai} r_{ci}/(\varepsilon + 1)$ and $r_{eqi} = r_{ai} + r_{ci}/(\varepsilon + 1)$ are defined from the surface properties of patch i.

The mixed-layer equation for $B(t)$ over the patchy surface is obtained from Equation (1) (with $C = B$) by summing the contributions from each patch to the total surface flux of B:

$$\frac{dB}{dt} = \left(\frac{\sum_i f_i F_{Bt}}{h}\right) + \left(\frac{B_+ - B}{h}\right)\frac{dh}{dt} . \qquad (25)$$

Using Equation (24), this becomes:

$$\frac{dB}{dt} = \left(\frac{\sum_i f_i F_{eqi}/r_{eqi}}{h}\right) - \left(\frac{\sum_i f_i/r_{eqi}}{h}\right) +$$

$$+ \left(\frac{B_+ - B}{h}\right)\frac{dh}{dt} , \qquad (26)$$

which is the equivalent for microscale inhomogeneity of Equation (17) for a homogeneous surface. In fact, Equations (17) and (26) are identical if average surface properties for the patchy surface (denoted by single brackets) are defined thus:

$$\frac{1}{\langle r_{eq} \rangle} = \sum_i \left(\frac{f_i}{r_{eqi}}\right), \qquad (27a)$$

$$\frac{\langle B_{eq} \rangle}{\langle r_{eq} \rangle} = \sum_i \left(\frac{f_i B_{eqi}}{r_{eqi}}\right). \qquad (27b)$$

This means that the entire development for homogeneous surfaces is also applicable to the case of microscale inhomogeneity, with averaged surface properties defined by Equation (27). The averaged $\langle r_{eq} \rangle$ is the area-weighted parallel sum of all the r_{eqi} values for the individual patches, but the mixed-layer equilibrium saturation deficit is not the area-weighted average of the individual B_{eqi} values; rather, $1/r_{eqi}$ appears as a weighting factor as well. Further, because the average of $r_{eqi} = r_{ai} + r_{ci}/(\varepsilon + 1)$ is a simple area-weighted parallel sum, it is not possible to write averages of r_{ai} and r_{ci} as area-weighted parallel sums. However, this is not a problem because (at least for energy partition purposes) it is unnecessary to define averaged bulk aerodynamic and bulk surface resistances at all.[1]

---

[1] In fact, from Equation (13), $\langle r_c \rangle = (\varepsilon + 1)\langle B_{eq} \rangle / \varepsilon\langle F_A \rangle$.

From the mixed-layer saturation deficit and the surface properties, one can find the latent and sensible heat fluxes from each patch, and the averaged flux densities across the whole surface, by using the combination equation – most conveniently in the form of Equation (16b). This gives, for each patch

$$\lambda E_i = \frac{\varepsilon F_{Ai}}{\varepsilon + 1} + \frac{B - B_{eqi}}{(\varepsilon + 1)r_{eqi}} , \qquad (28a)$$

and for the average over the whole surface,

$$\langle \lambda E \rangle = \frac{\varepsilon\langle F_A \rangle}{\varepsilon + 1} + \frac{B - \langle B_{eq} \rangle}{(\varepsilon + 1)\langle r_{eq} \rangle} , \qquad (28b)$$

where both $B(t)$ and the surface properties (available energies and bulk resistances) are time-dependent in general, and where $\langle F_A \rangle$, like $\langle F_B \rangle$ and $\langle \lambda E \rangle$ itself, is a simple summation over the flux contributions from each patch:

$$\langle F_A \rangle = \sum_i f_i F_{Ai}, \qquad \langle F_B \rangle = \sum_i f_i F_{Bi},$$

$$\langle \lambda E \rangle = \sum_i f_i \lambda E_i . \qquad (29)$$

Figure 1 shows how these solutions work out in an idealised case, by using the analytic solution

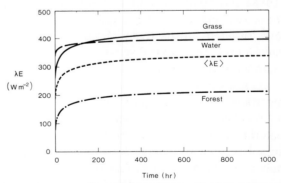

Fig. 1. Latent heat flux densities $\lambda E_i(t)$ and $\langle \lambda E \rangle (t)$ from a region with microscale heterogeneity, consisting of 50% grass, 40% forest and 10% water. Calculations use analytic solution for CBL depth proportional to $t^{1/2}$, Equations (20), (27), (28). Bulk aerodynamic and surface resistance values are from Table 1; available energy for all surfaces is $F_{Ai} = 500$ W m^{-2}, independent of time. Parameter values: $\varepsilon = 2.20$, $\rho = 1.20$ kg m^{-3}, $\lambda = 2.5 \times 10^{-6}$ J kg^{-1}, $K = 46$ m^2 s^{-1}, $\gamma_B/\lambda = 10^{-6}$ (kg m^{-3}) m^{-1}.

Table 1. The resistance r_{eq} and time scale T_{eq} (for a closed-box CBL) and T_g (for a growing CBL with $h(t) = (Kt)^{1/2}$). Assumed values were $\varepsilon = 2.2$ (appropriate for $T = 20\ C$), $h = 1$ km for the closed box CBL, and $h = 1$ km at $t = 6$ for the growing CBL, giving $K = 46$ m^2 s^{-1}.

Surface	r_a (s m^{-1})	r_c (s m^{-1})	r_{eq} (s m^{-1})	T_{eq} (hr)	T_g (hr)
Grass	50	50	66.6	18	36
Forest	20	200	82.5	23	55
Water	200	0	200	56	324

developed for Case 2 in the previous section. A heterogeneous region with $L_x \ll X$ (microscale heterogeneity) was assumed to be covered with 50% grass, 40% forest and 10% water; the bulk resistances for these three surfaces were taken as the values given in Table 1. For all surfaces, the available energy was taken as $F_{Ai} = 500$ W m^{-2}. The analytic solution does not include time dependence of the available energy or bulk resistances, so there is no diurnal cycle in the results, the time dependence being caused entirely by the evolution of the mixed-layer saturation deficit $B(t)$ according to Equation (20). The calculations show that there is a substantial difference in the evaporation rates between the three surfaces, with the forest evaporating about half as fast as the other two surfaces. This is a consequence of microscale advection associated with transport of saturation deficit in the mixed layer: all surfaces are exposed to a common, fairly low value of $B(t)$, so the forest (with high surface resistance r_c) evaporates more slowly than the other surfaces.

Mesoscale heterogeneity: The simplest assumption which can be made about a landscape which is heterogeneous on the mesoscale ($L_x \gg X$) is that the various surface patches ($i = 1$ to n) develop independent CBLs, each with its own depth $h_i(t)$ and mixed-layer saturation deficit $B_i(t)$. If the n patches appear within a single GCM grid cell, for example, then n independent mixed-layer models must be carried to determine $h_i(t)$ and $B_i(t)$. It is not possible to write a simple averaging scheme like Equation (27) to find the integrated behaviour of the patches from a single averaged CBL.

It must also be noted that dynamical motions

appear at the mesoscale, such as sea and lake breezes, which often have energies comparable with the resolved large-scale circulation in the GCM. These mean that the assumption of independent CBL development over different patches is often invalid, and can be considered as no more than an empirical averaging procedure. Even so, the assumption is probably much better than treating an entire grid cell with mesoscale heterogeneity as homogeneous.

Conclusions

Problems in modelling vegetation-atmosphere interaction, particularly energy, mass and momentum exchanges, fall into an 'upward' group concerned with the effect of vegetation on climate, and a 'downward' group concerned with the effect of a given climate on the vegetation microclimate. To describe processes at canopy scale, comprehensive, multilayer Canopy-Atmosphere Models (CAMs) are appropriate for the second task, while there is a role for single-layer Simplified Canopy-Atmosphere Models (SCAMs) in the first.

The presence of horizontal heterogeneity in most landscapes means that vegetation-atmosphere interaction can usefully be modelled at the PBL scale as well as the canopy scale, because the PBL is a well-mixed atmospheric region which naturally smooths out much of the surface-layer variability. A convective boundary-layer model for energy partition has been outlined, partially solved analytically under some idealisations, and then used to explore the effect of both microscale heterogeneity (with a terrain length scale

$L_x \ll X = hU/w_*$ where h is CBL depth, U wind speed and w_* the convective velocity scale) and mesoscale heterogeneity (with $L_x \gg X$).

The scheme outlined here offers a rational way of including microscale and mesoscale hetero-geneity into the CBL over flat terrain. Further pressing problems include the effects of hetero-geneity on the nocturnal boundary layer, and the complications caused by hilly terrain.

Appendix: a Simplified Canopy-Atmosphere Model (SCAM)

This illustrative SCAM is intended to describe the energy exchanges of a vegetated land surface within a GCM, incorporating the effects of albedo, vegetation type and soil moisture supply in the simplest possible way. The count of exter-nally prescribed land-surface parameters is kept as low as possible (to four). It is little more com-plicated than current practice in GCM land-surface description (Garratt 1990) but includes a more realistic vegetation treatment and avoids the need to iterate for surface temperature. The model is based on the bulk combination (Penman-Monteith) equation in the form

$$\lambda E = \frac{\varepsilon F_{A1} + \rho\lambda D_1/r_{aHR}}{\varepsilon + (r_c + r_{aE})/r_{aHR}} , \qquad (A1)$$

where the subscript 1 denotes values at a refer-ence level z_1, with r_{aH} and r_{aE} being the bulk aerodynamic resistances for heat and water vapour transfer between the surface and z_1, and r_{aHR} a modified resistance defined below.

The available energy flux density F_{A1} is defined by

$$F_{A1} = f_G[(1 - \alpha_c)R_S\downarrow + \varepsilon_c R_L\downarrow - \varepsilon_c \sigma T_1^4] . \qquad (A2)$$

Here, α_c is the canopy albedo, ε_c the canopy emissivity, σ the Stefan-Boltzmann constant and T_1 the air temperature at z_1 (not the surface: see below). To keep the SCAM parameter count as low as possible, it is reasonable to set ε_c to 1. The dimensionless number f_G accounts for the heat

flux density G into the ground; it is the fraction $(1 - G/R_n)$, where R_n is the net irradiance at the surface. Again to reduce free parameters, the universal value $f_G = 0.8$ may be adopted as a reasonable choice. The short-wave and long-wave downward irradiances at the surface, $R_{S\downarrow}$ and $R_{L,\downarrow}$ are assumed to be available from the GCM or from simple radiation models described (for example) by Paltridge and Platt (1976). With these assumptions, Equation (A2) determines F_{A1} from GCM computed variables T_1, $R_{S\downarrow}$ and $R_{L\downarrow}$ (though the last may be modelled) and the single surface parameter α_c.

The bulk resistances r_{aH} and r_{aE} for heat and water vapour transfer are defined by

$$r_{aH} = r_{aE} = \frac{\ln((z_1 - d)/z_0)\ln((z_1 - d)/z_{0HE})}{k^2 U_1} , \qquad (A3)$$

with k the van Karman constant (0.4), U_1 the mean wind speed at height z_1, d the canopy zero-plane displacement, z_0 the momentum roughness length and z_{0HE} the roughness length for heat and water vapour transfer. It is reasonable to approxi-mate d, z_0 and z_{0HE} as fractions of the canopy height h_c, by $d = 0.75h_c$, $z_0 = 0.1h_c$ and $z_{0HE} = z_0/5 = 0.02h_c$ (Thom 1971, 1972; Garratt & Hicks 1973).

The modified bulk resistance r_{aHR} for heat transfer includes a component accounting for the difference between the upward long-wave radiant flux density from the surface, $\varepsilon_c \sigma T_0^4$ (where T_o is the actual surface temperature), and the corre-sponding term used in Equation (A2), $\varepsilon_c \sigma T_1^4$. By linearizing the difference between these two flux densities, it is possible to eliminate the need to carry T_0 as a primary dependent variable in the SCAM. This is consistent with the overall philosophy of eliminating surface temperatures and humidities which underlies the bulk combi-nation equation, and also eliminates the need to iterate for T_0, though of course, T_0 can be calcu-lated *post facto* if required. Writing $\varepsilon_c \sigma(T_0^4 - T_1^4) \approx 4\varepsilon_c \sigma T_1^3(T_0 - T_1)$, one obtains the bulk combi-nation equation in the modified form of Equation (A1); compare with the more familiar form of

Equation (7). The modified bulk resistance r_{aHR} is defined by

$$r_{aHR} = (r_{aH}^{-1} + r_R^{-1})^{-1}, \qquad (A4)$$

where $r_R = \rho c_p / (4\varepsilon_c \sigma T_1^3)$ is a 'radiative resistance', typically about 230 s m^{-1}. Hence r_{aHR} is reduced below the actual bulk aerodynamic resistance for heat, r_{aH}, by parallel combination with the 'radiative resistance' r_R.

Finally the canopy (bulk surface) resistance must be specified. The simplest reasonable possibility appears to be

$$r_c = \frac{r_{c(\max)} f_I(I)}{f_M(M) f_{LAI}}, \qquad (A5)$$

where $r_{c(\max)}$ is an optimal canopy resistance dependent on the surface type (typically 0, 40, 100 s m^{-1} for water, grasses/crops and forests, respectively). Two 'physiological stress factors' f_M and f_{LAI}, both between 0 and 1, are included in Equation (A5) to account for environmental stresses which increase r_c above its optimal value (though more complicated models of r_c often include five or six stress factors: see Shuttleworth (1989) for examples). Here, $f_M(M)$ is a water stress factor dependent on the soil moisture deficit M, and f_{LAI} is a leaf area index factor accounting for incomplete canopy cover. Reasonable choices, guided by field data (Denmead & Shaw 1962; Dunin & Costin 1970; Dunin & Aston 1984) are:

$$
\begin{aligned}
f_M(M) &= 1 & (0 \le M < 5) \\
&= (20 - M)/15 & (5 \le M < 20) \\
&= 0 & (20 \le M) \qquad (A6)
\end{aligned}
$$

with M in centimetres of water, and

$$
\begin{aligned}
f_{LAI} &= LAI/3 & (LAI < 3) \\
&= 1 & (LAI \ge 3) \qquad (A7)
\end{aligned}
$$

The other factor in Equation (A5) is an interception factor $f_I(I)$, equal to 0 when the store I of intercepted water is greater than zero, and 1 when $I = 0$. This factor reduces r_c to zero when the canopy is wet. The soil moisture deficit M and

interception store I are both computed variables in the GCM, derived from simple canopy and soil water balances. If I_c is a canopy interception storage capacity (typically 1 mm), rainfall wets the canopy when $I < I_c$, otherwise entering the soil as throughfall; thereafter, runoff occurs when $M = 0$. In a GCM allowance must also be made for a possible patchy convective rainfall distribution across a grid cell, by specifying a rainfall intensity distribution (Shuttleworth 1989).

There are four surface parameters in this SCAM: α_c, h_c, $r_{c(\max)}$ and LAI. It is clear that a great deal of sophistication has been omitted to gain economy; the simplifications partly involve the omission of physics or plant physiology, and partly the assignment of global values to parameters which are not critical. Between this SCAM (which is probably close to the minimum possible level of description) and a fully-fledged multilayer CAM, there lies a large spectrum of possibilities.

References

Ball, F. K. 1960. Control of inversion height by surface heating. Quart. J. Roy. Meteorol. Soc. 86: 483–494.

Choudhury, B. J. & Monteith, J. L. 1988. A four-layer model for the heat budget of homogeneous land surfaces. Quart. J. Roy Meteorol. Soc. 114: 373–398.

Cowan, I. R. 1972. Mass and heat transfer in laminar boundary layers with particular reference to assimilation and transpiration in leaves. Agric. Meteorol. 10: 311–329.

Deardorff, J. W. 1972. Numerical investigation of neutral and unstable planetary boundary layers. J. Atmos. Sci. 29: 91–115.

Denmead, O. T. & Shaw, R. H. 1962. Availability of soil water to plants as affected by soil moisture content and meteorological conditions, Agron. J. 54: 505–510.

Dickinson, R. E. 1983. Land surface processes and climate – surface albedos and energy balance. In: B. Saltzmann (ed.) Theory of climate. Adv. Geophys. 25: 305–353.

Dickinson, R. E. 1984. Modeling evapotranspiration for three-dimensional global climate models. In: Hansen, J. E. (ed.) Climate processes and climatic sensibility. Geophysical Monograph 29: 58–72. American Geophysical Union.

Dickinson, R. E. & Henderson-Sellers, A. 1988. Modelling tropical deforestation: a study of ECM land-surface parameterizations. Quart. J. Roy. Meteorol. Soc. 114: 439–462.

Dickinson, R. E., Henderson-Sellers, A., Kennedy, P. J. & Wilson, M. F. 1986. Biosphere-atmosphere transfer scheme (BATS) for the NCAR Community Climate

Model. National Center for Atmospheric Research, Boulder, CO. Tech. Note TN 275 + STR.

Dunin, F. X. & Aston, A. R. 1984. The development and proving of models of large-scale evapotranspiration: an Australian study. Agric. Water Management 8: 305–323.

Dunin, F. X. & Costin, A. B. 1970. Analytical procedures for evaluating the infiltration and evapotranspiration terms of the water balance equation. Proc. IASH-UNESCO Symposium on the results of research on representative and experimental basins, Wellington, New Zealand, December 1970, pp. 39–54.

Finnigan, J. J. & Raupach, M. R. 1987. Transfer processes in plant canopies in relation to stomatal characteristics. In: Zeiger, E., Farquhar, G. & Cowan, I. (eds.), Stomatal Function. pp. 385–429. Stanford University Press, Stanford, CA USA.

Finnigan, J. J., Raupach, M. R., Bradley, E. F. & Aldis, G. K. 1990. A wind tunnel study of turbulent flow over a two dimensional ridge. Boundary-Layer Meteorol. 50: 277–317.

Garratt, J. R. 1990. The sensitivity of large-scale models to PBL and land-surface parameterizations. Proc. WRCP Workshop on PBL model evaluation, Reading, England, August 14–15 1989.

Garratt, J. R. & Hicks, B. B. 1973. Momentum, heat and water vapour transfer to and from natural and artificial surfaces. Quart. J. Roy. Meteorol. Soc. 99: 680–687.

McNaughton, K. G. 1976. Evaporation and advection I: evaporation from extensive homogeneous surfaces. Quart. J. Roy. Meteorol. Soc. 102: 181–191.

McNaughton, K. G. 1987. Comments on 'Modeling effects of vegetation on climate'. In: Dickinson, R. E. (ed.), The Geophysiology of Amazonia, pp. 339–342. Wiley, New York.

McNaughton, K. G. & Jarvis, P. G. 1983. Predicting the effects of vegetation changes on transpiration and evaporation. In: Water deficits and plant growth, VII. pp. 1–47. Academic Press, London.

McNaughton, K. G. & Spriggs, T. W. 1986. A mixed-layer model for regional evaporation. Boundary-Layer Meteorol. 34: 243–262.

Noilhan, J. & Planton, S. 1989. A simple parameterization of land surface processes for meteorological models. Mon. Weather Rev. 117: 536–549.

Paltridge, G. W. & Platt, C.M.R. 1976. Radiative processes in meteorology and climatology. Elsevier Scientific Publishing Company, Amsterdam.

Perrier, A. 1980. Etude microclimatique des relations entre les proprietes de surface, et les caracteristiques de l'air: application aux echanges regionnaux. Meteor. Environ., EVRY (France).

Philip, J. R. 1959. The theory of local advection. Int. J. Meteorol 16: 535–547.

Philip, J. R. 1987. Advection, evaporation and surface resistance. Irrig. Sci. 8: 101–114.

Raupach, M. R. 1989a. A practical Lagrangian method for relating scalar concentrations to source distributions in vegetation canopies. Quart. J. Roy. Meteorol. Soc. 115: 609–632.

Raupach, M. R. 1989b. Applying Lagrangian fluid mechanics to infer scalar source distributions from concentration profiles in plant canopies. Agric. For. Meteorol. 47: 85–108.

Raupach, M. R. 1989c. Stand overstorey processes. Phil. Trans. Roy. Soc. Lond. B. 324: 175–190.

Raupach, M. R. & Finnigan, J. J. 1988. Single-layer models of evaporation from plant canopies are incorrect but useful, whereas multilayer models are correct but useless: discuss. Aust. J. Plant Physiol. 15: 715–726.

Sellers, P. J. 1987. Modelling effects of vegetation on climate. In: Dickinson, R. E. (ed.), The Geophysiology of Amazonia. pp. 297–339. Wiley, New York.

Sellers, P. J., Mintz, Y., Sud, Y. C. & Dalcher. A. 1986. A Simple Biosphere Model (SiB) for use within General Circulation Models. J. Atmos. Sci. 43: 505–531.

Shuttleworth, W. J. 1988. Macrohydrology – the new challenge for process hydrology. J. Hydrol. 100: 31–56.

Shuttleworth, W. J. 1989. Micrometeorology of temperate and tropical forest. Phil. Trans. Roy. Soc. Lond. B. 324: 299–334.

Slatyer, R. O. & McIlroy, I. C. 1961. Practical microclimatology, CSIRO, Australia.

Suarez, M. J., Arakawa, A. & Randall, D. A. 1983. The parameterization of the planetary boundary layer in the UCLA general circulation model: formulation and results. Mon. Weather Rev. 111: 2224–2243.

Tennekes, H. 1973. A model for the dynamics of the inversion above a convective boundary layer. J. Atmos. Sci. 30: 558–567.

Tennekes, H. & Driedonks, A. G. M. 1981. Basic entrainment equations for the atmospheric boundary layer. Boundary-Layer Meteorol. 20: 515–531.

Thom, A. S. 1971. Momentum absorption by vegetation. Quart. J. Roy. Meteorol. Soc. 97: 414–428.

Thom, A. S. 1972. Momentum mass and heat exchange of vegetation. Quart. J. Roy. Meteorol. Soc. 98: 124–134.

Thom, A. S. 1975. Momentum, mass and heat exchange of plant communities. In: Monteith, J. L. (ed.), Vegetation and the atmosphere. Vol. 1. pp. 57–109. Academic Press, London.

Wilson, M. F., Henderson-Sellers, A., Dickinson, R. E. & Kennedy, P. J. 1987. Sensitivity of the biosphere-atmosphere transfer scheme (BATS) to the inclusion of variable soil characteristics. J. Clim. Appl. Meteorol. 26: 341–362.

Vegetatio **91**: 121–134, 1991.
A. Henderson-Sellers and A. J. Pitman (eds).
Vegetation and climate interactions in semi-arid regions.
© 1991 *Kluwer Academic Publishers. Printed in Belgium.*

Sensitivity of the land surface to sub-grid scale processes: implications for climate simulations

A. J. Pitman
School of Earth Sciences, Macquarie University, North Ryde, 2109, Australia

Accepted 24.8.1990

Abstract

Using a state-of-the-art land surface model in a 'stand-alone' mode with prescribed atmospheric forcing, a method for retaining the spatial extent and intensity of precipitation in Atmospheric General Circulation Models (AGCMs) is investigated. It is shown that the surface climatology simulated by this model is strongly dependent upon the fraction of the grid square, μ, receiving precipitation. It is also shown that fundamentally different hydrological regimes (one runoff dominated, the other evaporation dominated) are obtained for the otherwise identical situations. It is argued that the new generation of land surface models which explicitly incorporate vegetation may have to be 're-tuned' if differences between large-scale and small-scale precipitation events are accommodated in AGCMs. If precipitation is always assumed to fall uniformly over the grid square, the precipitation intensity will generally be underestimated. This will lead to an overestimation of canopy interception and interception loss. Furthermore, too little precipitation will reach the soil surface, and therefore surface runoff and the soil moisture will be underestimated. With too much interception loss, and too little soil water (and soil evaporation) there will be a tendency to re-cycle precipitation back to the atmosphere too quickly, leading to the unrealistic simulation of surface-atmosphere interactions. These results, if reproduced within an AGCM, would invalidate previous simulations of the effects of changing the state of the land surface on the atmosphere.

Introduction

A common method for investigating the effects of large scale soil or vegetation change on the climate is to use AGCMs. AGCMs were designed to simulate the Earth's climate. Recently, however, they have been coupled with advanced models of the land surface (incorporating both soil and vegetation) to simulate surface-atmosphere interactions at sub-continental resolutions. AGCMs are not ideal for sub-continental scale hydrological or meteorological prediction. Due to a lack of computational resources, AGCMs use a coarse resolution. The highest resolution currently used in AGCMs is around 2.8° latitude x 2.8° longitude (UK Meteorological Office model, Slingo 1985). At the other extreme is the Goddard Institute for Space Sciences (GISS) model (Hansen *et al.* 1983) which is commonly integrated with a horizontal resolution of 8° latitude × 10° longitude. The area of each grid 'element' is between approximately 10^4 and 10^6 km^2.

AGCMs predict a single value for each variable for each grid square at every timestep. At each time step, AGCMs predict whether precipitation occurred, and whether it was generated by large (e.g. frontal) or small scale (e.g. convective) processes. They also predict the quantity of precipitation which reaches the land surface. However, AGCMs currently simulate both large and small scale precipitation events as a uniform depth of water (a grid average) over the entire grid element (i.e. the spatial distribution of precipitation in AGCMs does not differ with different types of precipitation). In particular, precipitation formed in the model by convective processes (and therefore spatially restricted) is spread out uniformly over the entire grid square. Thus in the case of a small-scale precipitation event, such as a cumulus shower, the observed precipitation is localised and relatively intense (e.g. 10 mm h^{-1} over 100 km^2), in contrast to either a less intense event which occurs in the model (e.g. 10^{-2} mm h^{-1} over 100,000 km^2).

Actual large-scale precipitation events could conceivably fall over an area whose spatial extent is similar to the AGCMs entire grid element: cyclonic rainfall extends over considerable areas. More commonly, sub-grid-scale variability would exist so only a fraction of the area described by the grid element would receive rain or snow. Even if the entire area receives precipitation, the intensity, duration and hence total amount falling in various areas should vary. The assumption in current AGCMs of spatially averaged precipitation events implies that AGCMs must underestimate the surface precipitation intensity. However, it is also the case that AGCMs have been 'tuned' to produce reasonable amounts of precipitation (Hunt, pers. comm. 1990).

This paper examines a scheme which accounts for the lack of spatial heterogeneity in the precipitation simulated by AGCMs. In land surface schemes which include vegetation the intensity of precipitation is especially important. A dense canopy might intercept and re-evaporate virtually all the precipitation from the small scale event which is spread out over a full grid element. In contrast, if the precipitation were to be spatially restricted, the much higher precipitation intensity would result in fall-through or leaf-drip from the canopy reaching the soil moisture store or increasing runoff. The time difference between precipitation stored and re-evaporated from the canopy, and that stored and evaporated from the soil is considerable since canopies evaporate intercepted precipitation at the potential rate,

while soils offer a high resistance to it (Rutter 1975).

Below the current state of the parameterization of the land surface in AGCMs is briefly reviewed and the method for re-distributing precipitation over the grid element examined here is described and compared with the parameterization of hydrology in a state-of-the-art land surface model. Results from diurnal and seasonal scale simulations are discussed with the emphasis on the effects of incorporating the new precipitation distribution scheme.

Parameterizing land surface hydrology in AGCMs

The land surface in AGCMs

The land surface in AGCMs can be parameterized in a variety of ways. Methods range from the simple 'bucket model' described by Manabe (1969) which is still used in some AGCMs, to the recent generation of advanced parameterizations. The latter include an explicit representation of the effects of vegetation and soil on the surface energy and water balances.

There are now three land surface schemes which include a canopy explicitly and have been tested in AGCMs: The Biosphere Atmosphere Transfer Scheme (BATS, Dickinson et al. 1986), The Simple Biosphere Model (SiB) developed by Sellers et al. (1986) and Bare Essentials for Surface Transfer (BEST) developed by Pitman (1988). There are a series of other land surface schemes which, in terms of complexity, fall between Manabe's (1969) scheme and those discussed above. For instance, Warrilow et al. (1986) developed a model for the UK Meteorological Office 11-layer AGCM (Slingo, 1985). Although Warrilow et al. (1986) do not parameterize vegetation explicitly (although some of its effects are incorporated) they do provide a comparatively advanced soil model. Crucially, this model includes an attempt to incorporate a heterogeneous distribution of precipitation within the AGCM grid element. The model can account for the differences in the characteristics of large-scale and small-scale precipitation events only in terms of the simulation of surface runoff. Warrilow *et al.* (1986) suggest that the precipitation intensity simulated by an AGCM could be modified to account for its spatial variability by using a probability density function. Shuttleworth (1988) extended Warrilow *et al.*'s (1986) approach to include canopy interception, and it is Shuttleworth's (1988) scheme which is examined here. Sato *et al.* (1989) have recently suggested an alternative, although similar, scheme to Warrilow *et al.*'s (1986) and Shuttleworth's (1988) models (Sato *et al.* 1989, appendix 2). However, Sato *et al.* (1989) do not indicate how sensitive the AGCM is to their model, or how realistic their formulation is.

In this paper the sensitivity of BATS to incorporating the parameterization suggested by Warrilow *et al.* (1986) and Shuttleworth (1988) for distributing precipitation non-uniformly within the grid element for both the canopy and soil components of the model is investigated. In order to determine whether the type of approach suggested by Warrilow *et al.* (1986) and by Shuttleworth (1988) would produce a significant impact if incorporated into AGCMs, it is necessary to show whether current land surface models such as BATS are sensitive to changes in the spatial distribution of precipitation. Although these results are specific to BATS, the results are likely to be applicable to all models which incorporate a canopy parameterization.

Parameterization of land-surface hydrology in BATS

BATS is a state-of-the-art land surface model developed by Dickinson *et al.* (1986) and used, for example, by Wilson *et al.* (1987a,b) and Dickinson & Henderson-Sellers (1989). The model, by necessity, contains a number of simplifications. Here, the parameterization of surface runoff and canopy hydrology incorporated in BATS is reviewed. This model is fully explained by Dickinson *et al.* (1986), and only the relevant elements are discussed here.

Following Dickinson *et al.* (1986) surface runoff (R_{surf}) when the soil is unfrozen, is defined as

$$R_{surf} = \beta G \qquad R_{surf} \geq 0 \qquad (1)$$

where β is the soil wetness (kg kg^{-1}), and G is the net flux of water at the soil surface (kg m^{-2} s^{-1}).

In Equation (1), G is defined as

$$G = P - E_s + R_{drip} \qquad (2)$$

where E_s is the evaporation rate (kg m^{-2} s^{-1}), and R_{drip} is the canopy drip rate (kg m^{-2} s^{-1}).

The value of P used here is the grid box average rainfall rate. Thus, $R_{surf} > 0$ if $G > 0$, irrespective of the precipitation intensity. Thus in BATS no explicit infiltration capacity is defined, which suggests that BATS might overestimate R_{surf} particularly at low precipitation intensities as even a light shower will form some runoff (since $\beta = 0$ is extremely rare).

BATS parameterizes the canopy hydrology in an analogous way. In the version of BATS used here, the maximum water storage capacity (S) is given by

$$S = 0.1 L_{sai} \qquad (3)$$

where L_{sai} is the leaf and stem area index and the 0.1 factor is expressed in kg m^{-2}.

Then, if the amount of water intercepted by the canopy (C), which, in BATS is all the precipitation which falls over the vegetated fraction of the grid box, exceeds S

$$R_{drip} = (C - S)/\Delta t \qquad (4)$$

and the canopy storage is set to S. This scheme is conceptually similar to the parameterization of surface runoff used in Manabe's (1969) hydrology model, thus

$$R_{drip} = 0 \qquad \text{if } C < S \qquad (5)$$

The parameterization of R_{drip} and R_{surf} incorporated by BATS is probably a reasonable simplification if precipitation is assumed to fall homogeneously within the grid element. However, it will be shown later that if the parameterization described by Shuttleworth (1988) is included in BATS instead of Equation (1) to (5), a rather different set of results are obtained.

The Warrilow/Shuttleworth precipitation distribution scheme

In order to retain the spatial distribution and intensity of precipitation simulated by AGCMs, Warrilow et al. (1986) and Shuttleworth (1988, appendix 2), derived new expressions for surface runoff and canopy drip. The local precipitation rate, over the rain covered fraction of the AGCM grid box (μ) is assumed to follow a probability distribution (the expressions for R_{surf} and R_{drip} are fully derived by Warrilow et al. (1986) and Shuttleworth (1988) and are not repeated here).

When P_s (the net flux of water at the surface, in kg m^{-2} s^{-1}) exceeds F_s (the maximum surface infiltration rate (a function of soil texture and soil moisture content, in kg m^{-2} s^{-1}) the surface runoff rate (R_{surf}) over the fraction of the grid box not covered by vegetation can be defined following Warrilow et al. (1986) as (in kg m^{-2} s^{-1})

$$R_{surf} = P_s \exp\left[\frac{-\mu F_s}{P_s}\right] \tag{6}$$

When the precipitation intercepted by the canopy, P_c, (kg m^{-2} s^{-1}) exceeds the maximum canopy infiltration rate, F_c,(kg m^{-2} s^{-1}), the canopy drip rate (R_{drip}) over the fraction of the grid box covered by vegetation can be defined, following Shuttleworth (1988) as (in kg m^{-2} s^{-1}),

$$R_{drip} = P_c \exp\left[\frac{-\mu F_c}{P_c}\right] \tag{7}$$

In Equation (7) F_c can be defined following Shuttleworth (1988) as

$$F_c = (S - C)/\Delta t \tag{8}$$

where S is the maximum canopy storage capacity, (kg m^{-2}); C is the amount of water stored by the canopy, (kg m^{-2}) and Δt is the model time step.

Note that if the infiltration rate (F_s) is greater than P_s (i.e. the soil can absorb all the precipitation) there is no surface runoff. Similarly with the canopy, if the intercepted precipitation does not exceed F_c, then there is no canopy drip.

The main difficulty with this model is the definition of μ. Warrilow et al. (1986) define $\mu = 1.0$ for large scale precipitation and $\mu = 0.3$ for convective precipitation (precipitation type is predicted by the AGCM). Defining μ as a constant with respect to the precipitation type reduces the value of the approach. If μ can vary the advantages of incorporating Equations (6) and (7) increase considerably. AGCMs should be able to estimate the fraction of the grid element over which precipitation is occurring, either from fractional cloud cover (if this is a prognostic variable), or from the 'plume fraction of the grid element that is active during moist-convection events' (Entekhabi & Eagleson 1989). However, if μ cannot be calculated, defining it as Warrilow et al. (1986) proposed (i.e. distinguishing between large and small scales) may be better than current schemes.

Equations (6) through (8) provide a fundamentally different description of canopy and surface runoff than those currently incorporated in BATS. In this paper, the effects of incorporating these equations into BATS are described and the implications for re-modelling the land surface in this way in coarse resolution AGCMs are discussed.

Sensitivity experiments

Methodology

BATS was integrated for two model years in stand-alone mode (thereby preventing model feedbacks and sensitivity) in order to simulate the land surface climatology of a tropical forest using prescribed atmospheric forcing (summarised in Table 1) generated from World Weather Records

Table 1. List of the observations used in the prescription of atmospheric forcing for BATS.

	J	F	M	A	M	J	J	A	S	O	N	D
Air temperature (K)	300	300	301	301	300	299	298	298	299	299	300	300
Precipitation mm month^{-1}	40	96	159	270	319	402	440	279	392	276	164	34
Wind velocity	3 m s^{-1} (constant)											
Atmospheric pressure	1000 hPa (constant)											

(1957). The example of a tropical forest was chosen because, in BATS, this ecotype is prescribed with a dense canopy and high leaf area index which leads to a large interception capacity. Therefore using the example of a tropical forest should allow the sensitivity of the land surface to incorporating the μ type parameterization to be shown most clearly.

In the prescribed atmospheric forcing, air temperature and solar radiation are forced to show diurnal and seasonal cycles typical of a tropical forest biome typical of Calabar, Nigeria. The quantity of precipitation which occurs each month is realistic. However, the major weakness in the forcing is that the monthly total precipitation falls at regular intervals since this reduces the variability in the forcing which would be expected in natural systems. Although this is a simplification, this problem does not invalidate the results shown here as they are only sensitivity tests and not predictions. Wilson *et al.* (1987a,b) demonstrated that using BATS with prescribed forcing provided a valuable methodology to investigate the model response to forcing or parameter changes.

Four simulations were performed, a control simulation using a standard version of BATS, followed by three simulations using BATS, but incorporating the parameterization of R_{drip} and R_{surf} suggested by Shuttleworth (1988) for three different values of μ. In these three simulations, intended as sensitivity tests, the value of μ was changed, so that $\mu = 1.0$, $\mu = 0.5$ and $\mu = 0.1$. Only results from the second year will be discussed here as these are independent of initial conditions. The effect of different values of μ is shown in Fig. 1. Since μ describes the fraction of the grid element over which precipitation occurs, when $\mu = 1.0$ precipitation falls over the entire grid element, as is the case in most current AGCMs. However, when $\mu = 0.1$ precipitation only occurs over 10% of the grid element. As Fig. 1 shows, the resulting effective precipitation intensity at the surface is therefore strongly affected by the prescribed value of μ. The choice

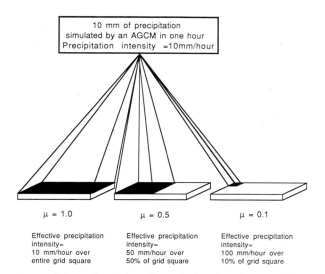

Fig. 1. Schematic of the effect of changing the prescribed value of μ on a hypothetical rainfall event. As μ decreases, the effective precipitation intensity increases since the area the rainfall occurs over is reduced.

126

of values for μ in these experiments was arbitrary but $\mu = 0.1$ is clearly not an extreme value for localised rainfall events. A selection of results will be shown which illustrate the effects on BATS of changing μ.

Diurnal simulations

The most important variable in terms of the daily results is the precipitation intensity. Table 1 shows that the amount of precipitation which falls in each month varies considerably. Figure 2 shows the precipitation forcing used by BATS for (a) January 15th and (b) July 15th. The prescribed precipitation forcing is seasonally variable (see

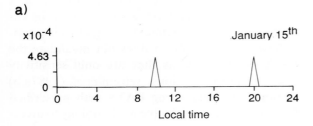

a)

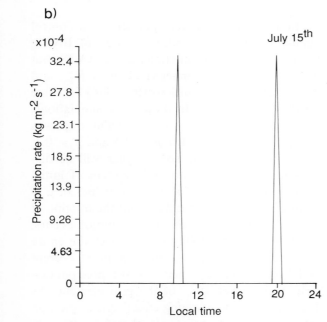

b)

Fig. 2. Precipitation forcing (kg m^{-2} s^{-1}) used by BATS for (a) January 15th and (b) July 15th.

Table 1), ranging from 40 mm month^{-1} in January to 440 mm month^{-1} in July. The precipitation forcing in Fig. 2 shows two showers occur each day. In January, each shower produces 1.1 mm while in July each shower produces 12.5 mm. There is only one curve in Fig. 2 because the precipitation forcing is identical for all values of μ. Although this precipitation forcing is rather schematic, the use of simple forcing helps to interpret these results. Further, the precipitation forcing used here is compatible with that used by Wilson *et al.* (1987b), and although more realistic forcing could be derived, remaining consistent with earlier experiments using BATS seemed advantageous.

The simulation of surface runoff in January 15th and July 15th is shown on Fig. 3. In Fig. 3a (January 15th) there is only one curve. The prescribed precipitation intensity did not exceed the infiltration capacity and hence surface runoff was zero for BATS including the μ parameterization for all the values of μ tested here. In contrast the standard version of BATS simulated some surface runoff from Equation (1) since $G > 0$. This surface runoff was about 4% of the incident precipitation at the surface. Therefore the quantitative differences between standard BATS and BATS incorporating the μ parameterization were relatively small in this case. Qualitatively however, Fig. 3a shows that at low precipitation intensities, standard BATS probably overestimates surface runoff as 1.1 mm of precipitation should not induce surface runoff unless the soil was saturated (which it was not in this case).

Figure 3b shows the surface runoff simulated for July 15th: it is strongly related to the value of μ. The standard version of BATS produced a runoff rate of 13.9 $\times$ 10^{-4} kg m^{-2} s^{-1}, while in the simulations using μ, surface runoff rates were 0 when $\mu = 1.0$, 16.2 $\times$ 10^{-4} kg m^{-2} s^{-1} when $\mu = 0.5$ and 29.5 $\times$ 10^{-4} kg m^{-2} s^{-1} when $\mu = 0.1$. Clearly the smaller the value chosen for μ, the higher the surface runoff. In the control case 41% of precipitation formed runoff, while when $\mu = 1.0$ no runoff occurred, when $\mu = 0.5$ 48% formed runoff and when $\mu = 0.1$ 88% formed runoff. The pattern of these results follow from

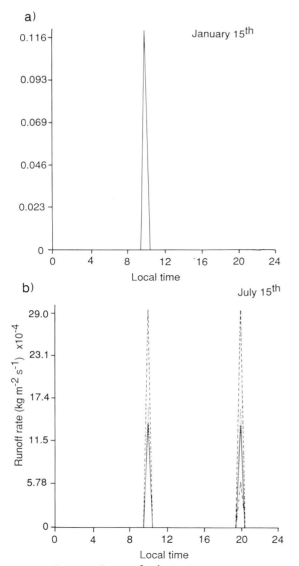

a)

January 15th

b)

July 15th

Runoff rate (kg m^{-2} s^{-1}) ×10^{-4}

Local time

Fig. 3. Surface runoff (kg m^{-2} s^{-1}) simulated by BATS for (a) January 15th 15th and (b) July 15th. In all figures, the solid line represents the control simulation, the dotted and dashed line represents $\mu = 1.0$, the dotted line represents $\mu = 0.5$ and the dashed line represents $\mu = 0.1$.

the fact that as precipitation intensity increased, the infiltration capacity was exceeded earlier and more often hence the surface runoff increased. In contrast, if $\mu = 1.0$, the precipitation intensity was not great enough to exceed the infiltration capacity and form runoff.

In the case of water intercepted by the canopy,

the results are similar to the those for surface runoff. Figure 4a shows the pattern for January 15th. There are three peaks because the precipitation intercepted the previous evening requires time to evaporate from the canopy. The midday peak of 0.4 mm is about half of the incident precipitation. However, this is not the amount of precipitation actually intercepted. Rather it is the amount of water remaining on the canopy at the end of a time step (equal to 30 minutes), and after evaporation for that time step has occurred. Hence the third peak of the day is higher (0.45 mm) since less evaporation occurred during this time step which was during the night. Figure

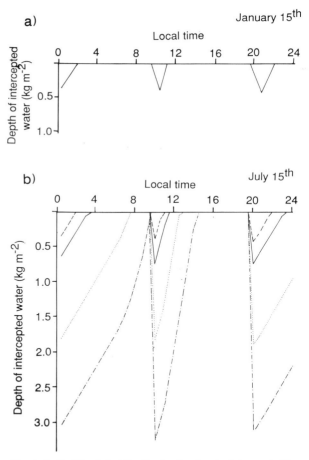

a)

January 15th

Local time

Depth of intercepted water (kg m^{-2})

b)

July 15th

Local time

Depth of intercepted water (kg m^{-2})

Fig. 4. As for Fig. 3, but for the depth of water intercepted by the canopy, and remaining at the end of the current time step (kg m^{-2}). The solid line represents the control simulation, the dotted and dashed line represents $\mu = 1.0$, the dotted line represents $\mu = 0.5$ and the dashed line represents $\mu = 0.1$.

128

4a shows that interception during January 15th was insensitive to the value of μ. It was pointed out earlier that the January 15th precipitation event shown here produced 1.1 mm of rainfall. Even when $\mu = 0.1$, the precipitation intensity was too low to affect the interception as the canopy could still intercept all the rainfall falling onto it.

Figure 4b shows a different pattern for July 15th. Although it is basically similar with evaporation of the previous day's interception occurring in the early morning, followed by two subsequent precipitation events refilling the canopy interception store, there are two major differences. In all the simulations discussed here, BATS has a maximum interception store of 0.64 mm (from Equation 3). With BATS incorporating the μ parameterization, more precipitation is intercepted (up to 3.2 mm) because the maximum interception store is used in a different way (e.g. Equation 8). In the standard version of BATS, when the maximum interception store is exceeded, all the excess water is assumed to drip to the ground surface in a single time step. In BATS incorporating the μ parameterization not all of the excess water necessarily drips to the surface in any given time step. Thus the maximum interception store can be exceeded, and remain for several time steps.

Figures 5a and 5b show that major differences occur in the soil water budget because of the differences in interception and runoff discussed above. With higher precipitation intensities, more precipitation forms runoff, leading to a relatively depleted soil moisture reservoir (curve $\mu = 0.1$). In contrast, precipitation at lower intensities does not run off, rather it adds to the soil moisture store (curve $\mu = 1.0$ and $\mu = 0.5$). Clearly less precipitation reached the surface at lower precipitation intensities because of higher interception rates. However, more of the precipitation which reaches the surface actually infiltrates into the soil. The differences between the four curves in Figs 5a and 5b are comparatively large. The percentage differences between the control and the perturbation simulations was 11% ($\mu = 0.1$), 1% ($\mu = 0.5$) and 4% ($\mu = 1.0$) for January 15th and 18%

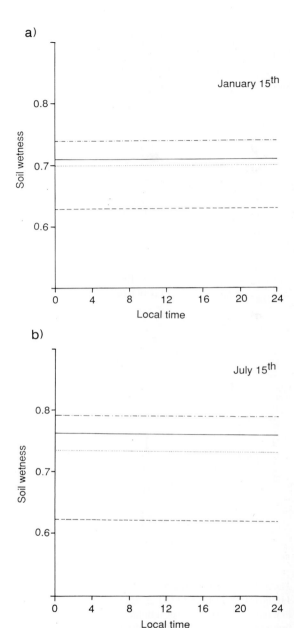

Fig. 5. As for Fig. 3, but for the soil moisture averaged over the total soil profile and expressed as a fraction of saturation where saturation = 1.0. The solid line represents the control simulation, the dotted and dashed line represents $\mu = 1.0$, the dotted line represents $\mu = 0.5$ and the dashed line represents $\mu = 0.1$.

($\mu = 0.1$), 4% ($\mu = 0.5$) and 4% ($\mu = 1.0$) for July 15th. At low values of μ (e.g. 0.1) BATS appears to be rather sensitive to how the precipitation forcing is prescribed.

BATS has been shown to be sensitive to the specification of μ. However, if the μ parameterization were to be incorporated into AGCMs the sensitivity of the variables which could feedback

into the climate simulation itself should also be examined. Figure 6a shows the simulation of evaporation for January 15th. These curves represent the area-average evaporation simulated by BATS

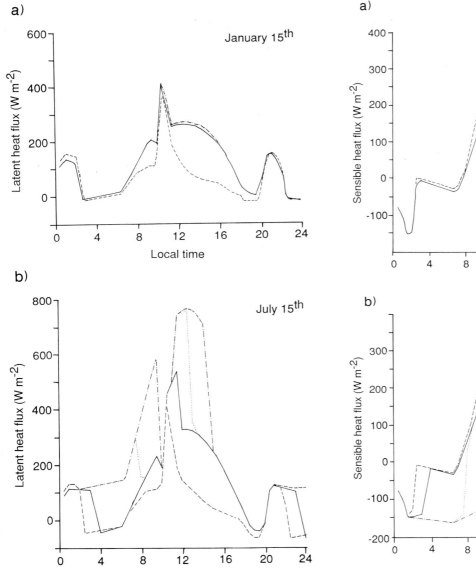

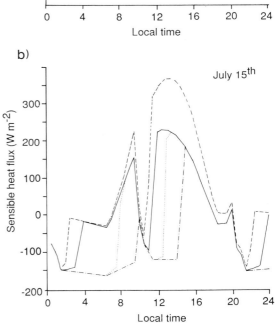

Fig. 6. As for Fig. 3, but for evaporation (W m^{-2}). This flux is the total latent heat flux between the land surface (soil and canopy) and the lowest model layer of the AGCM. The flux is directed upwards, so positive quantities cool and dry the surface. The solid line represents the control simulation, the dotted and dashed line represents $\mu = 1.0$, the dotted line represents $\mu = 0.5$ and the dashed line represents $\mu = 0.1$.

Fig. 7. As for Fig. 3, but for the sensible heat flux (W m^{-2}). This flux is the total sensible heat flux between the land surface (soil and canopy) and the lowest model layer of the AGCM. The flux is directed upwards, so positive quantities cool the surface. The solid line represents the control simulation, the dotted and dashed line represents $\mu = 1.0$, the dotted line represents $\mu = 0.5$ and the dashed line represents $\mu = 0.1$.

130

(evaporation includes soil and canopy fluxes plus transpiration). Three of the curves are almost identical, but curve $\mu = 0.1$ is quite different due to changes in the soil evaporation (since Fig. 4a shows that there was no difference in intercepted water). The soil moisture balance (Fig. 5a) for curve $\mu = 0.1$ was so different from the other three curves that it led to a lower soil evaporation rate. In contrast, Fig. 6b (July 15th) shows four different curves. Again curve $\mu = 0.1$ is lower than the other three curves, but otherwise, particularly around mid-day, the curves are different because of the variation in the canopy moisture store (Fig. 4b). The higher the quantity of intercepted water, the higher the resulting evaporation rate. Hence, while the control simulation predicted a peak daily evaporation rate of about 475 W m^{-2}, curve $\mu = 1.0$ and $\mu = 0.5$ predicted 670 W m^{-2}. Figures 7a and 7b shows an analogous and corresponding pattern for the sensible heat flux. In the drier $\mu = 0.1$ scenario the sensible heat flux was much higher (350 W m^{-2}), while lower fluxes (250 W m^{-2}) occur from the simulations where the soil and canopy were wetter.

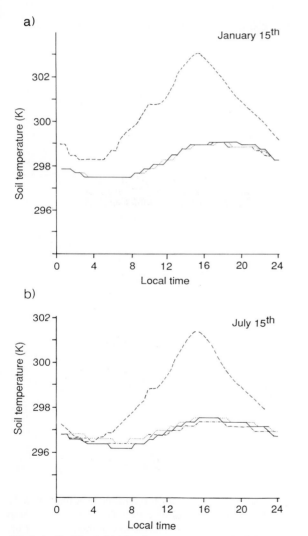

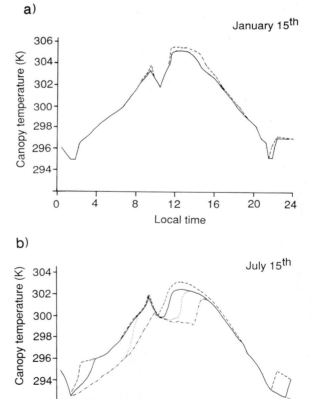

Fig. 8. As for Fig. 3, but for the temperature (K) of the upper soil layer (0.1 m deep). The solid line represents the control simulation, the dotted and dashed line represents $\mu = 1.0$, the dotted line represents $\mu = 0.5$ and the dashed line represents $\mu = 0.1$.

Fig. 9. As for Fig. 3, but for the temperature (K) of the canopy. The solid line represents the control simulation, the dotted and dashed line represents $\mu = 1.0$, the dotted line represents $\mu = 0.5$ and the dashed line represents $\mu = 0.1$.

The differences in the soil moisture and energy fluxes led to changes in the soil and canopy temperatures simulated by BATS. Figures 8a and 8b show the upper soil layer temperature. There are few differences between the control curve, and the curves when $\mu = 1.0$ and 0.5, but curve $\mu = 0.1$ shows a dramatic difference with an enhanced diurnal temperature cycle. The drier soil led to a reduced latent heat flux and thus higher temperatures. The different temperature in curve $\mu = 0.1$ partly explains the sensible and latent heat flux differences shown in Figs. 6 and 7. The canopy

temperature (Figs. 9a and 9b) is also different. The lower quantity of precipitation intercepted when $\mu = 0.1$ leads to less canopy evaporative cooling and a higher canopy temperature (Fig. 9a). The other three curves are identical. In contrast, Fig. 9b shows that the different interception and evaporation rates lead to canopy temperature differences of up to 4K around noon.

It must be re-iterated here that these simulations are not intended as predictions, rather they are sensitivity tests, hence the absolute value of the curves is much less important than the differ-

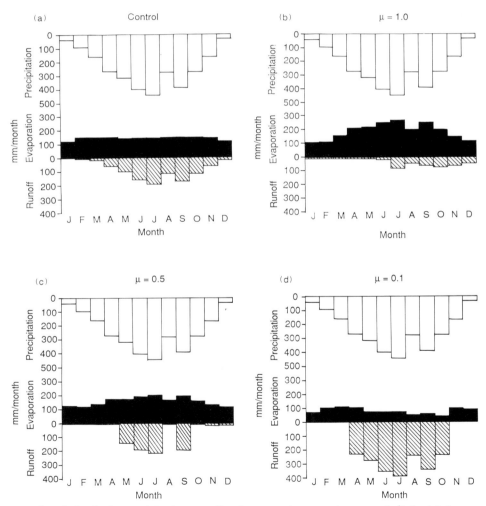

Fig. 10. The seasonal variation in the precipitation, runoff and evaporation rates (mm month^{-1}) for (a) the control simulation (BATS without the μ parameterization), 'b' for BATS with the μ parameterization with $\mu = 1.0$, 'c' with $\mu = 0.5$ and 'd' with $\mu = 0.1$. Precipitation, in each case, is identical, and is plotted downward from the top 'X' axis. Runoff is plotted downward form the lower 'X' axis, while evaporation is plotted upwards from the lower 'X' axis, and is shaded.

ences between the curves which require further discussion. It is clear that the amount of water intercepted and remaining on the canopy at the end of the time step is strongly related to the precipitation intensity, expressed here through μ. At the highest intensities ($\mu = 0.1$) most of the precipitation falls through a canopy leading to high surface runoff rates (Fig. 3b) or adding to the soil moisture store. The implications of this are fundamental to whether precipitation is stored for relatively long periods in the soil, or recycled back to the atmosphere rapidly through canopy evaporation. In current AGCMs, which do not consider the sub-grid-scale variability of precipitation (i.e. $\mu = 1.0$) and which include land surface models with a canopy, interception and re-evaporation would probably be overestimated. This would be at the expense of soil evaporation and runoff, since the effective precipitation intensity would be generally overestimated since the spatial distribution of precipitation is too large.

Seasonal simulations

Although Figs 2–9 show the basic pattern of the response of BATS to incorporating the parameterization suggested by Shuttleworth (1988) at a diurnal time scale, the analysis of AGCM simulations tends to be concentrated more on monthly and seasonal time scales. Thus results of an annual cycle for all four experiments are also examined.

Figure 10 shows four histograms, each with monthly total precipitation, evaporation and surface runoff. The standard BATS control simulation (Fig. 10a), can be compared with Figs 10b, c and d which shows results from BATS with the

μ scheme, with $\mu = 1.0, 0.5$ and 0.1 respectively. Despite strong seasonality in the precipitation forcing (Fig. 10a), the standard version of BATS shows little variation in the simulated evaporation flux which ranges between 115 mm month^{-1} (December) and 152 mm month^{-1} (April). Runoff, in contrast, mirrors the precipitation forcing with the maximum (197 mm month^{-1}) reached in July. Figure 10b shows a rather different pattern for BATS with $\mu = 1.0$. Here, evaporation is highly seasonal (116–269 mm month^{-1}), and closely mirrors precipitation. Runoff is less dependent on precipitation, and is negligible until June, after which about half the amount of runoff shown on Fig. 10a is simulated. Thus, in the control case, of the total precipitation, 37% formed runoff and 59% evaporated (see Table 2). When $\mu = 1.0$, only 15% formed runoff and 79% evaporated (note that the percentages do not sum to 100% because of soil moisture changes).

The simulation using BATS with $\mu = 0.5$ is shown in Fig. 10c. The seasonal pattern of evaporation is similar to, although higher than, the control case. Runoff, however, is quite different, with almost no runoff except in the summer months. Finally Fig. 10d shows BATS with $\mu = 0.1$. Here most of the precipitation forms runoff during the summer, with no runoff during the winter. In the case of $\mu = 0.5$, Table 2 shows that 65% of the precipitation evaporates with 29% as runoff. When $\mu = 0.1$, only 34% evaporates and 72% forms runoff.

It has been shown that BATS can simulate three quite different hydrological regimes simply by changing the prescribed value of μ (Fig. 10). It is clear that a considerable amount of care will have to be taken in the calculation of μ in AGCMs, otherwise the resulting simulations

Table 2. Variation of evaporation and runoff as a function of the prescribed value of μ.

	P	E	% of P	R	% of P
Control	2870	1687	59%	1062	37%
$\mu = 1.0$	2870	2274	79%	436	15%
$\mu = 0.5$	2870	1870	65%	822	29%
$\mu = 0.1$	2870	964	34%	2071	72%

could become meaningless. Figures 10b,c and d differ from one another both in terms of magnitude and shape. Evaporation peaks at different times of the year when $\mu = 1.0$ compared to the $\mu = 0.1$ case. The control simulation shows no seasonality in the evaporation, while with BATS incorporating the μ scheme, the seasonality in the simulated evaporation is decreased as μ decreases.

Discussion

A comparison of Figs 10b and 10d indicates how important it is to consider the sub-grid scale variability of precipitation. Fig. 10b shows a simulation of a hydrological regime dominated by evaporation, while Fig. 10d shows one dominated by runoff. It is true that the differences illustrated are maximised by the precipitation forcing used here (all large-scale ($\mu = 1.0$) or all small scale ($\mu = 0.1$) precipitation). Further, the choice of simulating a tropical forest ecotype means that these results can only be reliable for that ecotype. However, it would be expected that all land surface types dominated by vegetation would behave in similar ways to those discussed above.

In land surface models which incorporate canopy processes explicitly it is particularly important to incorporate a μ-type scheme. AGCMs should underestimate precipitation intensity because the simulated precipitation is spread out over the entire grid element, rather than only occurring over localised areas. Thus the importance of the canopy would become overstated since rather too much precipitation could be intercepted. Similarly, in AGCMs which do not incorporate a canopy, runoff may be underestimated since the AGCMs spread the precipitation over too large an area.

By changing the value of μ, very different hydrological regimes can be simulated, ranging from runoff to evaporation dominated systems. Despite the rather simple precipitation forcing, and the use of a constant value of μ in each simulation, the results described here do suggest that BATS is sensitive to the fraction of the grid element over which precipitation occurs. This finding has serious implications with respect to the use of AGCMs in studying the land surface-atmosphere interaction. Simulations using AGCMs which have tried to identify the causes and climatic consequences of deforestation and desertification, and yet lack either a realistic parameterization of the land surface or the μ type scheme discussed above are unlikely to produce similar results to those that do. The latest generation of land surface models do include vegetation (e.g. BATS) which leads to an increase in physical realism. However, this paper has shown that to improve the modelling of the land surface further requires improvements in the atmosphere-surface coupling, rather than simple changes to the land surface model itself.

Summary

According to Entekhabi & Eagleson (1989) μ can be calculated reasonably accurately. If this is not the case, it can easily be prescribed following Warrilow et al. (1986), hence it is important to incorporate the type of parameterization discussed above into AGCMs. Changing μ modifies the effective precipitation intensity. In the simulations reported here the Biosphere Atmosphere Transfer Scheme (BATS), exhibited considerable sensitivity to changes in precipitation intensity especially in the simulation of intercepted water and runoff. Changing μ altered the amount of intercepted water, leading to changes in the canopy temperature and canopy energy fluxes. Changing μ also affected the simulation of runoff, leading to changes in the soil wetness and soil temperature, and thus on the partitioning of available energy between latent and sensible heat fluxes. Varying μ also greatly affected the seasonal distribution of evaporation and runoff. Extreme values of μ (1.0 and 0.1) resulted in the simulation of fundamentally different hydrological regimes for otherwise identical input and forcing parameters.

However successfully a land surface model accounts for interception, runoff and evaporation, unless a reasonable precipitation intensity (i.e. appropriate to its spatial area) is simulated by the AGCM the calculation of interception, re-evaporation of water and runoff will be incorrect. If AGCMs do underestimate precipitation intensity, then interception and re-evaporation will be overestimated and runoff underestimated. The present treatment of the spatial distribution of precipitation in most AGCMs may therefore be unsatisfactory when coupled to state-of-the-art land-surface schemes.

Although these results are preliminary they do suggest that incorporating a μ type parameterization into AGCMs is a matter of some urgency. In particular, those AGCMs which incorporate a canopy may be extremely sensitive to the underestimation of precipitation intensity by current climate models. Rather than simply increasing the complexity of land surface models, more attention needs to be paid to how to incorporate sub-grid scale processes into AGCMs. Including the μ-type parameterization would seem to be a relatively easy (and computationally cheap) improvement. However, with respect to the canopy, some care has to be taken in the definition of the interception capacity otherwise excessive quantities of water can remain on the canopy for several time step.

Finally, rather than just assessing the amount of precipitation simulated by AGCMs, some attention needs to be paid to how well these models simulate precipitation intensity and frequency. Two AGCMs can both simulate 1000 mm of precipitation. However, if one simulates it all in one time step, while another simulates a series of showers, an analysis of monthly averages will not identify a problem with the simulated precipitation, and serious model deficiencies will remain hidden.

Acknowledgments

AJP holds a Macquarie University Research Fellowship. I would like to thank B.G. Hunt for a thorough and valuable review of this manuscript.

References

Dickinson, R. E., Henderson-Sellers, A., Kennedy, P. J. & Wilson, M. F. 1986. Biosphere Atmosphere Transfer Scheme (BATS) for the NCAR Community Climate Model, NCAR Technical Note, NCAR, TN275 + STR, 69 pp.

Dickinson, R. E. & Henderson-Sellers, A. 1988. Modelling tropical deforestation: a study of GCM land-surface parameterizations, Quart. J. Roy. Meteor. Soc. 114(b): 439–462.

Entekhabi, D. & Eagleson, P. S. 1989. Land surface hydrology parameterization for atmospheric general circulation models including subgrid scale spatial variability, J. Climate 2: 816–831.

Hansen, J. E., Russell, G., Rind, D., Stone, P. H., Lacis, A. A., Lebedeff, S., Ruedy, R., Travis, L., 1983. Efficient Three Dimensional Global Models for Climate Studies: Models I and II. Mon. Wea. Rev. 111: 609–662.

Manabe, S. 1969. Climate and the ocean circulation: 1. The atmospheric circulation and the hydrology of the earths' surface. Mon. Wea. Rev. 97: 739–305.

Pitman, A. J. 1988. The development and implementation of a new land surface scheme for use in GCMs. Unpublished Ph.D thesis, Liverpool University, 481 pp.

Rutter, A. J. 1975. The hydrological cycle in vegetation. In Vegetation and the Atmosphere, 1. Monteith, J. L., (ed.), 111–154, Academic Press, 278 pp.

Sato, N., Sellers, P. J., Randall, D. A., Schneider, E. K., Shukla, J., Kinter, J. L., Hou, Y.-T. & Albertazzi, E. 1989. Implementing the Simple Biosphere Model (SiB) in a General Circulation Models: Methodologies and Results, NASA Contractor Report 185509, August, 1989, 76 pp.

Sellers, P. J., Mintz, Y., Sud, Y.C. & Dalcher, A., 1986. A Simple Biosphere model (SiB) for use within general circulation models. J. Atmos. Sci. 43: 505–531.

Shuttleworth, W. J. 1988. Macrohydrology – the new challenge for process hydrology. J. Hydrology 100: 31–56.

Slingo, A. 1985. (ed.), Handbook of the Meteorological Office 11 layer atmospheric general circulation model, Volume 1: Model description. MET O 20 DCTN 29, Meteorological Office, Bracknell, Berkshire.

Warrilow, D. A., Sangster, A. B. & Slingo, A. 1986. Modelling of land surface processes and their influence on European climate, Dynamic Climatology Tech. Note No. 38, Meteorological Office, MET O 20. (Unpublished), Bracknell, Berks, 94 pp.

Wilson, M. F., Henderson-Sellers, A., Dickinson, R. E. & Kennedy, P. J., 1987a. Investigation of the sensitivity of the land-surface parameterization of the NCAR Community Climate Model in regions of tundra vegetation. J. Climatol. 7: 319–343.

Wilson, M. F., Henderson-Sellers, A., Dickinson, R. E. & Kennedy, P. J. 1987b. Sensitivity of the Biosphere-Atmosphere Transfer Scheme (BATS) to the inclusion of variable soil characteristics. J. Clim. Appl. Meteor. 26: 341–362.

World Weather Records, 1941–50. 1957. U.S. Dept. Commerce, Superintendent of documents. U.S. Govt. Printing Office, Washington, D.C.

Vegetatio **91**: 135–148, 1991.
A. Henderson-Sellers and A. J. Pitman (eds).
Vegetation and climate interactions in semi-arid regions.
© 1991 *Kluwer Academic Publishers. Printed in Belgium.*

Predicting evaporation at the catchment scale using a coupled canopy and mixed layer model

H. A. Cleugh
School of Earth Sciences, Macquarie University, North Ryde, NSW 2109, Australia

Accepted 24.8.1990

Introduction and rationale

Climate and vegetation are inextricably linked. Vegetation is sensitive to changes in the climate forcing and climate (at least at the regional scale) is sensitive to vegetation changes. Such modifications to the vegetation or climate may be either deliberate or inadvertent. Evaporation plays a key role in the interrelationship between climate and vegetation. The water balance shows that the difference between precipitation and runoff + evaporation determines the soil moisture, and plant growth and productivity strongly depend on this available moisture supply. The evaporation of water from the land surface provides the input of water vapour into the lower atmosphere for precipitation processes such as condensation, cloud formation and rain. All of these are of particular importance to everyday weather and the longer-term climate. This paper addresses the problem of predicting areal evaporation at large spatial scales.

A full understanding of the link between vegetation degradation and climate requires an ability to predict this exchange of moisture between the vegetation surface and the atmosphere. Climate and land use changes impact the water and energy balances at all spatial scales. While current general circulation models (GCMs) provide a view of climate changes at the global scale, planners and resource managers demand a knowledge of impacts on weather, climate and hydrology at the regional (10^2 km) or catchment (10^1–10^2 km) scales. Clearly there is a demand for prediction at

these mesoscales, the question is how can such predictions be made?

Areal evaporation can be measured using aircraft-based techniques. Although a number of studies (e.g. Schuepp *et al.* 1987; Hacker 1988; Hacker *et al.* 1989) demonstrate the viability of such methods, there is still considerable uncertainty regarding their accuracy. While they have considerable potential to provide validation data for evaluating regional and catchment scale evaporation models, aircraft-based measurements of turbulent fluxes are still not predictions.

A range of predictive surface evaporation models have been developed within the field of micrometeorology. One of the more popular is the Penman Monteith form of the Combination model. Such models are limited to local spatial scales (ca 1 km) and primarily one dimensional. The Penman Monteith model must therefore parameterize a complex, three dimensional vegetation canopy into two resistance terms which express the surface moisture availability (the surface resistance, r_s) and atmospheric turbulent mixing (r_a). The Penman Monteith model is forced using measured surface and atmospheric variables. These are typically measured within the surface layer where they are influenced by the underlying surface type. As a predictive model, this is a limiting feature. McLeod (1989) demonstrates the limitations of simulating potential evaporation rates using Penman Monteith (with a zero surface resistance) forced with variables measured 1 m above dry grass. This dry surface maintains a large near-surface vapour pressure deficit (D_s)

which in turn forces a high 'potential' evaporation rate. If the same surface was truly saturated (i.e. r_s close to zero) then D_s would be smaller, and the true potential evaporation rate lower. A prognostic form of the Penman Monteith model requires input data determined at a reference height where it is less influenced by the underlying surface. Without such inputs, it is limited as a predictive model. Any other form of the Combination model would suffer from the same limitations.

There are other models, for example the complementary approach, which endeavour to predict large scale evaporation. Some authors have found flaws with this model (McNaughton & Spriggs 1989), whilst others have found that it simulates evaporation rates realistically – although it is not always used in a truly predictive mode (e.g. Byrne *et al.* 1988).

This brief summary illustrates that predicting regional scale evaporation is difficult, despite the very real need for such estimates. The object of this paper is to present and evaluate an approach, albeit a simple one, to predicting evaporation at the catchment or regional scale. It links a canopy evaporation with a PBL (Planetary Boundary Layer) growth model and so incorporates feedbacks between the surface and the atmosphere. The theoretical background to this modelling approach is presented, its implementation described and performance evaluated for predicting evaporation from a grassland catchment in NSW. Finally, the contribution of the modelling approach to predicting evaporation is assessed using sensitivity analyses.

The coupled boundary layer growth/Canopy evaporation model

Theory and model equations

The model comprises two sub-models: a Penman Monteith canopy evaporation model and a boundary layer growth model. The two are linked by the potential saturation deficit: D_s, defined as

$$D_m = q*(\Theta_c) + s(\Theta - \Theta_c) - q \quad (1)$$

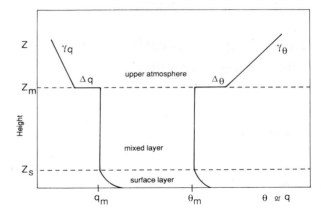

Fig. 1. A schematic representation of the PBL in the coupled mixed layer growth/canopy evaporation model. See text for explanation of symbols.

where q is the specific humidity; $q*$ is the saturation specific humidity; Θ is potential temperature and subscript c denotes a canopy value.

Figure 1 is a schematic view of the structure of the convectively-dominated PBL assumed by the 'slab' model of mixed layer growth. Observations have shown that this simplification is valid under clear sky, typically anticyclonic conditions. Further mention of these limiting synoptic conditions will be made later in the paper. Essentially the convectively-dominated PBL, or CBL, comprises a relatively thin surface layer where the gradients of potential temperature and humidity are steep. The fluxes of heat, moisture and momentum are considered constant with height in the surface layer – whose upper and lower limits are denoted as z_s and $z_o + d$, respectively. z_o is the roughness length which varies with heat, water vapour and momentum transfer; d is the zero plane displacement length. Above the surface layer (maximum depth *ca* 50–80 m) is the mixed layer. Here, as the name suggests, strong mixing eliminates gradients in potential temperature and humidity. The fluxes of water and heat therefore are assumed to linearly decrease with height. Above the mixed layer is the upper atmosphere whose properties are determined by synoptic scale processes. The upper limit of the mixed layer, z_m, is usually marked by a potential temperature inversion whose strength ($\gamma_\Theta = d\Theta/dz$) may be enhanced by anticyclonic subsidence. The humidity gradient (γ_q) is not only lapse in the upper atmosphere, but observations show that it is often slightly lapse throughout the mixed-layer. The interface between the mixed-layer and the upper atmosphere is parameterized as a 'jump' ($\Delta\Theta$) in the zero order slab model described here. This interface, in reality, is a layer – termed the interfacial layer or zone.

The mixed layer grows from sunrise as the input of sensible heat from the surface leads to encroachment into the nocturnal inversion. Plume bombardment at the base of the capping inversion also entrains air from aloft. Heat is transferred downwards into the mixed layer by encroachment and entrainment and there is thus a negative (downwards directed) heat flux from the upper atmosphere into the mixed layer. Water vapour, on the other hand, is transferred from the surface layer into the mixed layer and then upwards into the upper atmosphere.

This basic conceptual 'slab' model treats the mixed-layer as a slab or volume of air which is transparent to radiation and whose temperature and humidity depends on the input of heat and water vapour from the surface layer (which is sufficiently thin to have negligible storage) and the input (or removal) of heat and water vapour at the mixed layer top. The mixed layer temperature and humidity budgets change through time as a result of vertical divergence or convergence in heat and moisture fluxes. This model of the mixed layer enables the following set of equations to be derived which describe the time rate of change of PBL parameters:

$$z_m \frac{d\Theta_m}{dt} = \frac{Q_{HS}}{C_a} + \Delta\Theta \frac{dz_m}{dt} \tag{2}$$

$$z_m \frac{dq_m}{dt} = \frac{Q_{ES}}{\rho L_v} + \Delta q \frac{dz_m}{dt} \tag{3}$$

$$\frac{dz_m}{dt} = \frac{0.2\, Q_{HS}}{\Delta\Theta} \tag{4}$$

$$Q_{HI} = -0.2\, Q_{HS} \tag{5}$$

Boundary conditions:

$$Q^* - Q_G = Q_{HS} + Q_{ES} \tag{6}$$

$$Q_{ES} = \frac{s(Q^* - Q_G) + C_a D_m/r_{as}}{s + \gamma(1 + r_s/r_{as})} \tag{7}$$

where Q_H is the sensible heat flux density; Q_E is the latent heat flux density; Q^* is the net all-wave radiant flux density; Q_G is the soil heat flux density; s is the slope of the saturation specific humidity/temperature curve; L_v is the latent heat of vaporisation; ρ is the air density; γ is the psychrometric constant; and C_a is the heat capacity of air. D_m is the potential saturation deficit, defined in (1). Subscripts S and I refer to surface values and values at the base of the capping inversion, respectively.

Equation 7 is the Penman Monteith canopy evaporation model, using the potential saturation deficit of the mixed layer and the aerodynamic resistance (r_{as}) evaluated across the entire surface layer. Equations 4 and 5 describe the growth of the mixed layer. Driedonks (1982) has studied the dynamics of entrainment and developed more sophisticated models of this process. Sensitivity analyses (e.g. McNaughton & Spriggs 1986; Cleugh 1990) show that an accurate model of mixed layer depth is not critical to modelling evaporation.

Background

Such an approach is not a new idea. Brutsaert and his colleagues (Brutsaert & Mawdsley 1976; Mawsley & Brutsaert 1977), using a different model of the PBL, developed regional evaporation predictions for the United States based on upper atmospheric data from the U.S. radiosonde network. Subsequently de Bruin (1983) took advantage of some of the recent advances in modelling mixed layer growth, in particular the model for $d\Delta\Theta/dt$ developed by Tennekes (1973) and the progress made by Driedonks (1982) on modelling the dynamics of the entrainment process. De Bruin combined the slab model with a Penman Monteith canopy evaporation model to

138

examine the effects of changing synoptic conditions (through the humidity and temperature
gradients, γ_Θ and γ_q) and surface moisture availability on the Priestley and Taylor parameter α. De
Bruin assumed that the flow of moisture out of the
mixed layer (Q_{EI}) is directly proportional to the
moisture flux at the surface (Q_{ES}), thus including
moisture in the PBL equations. Critics of this
approach maintain that such an assumption is a
major flaw (e.g. McNaughton & Spriggs 1986).
The study was of considerable importance, however, as de Bruin showed that over a fairly large
range in soil moisture availability and synoptic
conditions, the value of α was close to unity.

McNaughton & Spriggs (1986) provided the
first really rigorous development and testing of the
coupled mixed layer growth/canopy evaporation
model, which they term a regional evaporation
model. Based also on the slab model, they test
several formulations for dz_m/dt and conclude that
modelled evaporation is relatively insensitive to
this term. They demonstrate that the value of the
Priestley and Taylor α can change as a result of
the entrainment of air from the upper atmosphere.
The direction and magnitude of change in α
depends on the relative saturation deficits of the
mixed layer and the upper air. The model is tested
against several days' data from the Cabauw
meteorological site in the Netherlands and they
find excellent agreement. It has since been used to
examine the validity of the complementary
approach (McNaughton & Spriggs 1989).

Limitations

The slab model is a simplified representation of
the atmosphere and this introduces a number of
limitations. Firstly, evaluating γ_Θ and γ_q is vital
to the success of the model. Figure 2(a) illustrates
potential temperature and mixing ratio profiles
measured at an inland site in NSW where the
mixed layer is 1000 m deep. Probing the atmosphere to this height using instrumented, tethered
balloons is often limited by Civil Aviation regulations. Figure 2(b) shows a typical situation
where the tethered instrumentation (which can

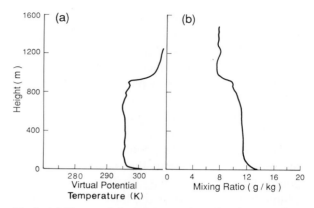

Fig. 2. (a) Measured, instantaneous boundary layer profiles
of (a) virtual potential temperatures and (b) mixing ratio at
Goulburn, NSW, December 2, 1988.

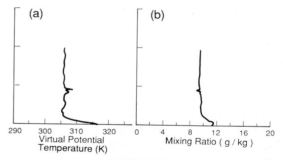

Fig. 2. (b) Measured, instantaneous boundary layer profiles
of (a) virtual potential temperature and (b) mixing ratio at
Gouldburn, NSW, January 18, 1989.

only be flown at heights lower than 300 m) cannot
resolve the top of the mixed layer. The first limitation, therefore, is the logistical problem of
resolving the full PBL. Radiosonde and airsonde
packages attached to weather balloons can be
used in place of tethered systems but costs are far
greater if time-dependent profiles are sought and
this is often at the expense of high quality data.

The model equations as presented above do not
include horizontal advective effects which may
occur at sites with a limited fetch. This can lead
to horizontal divergence in the moisture and temperature fields further modifying the mixed layer
temperature, humidity and growth. A coastal
location is an example where the model, as
described above, would be inappropriate for

predicting mixed layer growth, temperature and humidity.

A term representing the subsidence velocity is also absent from Equations 2–8. In anticyclonic conditions subsidence may have a large effect upon the mixed layer temperature and depth (Garratt *pers comm*). Although it would be desirable to account for this effect, the task of determining values for the subsidence velocity to input to the model would be a difficult one especially given the limitations mentioned above.

Both of these limitations can be resolved by modifying the equations. An advective form of the slab model has been developed (Steyn & Oke 1982; Steyn 1989 & Cleugh 1990) for coastal boundary layers. Steyn & Oke (1982) also include a subsidence velocity in their form of the model.

As mentioned, the equation for dz_m/dt is much more simplified than others currently in the literature. This may introduce errors in modelling z_m, for example Manins (1982) found that under conditions of strong wind shear across the interfacial zone the closure assumption ($Q_{HI} = -Q_{HS}$) does not hold. On the basis of past studies, this should not limit the usefulness of the coupled model in predicting surface evaporation.

The simplified model means that its use is only justifiable under a limited set of atmospheric conditions: typically anticyclonic, cloudless days where radiative divergence in the PBL is minimal. Given the results of Manins (1982), it may be also limited under conditions of strong wind shear in the interfacial zone.

Advantages

Despite its limitations the coupled model has some important advantages over and above surface-based predictive models such as Penman Monteith. Firstly, it is truly prognostic because the feedbacks between the surface and atmosphere are modelled interactively. The impact of changed land-use or synoptic conditions on both the surface latent heat flux and the PBL parameters therefore can be appropriately simulated. Secondly it encompasses larger scale 'controls' upon the surface latent heat flux e.g. the impact of

advective enhancement or suppression on Q_{ES} as a result of synoptic conditions. These two features also mean that the dominant controls upon evaporation can be assessed for given synoptic and/or surface conditions. Such insight is crucial when developing simpler models which are appropriate for predicting evaporation from particular land-uses. The three papers by de Bruin (1984) and McNaughton and Spriggs (1986, 1989) are examples of the way in which coupled models can be used diagnostically.

The temperature and humidity of the mixed layer partly results from the input of heat, moisture and other scalars whose source is the land surface. The vertical extent of the mixed layer means that its atmospheric constituents are influenced by a large source area so the mixed layer is an integration of the underlying surface heterogeneity. The coupled model is therefore applicable to modelling areal evaporation rates at large spatial scales, such as catchments and regions. As previously discussed, there is a real need for modelling at these scales – the coupled model offers an approach to meet this need.

The overall study aim was to model evaporation from a grassland catchment located near Goulburn, NSW, Australia using the coupled model and to evaluate its performance. This has already been achieved at one site (McNaughton & Spriggs 1986) and an objective of this study is to confirm the results found by those authors.

The Lockyersleigh Creek (Fig. 3) catchment has an area of 27 km^2 with undulating topography ranging in elevation from 600 m to 762 m. The vegetation is a mixture of tussock grasses and native sedges with open woodland (eucalypt species) in the eastern and south eastern areas.

These preliminary modelling results are a part of a collaborative study with the Water Resources Division of CSIRO who maintain a number of climate stations and two micrometeorological observing sites within the catchment. The two micrometeorological sites – P and Q – are spaced approximately 8 km apart. One some occasions, Site S (Fig. 3) was used in place of Site P as the micrometeorological site. The PBL profiles were conducted within 300 m of Site P (see Fig. 3).

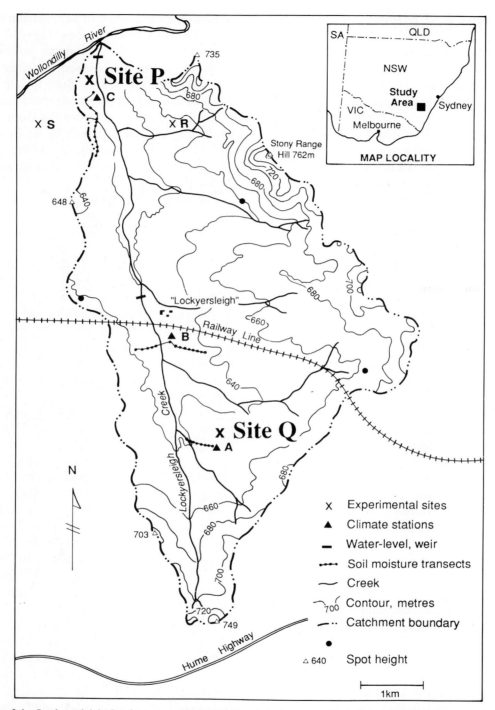

Fig. 3. Map of the Lockyersleigh Catchment and location of main micrometerological sites (P, Q); climate stations (A, B, C) and soil moisture transects. Sites S and R were used on occasions as main micrometeorological sites.

Implementation and results

Model implementation and inputs

Figure 4 presents a flow diagram of the coupled model: at each time step the surface latent heat flux (Q_{ES}) is computed from the (measured) available energy (Q^*-Q_G), aerodynamic and surface resistances and the modelled potential saturation deficit (D_m). The new Q_{ES} and energy balance define a surface sensible heat flux (Q_{HS}). The PBL grows as a result of the input of this sensible heat and entrains drier and warmer air from the upper atmosphere. This alters the mixed layer deficit and hence forces a new surface latent heat flux for the next time step. D_m provides a link between the three atmospheric layers – surface, mixed and 'synoptic' and the coupled model facilitates their interaction.

The set of differential equations are solved on a VAX computer, using a subroutine from the IMSL library which provides a numerical solution. All input data (see below) are linearly interpolated from hourly averages to six minute averages to achieve greater numerical stability.

The initial values required by the model are: Θ_m (initial mixed layer potential temperature), q_m (intial mixed layer specific humidity), γ_Θ, γ_q, $\Delta\Theta$ and Δq and mixed layer depth (z_m). These are obtained from measured profiles using a tethered balloon and sonde. Because of height restrictions, the measurements were supplemented with free flying weather balloons and airsondes. Profiles were measured throughout the day to provide validation data (z_m, Θ_m and q_m).

In addition time-dependent values of the available energy, surface and aerodynamic resistances are required by the canopy evaporation sub-model. Although net radiation and soil heat fluxes were measured at two sites (Fig. 3), only values from site P were used in the model simulation because they coincide with the times when PBL profiles were conducted.

The aerodynamic resistance to the top of the surface layer, r_{as}, was determined using wind speeds measured 2.5 m above ground and estimates of the roughness length from visual observations of vegetation height. A logarithmic wind profile form was assumed and adjusted for stability using the method of Van Ulden and Holtslag (1985).

The model was run using specified surface resistances. Following McNaughton and Spriggs (1986), these are derived by rearranging the Penman Monteith equation and solving for r_s using measured latent heat fluxes. They are referred to as 'effective' surface resistances. A second set of surface resistances are also available from the study of McLeod (1989). He developed an empirical relationship between effective surface resistance and measured soil moisture levels. Such a relationship could be used to estimate surface resistance – assuming that, for pasture, the soil moisture is an adequate predictor of the physiological control being exerted by the vegetation. At present, the uncertainties in this relationship are such that the use of these emperical resistances is not valid. This is an aspect that is

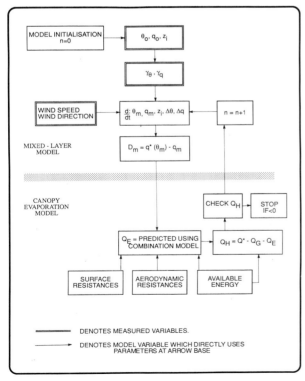

Fig. 4. Flow diagram of an implemented mixed layer growth/canopy evaporation model.

currently being examined for future use to further evaluate the coupled model.

The latent and sensible heat fluxes were determined at the two micrometeorological sites using a reversing differential psychrometer (ΔT_w and ΔT_d) and measured available energy to provide Q_{HS} and Q_{ES} via the Bowen ratio method. These measurements were used as validation data (see McLeod 1989 for further details).

Model validation

Three days have been selected to evaluate the hourly performance of the coupled model. These were the only time when all input and validation data were available and compatible. The three days are: December 1, 1988; January 18, 1989; January 20, 1989.

PBL parameters

The modelled and measured PBL parameters (Θ_m, q_m, z_m) are illustrated in Figures 5, 6 and 7. Combined they illustrate a more moist and shallow PBL than observed. The modelling of the mixed layer potential temperature is excellent,

particularly on 20 January. On 1 December, the model predicts a rapid increase in Θ_m, which levels out after 1400. The measurements indicate a more linear increase in potential temperature. This behaviour has been noted before (e.g. Cleugh 1990) and may result from the inadequacy of the closure assumption. The most serious errors, however, arise in modelling specific humidity. At the end of the model simulations for January 20, modelled q_m is 2.2 g kg^{-1} higher than that measured. The simulation for December 1 shows a similar relative difference. Both results are consistent with a modelled PBL which is much more shallow than the observed PBL. The discrepancy in mixed layer depths is also rather large (80–100 m) but it is the errors in q_m which are the most serious for modelling evaporation with accuracy.

Surface latent heat flux

Verifying the performance of the evaporation model is much more difficult for two reasons. The micrometeorological measurements of Q_{ES} represent a source area up to 1000 m^2 whereas the modelled estimates are for a much larger area. This issue is addressed later. Secondly, inte-

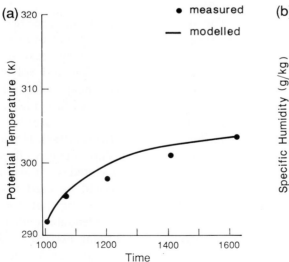
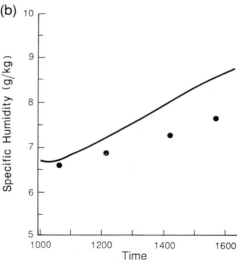

Fig. 5. Modelled and measured mixed layer (a) potential temperature and (b) specific humidity for December 1, 1988. Note: modelled values are interpolated from 6 min values; measured vaues are instantaneous and averaged over mixed layer depth.

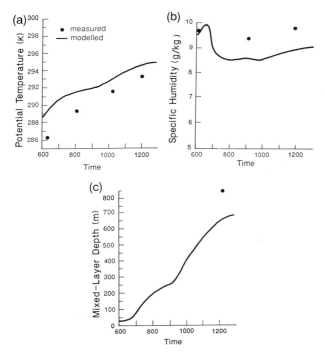

Fig. 6. Modelled and measured mixed layer (a) potential temperature, (b) specific humidity and (c) depth for 18 January, 1989. Note: modelled values are interpolated from 6 min values; measured values are instantaneous and averaged over mixed layer depth.

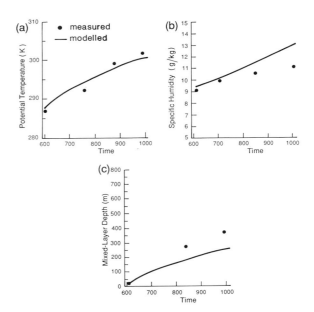

Fig. 7. Modelled and measured mixed layer (a) potential temperature, (b) specific humidity and (c) depth for 20 January, 1990. Note: modelled values are interpolated from 6 min values; measured values are instantaneous and averaged over mixed layer depth.

grating individual, measured stomatal resistances for a grassland canopy is not only a laborious proedure, but may not even be appropriate. In summary, finding accurate values of r_s is difficult *a-priori*. As mentioned above, the model is therefore initially evaluated using effective surface resistances.

The modelling simulations and statistics presented in Fig. 8 and Table 1 illustrate reasonable agreement between measured and modelled Q_{ES}. Note that averaging the six minute values up to hourly averages does not alter the statistics. The correlations are very high indicating that the diurnal variation is successfully simulated, however these don't reflect the consistent overestimate of the latent heat flux evident in Figure 8 for all three days. The differences are fairly small (ratio of means *ca* 1.11; relative difference between means *ca* 13–15%) for the January days, but much higher (1.18 and 17%) for December 1. The higher modelled latent heat flux values are consistent with the modelled PBL parameters *vis* a large specific humidity arising from a large surface evaporation. This limits sensible heating from the surface, and thus mixed layer depth.

Table 1. Statistics comparing modelled and measured latent heat fluxes.

Day	Q_{ES} (measured)	Q_{ES} (modelled)	r^2	Slope	Intercept
December 1, 1988	277.4	334.9	0.97	1.18	8.8
January 18, 1989	116.1	134.7	0.99	1.11	5.6
January 20, 1989	199.6	235.3	0.99	1.12	12

Note: Q_{ES} values are means (W m^{-2}) calculated for the model day length.

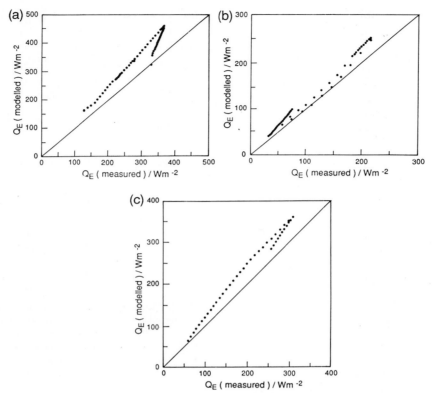

Fig. 8. Scatterplots of measured vs modelled surface latent heat fluxes for December 1, January 18 and January 20. All values are 6 min values – interpolated from hourly mean values.

The effective surface resistances were derived from the Penman Monteith equation using near-surface values Q_E, D_S and r_a. The increase in modelled Q_{ES} when a larger aerodynamic resistance is used (i.e. r_{as} which extends to the top of the surface layer, rather than r_a which only extends to sensor height) is consistent with the sensitivity of the Penman Monteith equation to aerodynamic resistance. As demonstrated by Monteith (1973), if the Bowen ratio (β, ratio of sensible to latent heat fluxes) is greater the γ/s then increasing the aerodynamic resistance may reduce the transfer of sensible heat and hence increase evaporation.

One of the problems highlighted above is the availability of appropriate *measured* surface heat fluxes. This coupled model is evaluated using measurements of sensible and latent heating which are representative of an area *ca* 100–1000 m². The PBL moisture and tempera-

ture budgets have a much larger source area as discussed previously. At the Lockyersleigh catchment a second set of measurements are available from Site Q (Fig. 3), located 8 km to the south of Site P. Figure 9 illustrates the diurnal variation of latent heat fluxes at both sites for four days in the interval of model simulations. The measured spatial differences in fluxes for December 1st (and the morning of December 2nd) are quite large. Although these differences may arise from instrument errors, surface observations at the time would suggest that they reflect the spatial variation in soil moisture between sites P and Q (see later). This, plus the poorer model performance on this day, suggests that an adequate simulation of Q_{ES} would require a better method for parameterizing an 'areal' r_s. For the other two model days, the spatial variation of fluxes is less than the modelled-measured differences.

Although the r² values and scatterplots (Fig. 8)

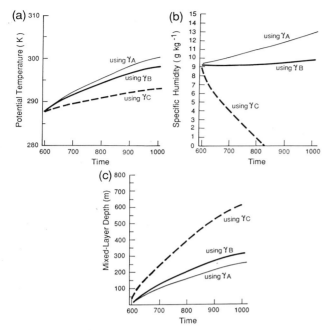

are indicative of good agreement, there are still systematic differences between measured and modelled evaporation. These arise partly from the effects of spatial variability, however the preceding discussion eliminated that possibility at least for the January days. The next section uses sensitivity analyses to examine the influence of errors in the input data set on the modelling performance.

Model sensitivity

The practical difficulties of obtaining reliable measurements of the humidity and temperature gradients (γ_Θ and γ_q) were demonstrated earlier. Together with the dilemma of determining an appropriate surface resistance, it is these three aspects that pose the greatest limitation to the model's success.

It is thus instructive to look at the effects of errors in these profiles upon modelled latent heat fluxes, using January 20 as a case study. The measured profiles on this day suggested differing

Fig. 9. A comparison between the measured latent heat flux (hourly averages) at sites P (or S) and Q for: (a) December 2, 1988; (b) December 1, 1988; (c) January 18, 1989; (d) January 20, 1989.

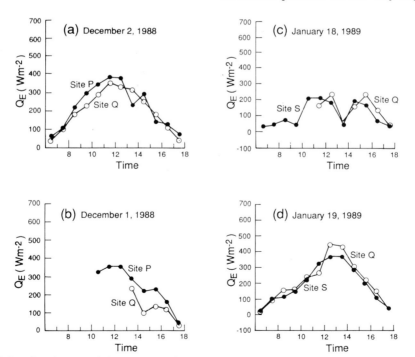

Fig. 10. The effect of changing the potential temperature inversion strength and the humidity lapse rate (above the mixed layer) on mixed layer potential temperature, specific humidity and depth for January 20, 1989. See text for definition of profiles A, B and C.

Table 2. Effects of changing surface resistance upon modelled latent heat flux.

Day	Q_{ES} (r_s varying)	Q_{ES} ($r_s = 100$ s m^{-1})	Q_{ES} ($r_s = 300$ s m^{-1})	Q_{ES} (measured)
December 1, 1988	334.9	355.7	278.3	277.4
January 18, 1989	134.7	130.7	97.7	116.1
January 20, 1989	235.3	230.3	200.4	199.6
	W m^{-2}	W m^{-2}	W m^{-2}	W m^{-2}

gradients for the temperature and humidity in the upper atmosphere. This uncertainty arose from difficulties in locating the top of the mixed layer. The values finally used in the simulation were believed to be the most representative. In order to determine the effect of errors, the other gradients were used as input for January 20 simulations. In the following discussion,

A $= \gamma_\Theta = 0.0622$ Km^{-1}; $\gamma_q = -0.0075$ g kg^{-1} m^{-1} (used in initial model runs)

B $= \gamma_\Theta = 0.0411$ Km^{-1}; $\gamma_q = -0.02$ g kg^{-1} m^{-1}

C $= \gamma_\Theta = 0.0110$ Km^{-1}; $\gamma_q = -0.05$ g kg^{-1} m^{-1}

The mixed layer becomes progressively drier, cooler and deeper (Fig. 10) as the upper atmosphere approaches neutrality and γ_Θ is reduced. Increasing the lapse rate in humidity means that progressively drier air is entrained into the mixed layer. Model simulations using profiles A and B lead to fairly small differences in modelled temperatures, but larger differences in humidity an mixed layer depth. Profile C, however, gives not only widely differing results, but humidities that are obviously incorrect. Clearly the PBL is sensitive to γ_q and γ_Θ, and so errors in these terms will be important. However the coupled model is intended primarily as a regional evaporation model, hence a more relevant question is – how sensitive is Q_{ES} to errors in γ_q and γ_Θ? This is not to say that errors in modelling the PBL parameters are unimportant. Rather, sensitivity analyses should be used to identify those input variables which most affect modelled evaporation. It is these variables which must be determined with the greatest accuracy.

The model was thus used to assess the effect of these differences in γ_Θ and γ_q upon modelled Q_{ES} for January 20 alone. The modelled evaporation using Profiles A and B only differed by less than 5%, however this should be treated with caution as further analyses (see below) illustrate that this sensitivity has a strong dependence on the surface and aerodynamic resistances. However for this day, errors in measuring the upper atmospheric humidity and temperature gradients should not lead to significant errors in modelled evaporation.

Applications of the coupled model require input values of r_s, *a priori*. This is probably the most difficult term to evaluate accurately. Table 2 compares Q_{ES} modelled with an effective surface resistance to Q_{ES} modelled using two diurnally constant surface resistances. The relationships identified by McLeod (1989) indicate that a surface resistance of 100 s m^{-1} would be appropriate, based on the soil moisture, for each model day. Table 2 shows that this leads to greater errors (28% difference) between measured and modelled on December 1; and a very slight improvement on January 18 and 20. For January 18 and 20, the small errors in using an a-priori surface resistance based on soil moisture do not lead to significant errors in modelled evaporation. The sensitivity of modelled evaporation to a change in surface resistance for these days varies from 0.15 to 0.38 W m^{-2}/s m^{-1}. If estimates of surface resistance can be made which depart from actual values by 100 s m^{-1}, then errors in modelled evaporation will be of the order of 10–15%. This suggests that, on these days and for this site, modelled evaporation is slightly more sensitive to surface resistance than to the PBL parameters. Hence errors in modelled evaporation may tend to arise more from errors in specifying surface resistance than the upper atmosphere temperature and moisture gradients.

December 1 is an interesting example of the difficulty of determining an 'areally-averaged' surface resistance. Based on soil moisture measurements, a value of 100 s m^{-1} would be selected, however the modelling results illustrate that this is too low. In fact the soil moisture observations across the Lockyersleigh catchment at this time indicate quite large spatial differences. Site Q was much drier than site P, and had a lower (measured) latent heat flux (recall Figure 9). Using effective resistances from Site P, and validating the model using flux measurements from Site P, may be inappropriate because of the mismatch between local scale measurements at site P and areal prediction from the coupled model.

Discussion and conclusions

The sensitivity results presented earlier showed the influence of some errors in input variables upon Q_{ES}. However, the sensitivities found were relevant only to those particular days with their specific forcing. It is useful to examine the general sensitivity of evaporation to changing atmospheric and surface conditions. Table 3 illustrates the sensitivity of Q_{ES} to changes in γ_Θ and γ_q. An equivalent change in γ_Θ and γ_q lead to an 18% difference in evaporation when r_{as} is large, compared to 30% when r_{as} is halved. This illustrates the decoupled evaporation regime over surfaces that are aerodynamically smooth. In more aerodynamically rough environments, the latent heat flux is sensitive to changes in the upper atmospheric forcing and also changes in surface resistance. The latter is evident in Table 3, the sensitivity of Q_{ES} to surface resistance is 0.38 for rough and 0.24 for smooth surfaces.

The implications of these results for the Lockyersleigh catchment are that, in general, errors in specifying PBL variables (i.e. γ_q and γ_Θ) are less important for an accurate evaporation prediction. This is supported by McLeod's (1989) finding that an equilibrium evaporation model (based primarily on available energy) provided the most accurate estimate of evaporation for this cathment. Nonetheless, the sensitivity results demonstrate that the variations in the performance of the model between days may be due to changing surface or upper atmosphere conditions.

These sensitivity analyses illustrate the importance of the coupling between atmospheric layers that occurs with efficient mixing. It confirms the results of McNaughton and Spriggs (1986) and Jarvis and McNaughton (1985) that aerodynamically rough surfaces will tend to have an evaporation regime driven by advective influences, and sensitive to surface resistance.

In conclusion, the results of these model simulations support those of other similar studies in the literature. A coupled mixed layer/canopy evaporation model has been shown to predict hourly values of the latent heat flux with reasonable accuracy. This model was also able to predict mixed layer potential temperature well, but there were large errors in the specific humidity and the mixed layer depth.

Other studies have shown that errors in modelling the mixed layer depth are unimportant for modelling evaporation. This is confirmed here. There are two major barriers to applying the model: (a) determining γ_Θ and γ_q and (b) specifying accurate values for the surface resistance. At

Table 3. Sensitivity of modelled latent heat fluxes to changes in surface and aerodynamic resistances; and upper atmosphere moisture and temperature gradients.

r_s (s m^{-1})	Low r_{as} (s m^{-1})	High r_{as} (s m^{-1})
	Q_{ES} (modelled/W m^{-2})	
60	360	357
120	323	337
480	199	258
$\Delta Q_{ES}/\Delta r_s$	0.38	0.24
γ_Θ, γ_q (units in Km^{-1} and g kg^{-1} m^{-1})		
0.062, -0.0075	323	337
0.011, -0.0500	239	283
ΔQ_{ES} W m^{-2}	143	87
% increase	30%	18%

this site errors in γ_q and γ_Θ were slightly less important than errors in the surface resistance terms. However this sensitivity strongly depends on the efficiency of turbulent mixing and the availability of moisture at the surface.

Modelled fluxes are likely to represent a much larger source area than micrometeorological measurements, and this leads to problems when evaluating a catchment scale model using local scale measurements. Fortunately, the spatial variation in the turbulent fluxes was not great for two of the three days modelled.

Therefore the results of this study raise a number of interesting questions that should be pursued as further research. The first is the validity of comparing local-scale measurements of surface fluxes with modelled fluxes which may represent a much greater source area. Of course this 'integrative' aspect of the coupled model is one of its advantages, but the problem of verifying model output remains. A second question raised is that of providing input data to these mixed-layer evaporation models. The problems of providing this information have been addressed. The fact that sensitivity analyses indicate that errors in γ_Θ and γ_q may lead to errors in modelled latent heat fluxes simply reinforces this problem.

Such models are particularly useful in a predictive mode as they are free of the constraints of providing input data measured near to the land surface. This makes them a very powerful tool for investigating effects of climate change on evaporation. The resolution of these two aspects is therfore of paramount importance.

References

Brutsaert, W. & Mawdsley, J. A. 1976. The appplicability of planetary boundary layer theory to calculate regional evapotranspiration. Water Resources Research 12: 852–857.

Byrne, G. F., Dunin, F. X. & Diggle, P. J. 1988. Forest evaporation and meteorological data: A test of a complementary theory advection-aridity approach. Water Resources Research 24: 30–35.

Cleugh, H. A. 1990. Development and evaluation of a suburban evaporation model: A study of atmospheric and surface controls on the suburban evaporation regime. Unpublished Ph.D. Thesis, 1990, 249 pp.

de Bruin, H. A. R. 1983. A model for Priestley-Taylor α. J. Clim. Appl. Meteorol. 22: 572–578.

Driedonks, A. G. M. 1982. Models and observations of the growth of the atmospheric boundary layer. Boundary-Layer Meteorol. 23: 283–307.

Hacker, J. M. 1988. The spatial distribution of the verical energy fluxes over a desert lake area. Aust. Met. Mag. 36: 235–243.

Hacker, J.M., Shao, Y., Byron-Scott, R. A. D. & Schwerdtfeger, P. 1988. The vertical sensible and latent heat transport in the boundary layer of the upper Spencer Gulf area – a case study. Beitr. Phys. Atmos. 61: 56–61.

Jarvis, P. G. & McNaughton, K. G. 1985. Stomatal control of transpiration: Scaling up from leaf to region. Advances in Ecological Research 15: 1–49.

Manins, P. C. 1982. The daytime planetary boundary layer: A new interpretation of Wangara data. Quart. J. R. Met. Soc. 108: 689–705.

Mawdsley, J. A. & Brutsaert, W. 1977. Determination of regional evapotranspiration from upper air meteorological data. Water Resources Research 13: 539–548.

McLeod, K. J. 1989. An analysis of evaporation from pasture. Unpublished Honours Thesis, 113 pp.

McNaughton, K. G. & Spriggs, T. W. 1986. A mixed layer model for regional evaporation. Boundary-Layer Meteorol. 34: 243–262.

McNaughton, K. G. & Spriggs, T. W. 1989. An evaluation of the Priestley and Taylor equation and the complementary relationship using results from a mixed-layer model of the convective boundary layer. Estimation of Areal Evapotranspiration, IAHS Publ. no. 177.

McNaughton, K. G. & Jarvis, P. G. 1983. Predicting effects of vegetation changes on transpiration and evaporation. In: Water Deficits and Plant Growth, Vol VII, edited by T. T. Kozlowski, 1–47, Academic New Tork.

Monteith, J. L. 1973. Principles of Environmental Physics, 241 pp., Edward Arnold, New York.

Schuepp, P. H., Desjardins, R. L., Macpherson, J. L., Boisvert J. & Austin, L. B. 1987. Airborne determination of regional water use efficiency and evapotranspiration: present capabilities and initial field tests. Agric. and For. Meteorol. 41: 1–19.

Steyn, D. G. 1989. An advective mixed-layer model for heat and moisture incorporating an analytic expression for moisture entrainment (in press).

Steyn, D. G. & Oke, T. R. 1982. The depth of the daytime mixed layer at two coastal sites: A model and its validation. Boundary-Layer Meteorol. 24: 161–180.

Tennekes, H. 1973. A model for the dynamics of the inversion above a convective boundary layer. J. Atmos. Sci. 30: 558–567.

Van Ulden, A. P. & Holtslag, A. A. M. 1985. Estimation of atmospheric boundary layer parameters for diffusion applications. J. Clim. Appl. Meteorol. 24: 1196–1207.

Vegetatio **91**: 149–166, 1991.
A. Henderson-Sellers and A. J. Pitman (eds).
Vegetation and climate interactions in semi-arid regions.
© 1991 *Kluwer Academic Publishers. Printed in Belgium.*

149

Developing an interactive biosphere for global climate models

A. Henderson-Sellers
Macquarie University, North Ryde, 2109, NSW, Australia

Accepted 24.8.1990

Abstract

A highly generalised (five classes) grouping of Holdridge life zones, has been used to derive predicted 'natural' ecosystems for the present day climate using temperature and precipitation derived from two experiments undertaken with the NCAR Community Climate Model. These predictions differ from one another and both differ significantly from the prescribed classification groupings of ecosystem complexes used with a state-of-the-art land surface parameterization submodel, the Biosphere-Atmosphere Transfer Scheme. The highly generalised groupings show relatively little sensitivity to the temperature changes induced by doubling atmospheric CO_2 but greater response when precipitation is also modified. All the doubled-CO_2 scenarios predict increased 'desert' areas although these future climatically-induced changes to the global-scale ecology are very much smaller than the extensive disturbances already caused by mankind's land clearance and poor agriculture. Land-use change rendered 13% of the Earth's land surface 'desert' whereas the most pessimistic doubled-CO_2 result gives rise to only a 2% increase in 'desert' area.

Introduction

At present, strenuous efforts are being made to try to understand the differences between sensitivities exhibited by different climate models (e.g. Schlesinger & Mitchell 1985 and Meehl & Washington 1988 cf. Manabe & Wetherald 1987). It seems clear that, ideally, these gross differences should be reduced and harmonised before (i) more detailed 'predictions' of the distribution of ecosystems are attempted and before (ii) the direct effects on the plants of the increased atmospheric CO_2 are contemplated within the atmospheric general circulation model (AGCM) environment. However, recent initiatives to study global systems such as the International Geo-

sphere Biosphere Programme (IGBP) and the Human Response to Global Change programme (HRGC) prompt the evaluation reported here of the accuracy with which a simple, generalised grouping of ecosystems can be predicted from surface parameters generated by current climate models. Thus this study underlines that improvements must occur in numerical models before an interactive land biosphere can be included in global climate models.

The practical importance of studying climate derives from the dependence of mankind upon the processes that occur at the atmosphere-land interface; in particular, the availability of fresh water and the growth of vegetation for food and fibre. Despite this crucial dependence, relatively little attention has yet been paid by the climate modelling community to the accurate prediction of land surface climates and we are still uncertain about both the sensitivity of the overall climate system to land surface processes and the sensitivity of land surface climates to perturbations in the overall climate (for example, the impact of doubling atmospheric carbon dioxide).

The physical characteristics of vegetation, and also the soil, control the absorption of solar radiation and emission of thermal infrared radiation, the exchanges of sensible and latent heat with the atmosphere and, to a lesser extent, the exchange of momentum between the atmosphere and the surface. Most modelling groups have acknowledged similar sensitivities to gross, global-scale changes in specified surface albedo, typically the impact of a 5% increase in albedo being a reduction in precipitation of between 5% and 20% (the Charney effect e.g. Charney 1975; Charney et al. 1977). More recently it has been recognised that changes in soil moisture have at least comparably significant effects (e.g. Shukla & Mintz 1982; Meehl & Washington 1988).

A much more comprehensive parameterization of vegetation and soil is required before more subtle, but perhaps equally important, interactions can be captured. For example, over forested areas, interception of rainfall by the canopy and re-evaporation can stop most of a light drizzle ever reaching the soil surface. Additionally, much of the transfer of moisture from the ground to the air is through leaf transpiration rather than directly from soil evaporation. The presence of snow and the occurrence of frozen soil moisture are also significantly affected by vegetation cover.

Recently, attempts have been made to include aspects of soil hydrology and canopy energy and moisture exchanges into GCMs. The European Centre for Medium Range Weather Forecasting model now incorporates a canopy resistance term which parameterizes stomatal and root resistance and the interception of moisture by the canopy without calculating a leaf or canopy temperature and hence neglecting the canopy energy budget. Three models of land surface processes explicitly include two or more soil layers and a canopy parameterization. These are the Simple Biosphere model (SiB) of Sellers et al. (1986), the Biosphere-Atmosphere Transfer Scheme (BATS) of Dickinson et al. (1986) and the Bare Essentials model of Pitman (1988) which have been incorporated into versions of the Goddard Laboratory for Atmospheric Sciences, the National Center for Atmospheric Research and the Canada Climate Centre AGCMs respectively. These parameterization packages treat land surface processes in a much more elaborate fashion than do most conventional GCMs, but they remain conceptually very much simpler than the most elaborate models available for individual surface processes; in particular, soil hydrology, plant water budgets and snow physics.

Vegetation is also crucially important in determining some chemical properties of the climate system, the most obvious example being partial control by the biosphere of the global carbon dioxide budget. Attempts have been made to relate the satellite-derived normalised differenced vegetation index (NDVI) to seasonal cycles of atmospheric CO_2 (Tucker et al., 1986). More recently, it has been recognised that the distribution of ecosystems also determines to some extent the methane budget of the atmosphere and perhaps affects other chemical constituents (e.g. Ehhalt 1988).

In this investigation results from two versions of one particular global climate model are used. The climate model, which is described in Section 2, was selected because it has been used both in conjunction with one of the current 'big leaf' land-surface parameterization schemes and for doubled CO_2 experiments, and readily accessible results from these studies were available.

Two experiments using the NCAR CCM which include (i) a mixed layer ocean model and (ii) the Biosphere-Atmosphere Transfer scheme (BATS)

The National Center for Atmospheric Research (NCAR) Community Climate Model (CCM) is a three-dimensional AGCM based in part on the

Australian spectral model first described by Bourke *et al.* (1977). The results were derived in the two experiments used here from model integrations truncated at wavenumber 15 (a rhomboidal spherical harmonic truncation) producing horizontal grid elements approximately 4.5° of latitude by 7.5° of longitude. A number of different versions of the CCM exist; here we use the results of two separate experiments, the first of which uses the older (nine-layer) version of the CCM, termed CCM0, to which a simple mixed layer ocean model has been coupled. The second experiment uses a more recent (twelve-layer) version of the model termed CCM1, but in this case only the atmospheric component of the climate is computed, the ocean being represented by specified monthly sea surface temperatures and sea ice extent.

CCM0 plus a simple mixed layer ocean model

In this version of the CCM, the land surface hydrology is modelled by a simple bucket (or Budyko) parameterization in which the bucket depth (or the 'field capacity') is a uniform 15 cm over the globe (e.g. Meehl & Washington 1988). The land surface albedos are specified as 0.13 for all non-desert and snow-free areas, 0.25 for deserts and 0.80 for snow. Cloud formation is determined interactively using the Ramanathan *et al.* (1983) radiation model. This version of the CCM has been coupled to a simple mixed layer ocean model that computes seasonal heat storage based upon the assumption of an isothermal layer of fixed ocean depth equivalent to 50 m. The ocean model, described in Washington & Meehl (1984), does not include horizontal or vertical heat transport or changes in salinity.

Two experiments were performed using this coupled atmosphere plus mixed layer ocean model: a control climate simulation (also termed $1 \times CO_2$) and a doubled CO_2 simulation ($2 \times CO_2$). In each case, the experiment comprised three parts: a two phase 'spin-up' followed by simulation of 11 full annual solar cycles. The results used here, described in full in Washington

& Meehl (1984), represent three-year averages of the surface air temperature (i.e. temperature of the lowest model σ level layer, $\sigma = 0.991$) and the averaged annual total precipitation from the final three years of the 11 year simulation.

Washington & Meehl (1984) comment that the modelled equator to pole temperature gradient is greater than observed in both hemispheres for the model December/January/February period, during which time tropical latitudes are warmer than observed and middle to high latitudes are cooler. Some of these discrepancies are probably the result of the lack of ocean currents in the simple mixed layer model. Generally, however, the control ($1 \times CO_2$) simulation is accepted as a reasonable representation of the present-day global climatic patterns (Washington & Meehl 1984; Marshall & Washington 1989). Uncertainties about regional details of present-day and perturbed climates and different sensitivities between models have been discussed by Schlesinger & Mitchell (1985) and Jenne (1989).

CCM1 plus the Biosphere-Atmosphere Transfer Scheme (BATS)

The CCM1, now incorporating an updated radiative transfer package (Kiehl *et al.* 1987), has been coupled to the Biosphere-Atmosphere Transfer Scheme of Dickinson *et al.* (1986) in which prognostic equations are solved for the temperatures and water contents of a surface and deep soil layer and a vegetation canopy (e.g. Wilson *et al.* 1987b). Foliage and canopy air temperatures are calculated diagnostically using an energy balance equation which includes stomatal transpiration and direct evaporation of moisture from the leaves. The BATS scheme also calculates a 'Stevenson's screen' temperature which is the air temperature at a height of 1.5 m above the soil surface. It is this temperature, averaged over each 24 hour diurnal cycle, which has been used here, together with the computed total precipitation.

The BATS package uses a vegetation/land cover classification derived from Matthews (1984) and Wilson & Henderson-Sellers (1985), containing 16 land use classes plus two classes for oceans and large inland water bodies, each of the grid blocks having 4.5° by 7.5° resolution and being associated with one of these 18 classes. The depth of the upper and total soil layers, the percentage vegetation cover, leaf area index and hence the surface radiative and hydrological properties are all dependent upon the land surface class prescribed within the BATS submodel. Table 1 lists these 18 classes and associates each with a more general ecotype classification which is to be used in the following section. Note that the generalised ecotype class termed 'woods' here

Table 1. Groupings of the eighteen vegetation/land cover types specified in the BATS plus CCM experiment into the five generalized life zones and implied ecosystem groups.

Specified vegetation type	Generalized life zones
evergreen broadleaf tree	'Rainforest' (T ≥ 6 and P ≥ 2000)
deciduous broadleaf tree deciduous needleleaf tree evergreen needleleaf tree mixed woods	'Woods' (seasonal woodland and forests) (T ≥ 3 & P ≥ 500 & P < 2000 or T ≥ 3 & T < 6 & P ≥ 250 & P < 500)
crop short grass tall grass irrigated crop evergreen shrub deciduous shrub	'Grass and shrub' (sometimes including interrupted woods) (T ≥ 3 & T < 10 & P ≥ 125 & P < 250 or T ≥ 6 & P ≥ 250 & P < 500)
desert semi-desert	'Desert' (T ≥ 3 & P < 125 or T ≥ 10 & P < 250)
tundra ice cap/glacier	'Tundra' (T < 3)
bog or marsh	represented by 'tundra' or 'woods'
[inland water	unclassified]
[oceans	unclassified]

* specification of classes used in Section 3 given in parentheses as a function of
T = mean annual biotemperature derived from seasonal means (°C) [biotemperature is defined in the text]
P = total annual precipitation (mm)

includes non-tropical forest which is evergreen. Figure 1 shows the distribution of these five generalised groups; it is important to recognise that this ecological classification does not represent a climax vegetation for the present-day climate but rather it incorporates the effect of mankind's impact on the land surface vegetation, particularly the clearance of trees and extensive agricultural activities. In this specified classification, 'grass and shrub' occupies the largest surface area (58.66×10^6 km^2) while 'rainforest' occupies the smallest (15.21×10^6 km^2). It must be anticipated that this distribution will differ from categories predicted using the temperatures and precipitation generated by the CCM experiments which cannot include mankind's disturbance of the natural ecosystem.

The results described here are taken from a three year simulation in which sea surface temperatures and sea ice extent are specified from monthly-averaged observed values. The predicted present-day climate is therefore much more closely constrained than in the case of CCM0 coupled to the mixed-layer ocean. On the other hand, the surface air temperature is at a more appropriate height and both the temperature and the precipitation will have been modified by the incorporation of a canopy and soil submodel. For example, continental precipitation is affected by canopy re-evaporation (e.g. Dickinson & Henderson-Sellers 1988).

Predictions of generalised ecosystem groups

The Holdridge life zone classification

The Holdridge life zone classification system is a scheme which replaces the general character of natural vegetation with simple climatic indices: the total annual precipitation, the mean annual biotemperature and the potential evapotranspiration ratio. Figure 2(a) illustrates the Holdridge diagram (Holdridge 1947) which contains 37 named life zones. Note that only two primary variables: for example, total annual precipitation and the biotemperature (which is a representation of the growing season temperature) are required

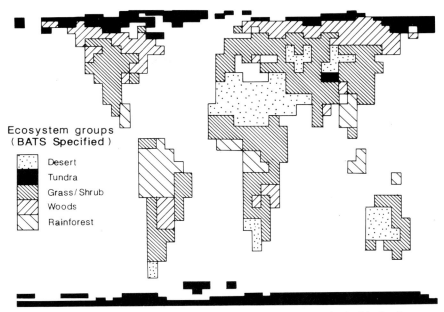

Fig. 1. Specified vegetation used in the BATS submodel (Dickinson *et al.* 1986) matched with the five generalized ecosystem groups following Table 1.

to define a location within the life zone triangle (e.g. Holdridge 1964). Emanuel *et al.* (1985a) showed that this life zone classification produced a reasonable representation of the undisturbed present-day patterns of ecosystem types when the input parameters were climatological temperatures and precipitation. Indeed it is this agreement between the observed and predicted life zones (based on *observations* of temperature and precipitation) and the relative simplicity and availability of the input parameters which has prompted this analysis based upon the Holdridge scheme rather than more recent relationships between climate and plant distributions which take account of, for example, soil water deficit and heating degree days (Woodward 1987).

In this study these 37 life zones have been reduced to a very much simpler classification of only five generalised groups termed 'desert', 'tundra', 'grass and shrub', 'woods' and 'rainforest' (Fig. 2(b)). The range of biotemperature and mean annual precipitation for each group is given in Table 1. The 'woods' group can be considered as representing seasonal woodland areas affected both by intermittent droughts and by

cold. Thus this class includes non-tropical forest areas, such as the 'rainforest' hexagon in the boreal/sub-alpine belt in keeping with the classification of the BATS land surface types which includes an evergreen needleleaf tree category. This gross simplification of the Holdridge classification is used here to predict generalised ecotypes from the CCM experiments previously described. It must be recalled throughout the following discussion that this prediction scheme gives rise to a life zone classification (on the basis of temperature and precipitation alone) which is taken as a surrogate for the more interesting prediction of ecosystems (i.e. vegetation plus soils plus climate). The highly generalised nature of the life zone classes used will continue to be emphasised by the use of quotation marks around the names of the groups.

Ecosystem prediction using the CCM1 plus BATS simulation

The CCM1 plus BATS three year experiment gave rise to Stevenson screen air temperatures

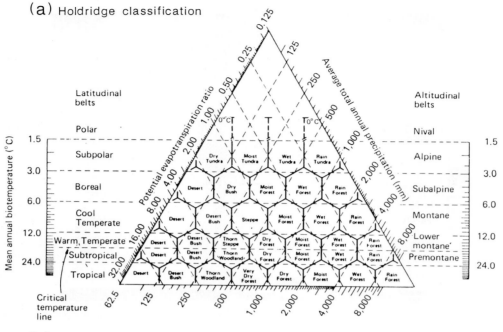

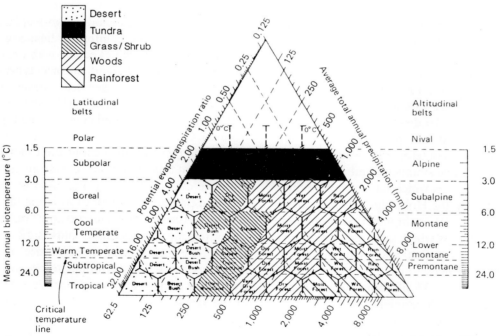

Fig. 2. (a) Holdridge life zones; (b) Five derived generalized life zones relabelled with typical vegetation specified by climate parameters of precipitation (P) and biotemperature (T).

'Rainforest' ($T \geq 6$ and $P \geq 2000$)

'Woods' ($T \geq 3$ & $P \geq 500$ & $P < 2000$ or $T \geq 3$ & $T < 6$ & $P \geq 250$ & $P < 500$)

'Grass and shrub' ($T \geq 3$ & $T < 10$ & $P \geq 125$ & $P < 250$ or $T \geq 6$ & $P \geq 250$ & $P < 500$)

'Desert' ($T \geq 3$ & $P < 125$ or $T \geq 10$ & $P < 250$)

'Tundra' ($T < 3$)

and total precipitation values as monthly means and monthly totals respectively. On the other hand, the CCM0 plus mixed layer ocean results were available only in the form of seasonal averages of temperature at $\sigma = 0.991$ (i.e. ~ 70 m above the surface) and precipitation totals. The average 'biotemperature' represents conditions important for the growing plant. Thus the effects of temperatures below (and sometimes above) prescribed thresholds are modified in the calculation of average conditions. Biotemperature is defined as the sum of the temperatures over a year with each temperature value (daily, weekly, monthly or seasonal) set to 0 °C if it is less than or equal to 0 °C and this sum then divided by the total number of values (i.e. 12 for monthly temperatures) (Holdridge 1947). For the 3 month seasonal calculation, the calculated biotemperature will be modified only when the season contains months with temperatures below zero and months with above zero temperatures. The contribution of these transition seasons to the annual biotemperature will be smaller for the computation using seasonal means than for the computation using monthly means. The sensitivity of the generalised ecotype classification to the use of seasonal temperatures in the construction of annual mean biotemperature was examined using the monthly temperatures from the CCM1 plus BATS experiment.

Figure 3(a) identifies the grid elements for which the month and three monthly (seasonal) computation of annual mean biotemperature gave rise to a different generalised ecotype class. There were only seven grid elements, 4 in northern Canada and 3 in northern Siberia, all of which were classified as 'tundra' when seasonally averaged temperatures were input to the biotemperature calculation, but as 'woods' when monthly average temperatures were input to the biotemperature calculation; and one grid element in the Nepal/southern Tibet region which was classified as 'rainforest' on a monthly basis and 'woods' using seasonal data. In view of these very small differences, it was deemed adequate to use seasonal values of temperature to compute the biotemperature for the rest of the analysis. It must

be recalled that this seasonally-based calculation results in slightly lowered biotemperatures in a few locations. For ease of identification of high latitude locations, Fig. 3(a) has been drawn using a cylindrical equi-distant projection in which each 4.5° by 7.5° grid element appears as the same size (on the map) as every other. It is very important to recognise that although this map offers many advantages, it is not an equal area representation of the globe. Figure 3(b) (and all other maps in this paper) are based on an equal area map projection (i.e. one in which equal areas on the ground are represented by equal areas on the map, thus, for example, rendering the island of Greenland significantly smaller than the continent of Australia as indeed is the case in the real world, their relative sizes being $2\,175\,600$ km^2 as compared with $7\,704\,135$ km^2). The content of Fig. 3(b) is discussed more fully in Section 3.4.

Figure 4 shows the distribution of the five generalised ecotypes based on the Holdridge life zone classification computed from the (D/J/F, M/A/M, J/J/A, S/O/N) averaged screen temperatures and total precipitation. There is, as was anticipated, a very much larger area covered by the 'woods' class as compared with the 'grass and shrub' class if this distribution is compared with the specified vegetation classification shown in Fig. 1. This is, at least partly, due to the difference between the basis for classification: the predicted life zones (and, by implication, ecosystems) are of 'natural' (i.e. dependent only upon climatic conditions) vegetation whilst the specified ecotypes include mankind's clearance of forest and agricultural activity. The surface area covered by 'grass and shrub' has dropped from 58.66×10^6 km^2 in the specified classification (Fig. 1) to only 19.80×10^6 km^2 in this predicted life zone scheme. The other obvious differences are a very much smaller area of 'desert' and a somewhat smaller area of 'tundra' in Fig. 4 as compared with Fig. 1.

Table 2 lists the percentage coverage of the land surface by the five generalised groups for this experiment and the others described in this section. The relative reduction in 'grass and shrub' is partly explicable in terms of agricultural

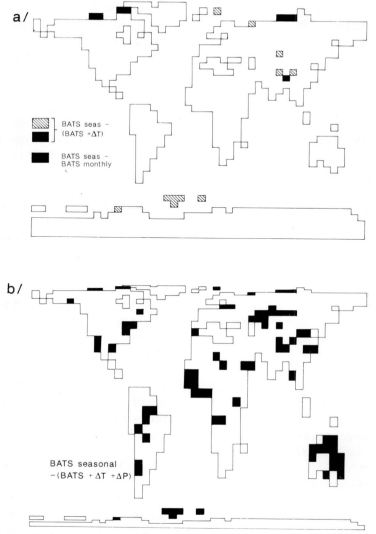

Fig. 3. (a) A cylindrical equidistant map identifying grid locations where differences occur in life zone predictions between two pairs of experiments: (i) BATS predicted (from seasonal values) – BATS predicted (from monthly values) and (ii) BATS predicted – (BATS + ΔT) predicted [i.e. between (i) Fig. 4 and the monthly-value based prediction and (ii) Fig. 4 and Fig. 8]. Note that grid areas with *both* types of shading fall into the second group (ii); (b) As (a) except for an equal area map projection and identifying locations where differences occur between BATS predicted and (BATS + ΔT + ΔP) predicted life zones [i.e. between Fig. 4 and Fig. 9].

activities; especially in the USA, Europe and China. The relative lack of 'tundra' is in part due to the fact that the CCM predicts summer air temperatures too warm in the northern high latitudes (cf. Wilson *et al.* 1987a). The relative reduction in the area of 'deserts' is the result of excessively large precipitation totals in continental desert regions. Figure 5, for example, compares the CCM1 plus BATS prediction of total annual precipitation for the continent of Australia with the observed total. The central desert region is generally too wet, whilst the large rainfall totals in the narrow coastal strip around most of the continent are underpredicted. Figure 5 suggests that

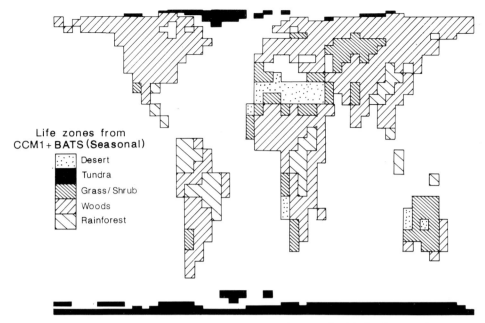

Fig. 4. Life zone groups computed from climate derived using CCM1 plus BATS with fixed sea surface temperatures.

not only the coarse block approximations of land but also the lack of orographic control of precipitation limit the realism of any model at the R15 resolution. For successful regional-scale prediction of precipitation, and hence of life zones, an improved spatial resolution is necessary. This improvement can be achieved either by truncation at a larger wavenumber (wavenumber 42 is to be used as the standard truncation for the next version of the CCM) or by embedding higher resolution mesoscale models into the current global-scale GCMs (e.g. Giorgi & Bates 1989).

Overall, the ecological classification produced from the biotemperatures and total precipitation resulting from the CCM1 plus BATS three year integration is reasonable but it must be recalled that this is in part due to the fact that this present-day climate simulation is fairly strongly controlled by specification of observed sea surface temperatures.

Table 2. Percentage coverage of land surface by the five generalized life zones.

Generalized Life zones (and implied) vegetation)	Prescribed	Present-day		Doubled CO_2		
	BATS specified	BATS predicted	$1 \times CO_2$ includes simple ocean	$2 \times CO_2$ includes simple ocean	BATS predicted $+ \Delta T$	BATS predicted $+ \Delta T + \Delta P$
'Rainforest'	10.0	10.4	7.4	7.6	11.1	13.9
'Woods'	16.2	58.5	62.4	68.6	59.2	56.9
'Grass & shrub'	38.7	13.2	12.9	8.9	13.0	10.6
'Desert'	15.5	5.2	2.0	2.3	5.4	7.6
'Tundra'	19.5	12.8	15.3	12.7	11.4	11.4

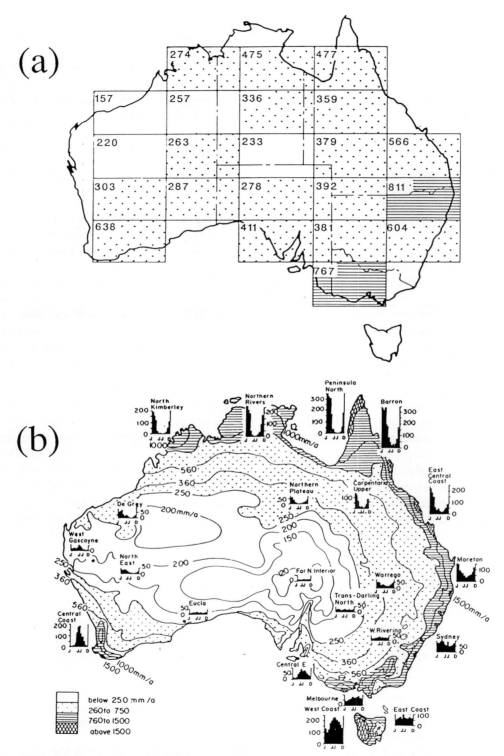

Fig. 5. (a) Annual precipitation for Australia derived using CCM1 plus BATS with fixed sea surface temperatures; (b) Observed annual precipitation for Australia (after Linacre and Hobbs, 1977).

Life zone prediction for 1 × CO_2 climate using the CCM0 plus mixed layer ocean model

The last three years of the 11 year experiment using the earlier version of the CCM coupled to a simple mixed layer ocean renders seasonally averaged air temperatures and total seasonal precipitation values. These have been used to construct mean annual biotemperatures and total precipitation from which the generalised classification shown in Fig. 6 has been drawn. This is a somewhat unsatisfactory representation of present-day ecosystem patterns since it is primarily 'woods'. (This class covers 93.84×10^6 km^2, as compared with 24.54×10^6 km^2 in the distribution specified for the present-day (Fig. 1)). The 'rainforest' region in South America is restricted to only four grid elements located on the west coast. The 'desert' area occupies only 8 grid elements all in northern Africa but none of which lie in the central Sahara. There is no 'desert' in Australia. (The 'desert' covers 3.03×10^6 km^2 compared with 23.48×10^6 km^2 in the specified land use classification).

The CCM0 plus mixed layer ocean model seems to be somewhat too wet. Indeed Washington & Meehl (1984) note that although the model correctly locates regions of high precipitation, the predicted totals are very much too high as compared with observations. This excess precipitation is, in some part, the result of warmer than observed predicted sea surface temperatures especially in equatorial regions. The other factor causing this disturbed ecotype classification is the tendency for the modelled continental surface temperatures to be higher than observed, especially in the latitude range $30°-70°$N. Washington & Meehl (1984) suggest that this may be in part due to an under prediction by the model of total cloud amount leading to an overestimate of the amount of direct insolation reaching the surface especially in the summertime. Note that in the highly simplistic surface parameterization package used in CCM0, the prescribed surface albedo values over the continents are somewhat lower than those which are observed.

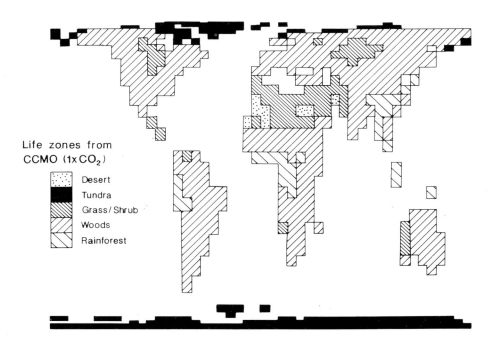

Life zones from CCM0 (1 × CO_2)

- :·.·: Desert
- ■ Tundra
- ⧄ Grass/Shrub
- ⟋ Woods
- ⟍ Rainforest

Fig. 6. Life zone groups predicted using CCM0 plus mixed layer ocean for $1 \times CO_2$.

Life zone groups predicted for a climate in which the atmospheric CO_2 concentration is doubled

The most straightforward method of establishing a generalised pattern which might result from the climatic perturbation caused by doubling atmospheric CO_2 can be obtained by using the seasonal temperatures and precipitation totals from the CCM0 plus mixed layer ocean model experiment in which CO_2 was doubled. Figure 7 shows the resulting distribution of groups. Note that the areal coverage is little changed, as compared with the $1 \times CO_2$ case (Table 2) for 'desert' and 'rainforest' but that there is about a 6% increase in cover by 'woods' compensated by $\sim 3\%$ decrease in both 'tundra' and 'grass and shrub'. The 'desert' region in north Africa has been relocated but the area is little changed (3.36×10^6 km^2 for $2 \times CO_2$ cf. 3.03×10^6 km^2 for $1 \times CO_2$). The area of 'rainforest' has decreased significantly in South America and increased slightly over the globe as a whole (11.38×10^6 km^2 for $2 \times CO_2$ cf. 11.14×10^6 km^2 for $1 \times CO_2$). This very small increase (0.2% of the land surface) is hard to explain since at the global

scale doubling the atmospheric concentration of CO_2 results in a 3.5 °C increase in temperature and a 7.1% increase in total precipitation as compared with the $1 \times CO_2$ experiment. Generally, increasing temperature and precipitation might be expected to increase significantly the area classified as 'rainforest' (see Fig. 2). The 'grass and shrub' area has decreased from 19.46×10^6 km^2 ($1 \times CO_2$) to 13.36×10^6 km^2 ($2 \times CO_2$) presumably in response to the warmer, wetter conditions.

Overall, the distribution of groups looks somewhat similar to that predicted from the present-day climate simulation using CCM0 plus a mixed layer ocean (Fig. 6 and Table 2) except for the replacement of the 'grass and shrub' areas in northern mid-latitudes by 'woods' and for the unexpected reduction in the area of 'rainforest' in South America in a doubled CO_2 world. Moreover the differences between Figs. 6 and 7 are very similar in extent and degree as the differences between Figs. 4 and 6. Compare, for example, the continents of Australia and South America in Figs. 4, 6 and 7.

Since the control climate (i.e. $1 \times CO_2$) of the

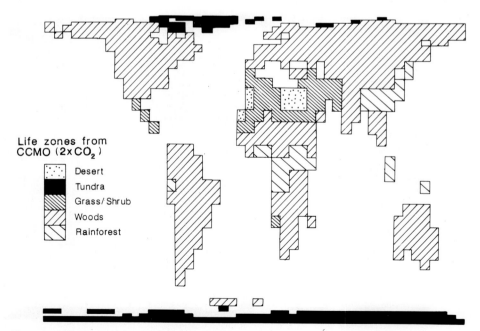

Fig. 7. Life zone groups predicted using CCM0 plus mixed layer ocean for $2 \times CO_2$.

NCAR CCM0 plus mixed layer ocean model gave rise to a somewhat unrealistic land pattern, especially in terms of the occurrence and location of 'desert', an alternative method of examining the impact on a prediction of ecosystem types of doubling atmospheric CO_2 should perhaps be sought. One such method is to compute the changes in temperature and precipitation, derived from subtracting the $1 \times CO_2$ from the $2 \times CO_2$ values, and to add these temperature and precipitation increments (or decrements) to the present-day predictions of climatic variables derived from the CCM1 plus BATS experiment. This has been done in two phases: (i) adding only the temperature increment and (ii) adding both the temperature and the precipitation differences. Although the latter of these two experiments is the more complete, the former has been included first in order to be directly comparable with an earlier estimation of the likely impact on a generalised ecological pattern of doubling atmospheric CO_2 undertaken by Emanuel *et al.* (1985a, b).

Figure 8 shows the ecological classification derived by adding only the ΔT values (derived from $2 \times CO_2$ minus $1 \times CO_2$). The difference

between this map and the present-day CCM1 plus BATS ecotype classification (Fig. 4) is very small. There is one new 'desert' grid element in the USSR, three 'woods' elements in China and these four single element changes comprise the only differences in low and middle latitudes. In addition, there are nine fewer elements classified as 'tundra' in the Arctic and six fewer classified as 'tundra' in the Antarctic. (Fig. 3(a) identifies all the locations at which differences occur between Fig. 8 and Fig. 4.) Note also in Table 2 that the percentage areas covered by each of the groups differs very little between these two classifications. The similarities between Fig. 8 and Fig. 4 are very much more impressive than the differences.

Adding the precipitation changes which are the consequence of doubling the atmospheric concentration of CO_2 to the predicted rainfall totals from the CCM1 plus BATS modifies the classification significantly more than the ΔT change alone (Fig. 9) (as compared with Fig. 4 rather than with Fig. 8). The most impressive changes occur in Australia where the eastern half of the country becomes 'desert' rather than 'grass and shrub'.

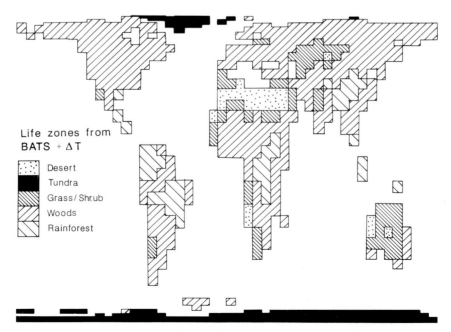

Fig. 8. Life zone groups predicted using CCM1 plus BATS climate with the addition of ΔT derived from $(1 \times CO_2 - 2 \times CO_2)$.

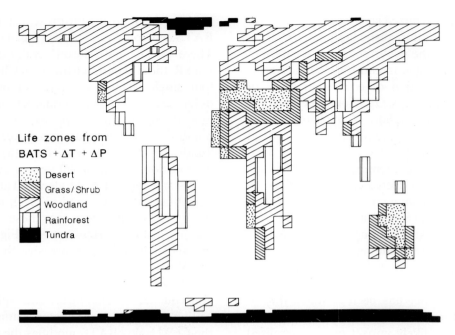

Fig. 9. Life zone groups predicted using CCM1 plus BATS climate with the addition of ΔT and ΔP derived from $(1 \times CO_2 - 2 \times CO_2)$.

There are new areas of 'rainforest' on the east coast of North America and joining two larger 'rainforest' areas in South America. In central Europe the area of 'grass and shrub' is reduced being replaced by 'woods' while on the central west African coast 'woods' are replaced by 'desert' and 'grass and shrub'. Figure 3(b) locates the grid elements which have had their classification altered as a result of adding both the temperature and precipitation increments to the CCM1 plus BATS present-day prediction (i.e. Figs. 4 and 9). The percentage coverage of the different groups for this doubled CO_2 as compared with the CCM1 plus BATS prediction shows ~3% more 'rainforest', but almost as large an increase in 'desert' compensated by ~3% less 'grass and shrub' and smaller reductions in 'tundra' and 'woods'.

What overall conclusion must be drawn from this exercise? Is the generalised ecosystem pattern well predicted from the seasonally-based mean annual biotemperature and the total annual precipitation? The ecosystem pattern seems to be relatively little affected when the effects of doubling the concentration of atmospheric carbon dioxide are considered only in terms of temperature changes. Even when the alterations in precipitation are included the differences derived are very much smaller than the impact that mankind's clearance of land for agriculture has had upon the natural distribution of ecosystems. Moreover, the impact on the predicted ecosystem groups by doubling CO_2 is no greater than the differences between two representations of ecosystem pattern based on present-day predictions of average temperature and precipitation totals.

Discussion

A highly generalised ecotype classification has been used to derive 'natural' predicted life zones for the present-day and for three different estimates of a $2 \times CO_2$ climate. The most important conclusion is that the highly generalised vegetation patterns derived show relatively little sensitivity to the climatic changes produced by doubling CO_2; the degree of sensitivity being about the same as the differences caused to the classification of the present-day (or future) ecosystems by using different predictions of seasonal temperature and precipitation (compare Fig. 4 with Fig. 6 and Fig. 7 with Fig. 9).

This result contradicts the earlier assertion of Emanuel *et al.* (1985a) resulting 'from increases in atmospheric CO_2 concentration. The indicated changes in global vegetation distribution are quite substantial' (Emanuel *et al.* 1985a, p41). In part this is, as was pointed out by Rowntree (1985), the result of the failure by Emanuel *et al.* (1985a) to compute correctly the biotemperatures resulting from doubling of atmospheric CO_2 (see also Emanuel *et al.* 1985b). It is also, in part, the result of the highly distorted representation of the eco-

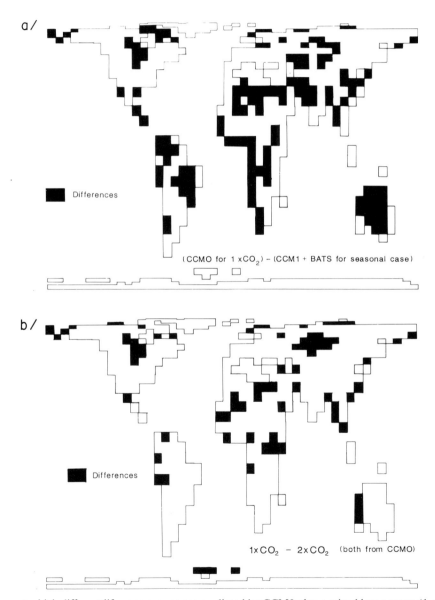

Fig. 10. (a) Locations at which different life zone groups are predicted by CCM0 plus a mixed layer ocean ($1 \times CO_2$) and CCM1 plus BATS also for the present day; (b) As for (a) except differences between $1 \times CO_2$ and $2 \times CO_2$ both predicted by CCM0 plus a mixed layer ocean (cf. Fig. 3(b)).

logical distributions offered by Emanuel *et al.* (1985a) in their figures 2 and 4 which, despite their assertion to the contrary, are maps not based on equal area projections (cf. Fig. 3). In this study there are more grid elements which have different predicted vegetation types when the two present-day simulations are compared (Fig. 10(a)) than when the $1 \times CO_2$ and $2 \times CO_2$ simulations are compared (Fig. 10(b)) or when the two CCM1 plus BATS simulations of present and future climates are compared (Fig. 3(b)).

This conclusion does not, of course, deny the fact that in certain areas the response of natural vegetation may indeed be strongly indicative of climatic change. All the present day/increased CO_2 pairs of maps indicate for example that 'tundra' will give way to 'woods' in high latitudes. However, the spatial resolution of most AGCMs (e.g. $4.5° \times 7.5°$ for the CCM used here) does not permit the type of predictions of transitional vegetation changes that such indicators would require.

The conclusion that a generalised land surface classification is relatively unaffected by doubling of atmospheric CO_2 is in agreement with the results of Marshall & Washington (1989), who found quite small differences between the areas defined under the five major Köppen climate classes for $1 \times CO_2$ and $2 \times CO_2$ respectively. Table 3(a) lists the differences between the control $(1 \times CO_2)$ percentage coverage and (i) observation [observed percentages also given in the table] and (ii) the control and the $2 \times CO_2$ climate. For all five classes the differences between the control (a representation of the present-day climate) and observations is greater than the difference between the control and $2 \times CO_2$.

Similar relationships exist for the differences derived from this study (Table 3(b)). For all five ecosystem groups the differences between the

Table 3. Relative differences between observed distributions, predicted distributions and doubled CO_2 distributions.
(a) Difference between percentage coverage of land surface by the five primary Köppen climate classes for CCM 'control' climate (i.e. CCM0 plus ocean model $1 \times CO_2$ results) and observations and for the control climate and the doubled CO_2 climate. (Observations after Haurwitz & Austin, 1944; differences derived from Marshall & Washington, 1989).

Köppen climate class	Control − observations	Control − $2 \times CO_2$	Observations (absolute % cover)
A (Tropical, rainy)	1.2	− 0.9	19.9
C (Temperate, rainy)	4.0	− 1.2	15.5
B (Arid)	− 9.7	− 1.9	26.3
D (Cold, snowy)	7.9	2.6	21.3
E (Polar)	− 3.4	1.3	17.0

(b) Differences between percentage coverage of land surface by the five generalized life zone classes for CCM 'control' climate (i.e. CCM0 plus ocean model) and specified vegetation; for the control climate and the doubled CO_2 climate and for the 'CCM1 plus BATS' predicted climate and the climate derived by incrementing ΔT and $\Delta T + \Delta P$ from $(2 \times CO_2 - 1 \times CO_2)$.

Generalized ecotype class	$(1 \times CO_2$ control) − (BATS specified)	$(1 \times CO_2$ control) − (BATS prediction)	$(1 \times CO_2$ control) − $(2 \times CO_2)$	(BATS predicted) − (BATS + ΔT	(BATS predicted) − (BATS + $\Delta T + \Delta P$)
'Rainforest'	− 2.6	− 3.0	− 0.2	− 0.7	− 3.5
'Woods'	46.2	3.9	− 6.2	− 0.7	1.6
'Grass & shrub'	− 25.8	− 0.3	4.0	0.7	2.6
'Desert'	− 13.5	− 3.2	− 0.3	− 0.2	− 2.4
'Tundra'	− 4.2	2.5	2.6	1.4	1.4

specified (observed) vegetation and the present climate represented by the $1 \times CO_2$ control (column one) exceed all other differences obtained. However, this is a somewhat unfair comparison since, as has been noted, the prescribed vegetation incorporates forest clearance and agriculture. A better comparison is probably between the $1 \times CO_2$ control and the CCM1 plus BATS predictions (column two) and the present-day and doubled CO_2 pairs of climates (columns three, four and five). For the two 'extreme' ecotypes ('desert' and 'rainforest'), the differences between the two simulations of the present-day climate (column two) are greater than the changes caused by doubling CO_2. This is because the CCM0 plus mixed layer ocean captures the areal extent and location of these two ecotypes so poorly. On the other hand, the CCM1 plus BATS significantly under-represents the present-day extent of northern tundra and this is the reason for the differences in percentage coverage of 'tundra' in columns two and three being so similar. The differences between the $1 \times CO_2$ and the $2 \times CO_2$ experiments do considerably exceed those in column two for 'grass and shrub' and 'woods'.

The same is not true when the alternative method of deriving the ecotype classification for a doubled CO_2 climate in which the increments of temperature and precipitation are added to the CCM1 plus BATS climate variables (column five cf. column two). In this comparison the differences are commensurate. The larger changes forced by including precipitation differences rather than the temperature changes alone are seen in column five cf. column four. These larger changes due to precipitation imply that the experiment of Emanuel *et al.* (1985a), who added only a predicted temperature increment on to a present-day climatology based on meteorological observations, might have produced an underestimate of ecosystem disturbance. The appropriate set of differences for comparison with their results is column four which incorporates only the effect of increasing temperatures. The temperature-only differences determined here are smaller than was suggested in the earlier study.

These results contrast with earlier suggestions

that ecological changes could be used as a sensitive monitor of trace-gas induced climatic warming. The large-scale disturbance of the natural ecosystems of the globe by mankind dominates almost all the shifts predicted as a result of greenhouse-induced climatic change (compare Fig. 1 with Figs. 4 and 9 or with Figs. 6 and 7). The exception to this might be in the high latitudes where tundra regimes may become able to support other ecosystems. Unfortunately the predictions of the present-day climate (especially CCM1 plus BATS) and the future climate (especially CCM0 plus a mixed layer ocean) are less successful here than elsewhere.

The highly simplistic 'predictions' of global pattern generated here are clearly only a first step towards incorporating the land-surface plants as an interactive component in a global climate model. The representations of the present-day vegetation differ from one another to about the same extent as the defining climate variables by the doubling of atmospheric CO_2. All the differences are much less than those caused by mankind's clearance of natural vegetation for agriculture and the extension of desert areas by overgrazing.

Acknowledgments

This project was funded by the Australian Research Council. Warren Washington & Gerald Meehl kindly provided the 3 year seasonal averages of temperature and precipitation from their $1 \times CO_2$ and $2 \times CO_2$ experiments and Linda VerPlank supplied these data on tape. Robert Dickinson kindly provided the 3 year monthly means of temperature and precipitation from his BATS experiment and Patrick Kennedy supplied these data on tape. Andrew Pitman & Jerry Olson read and commented constructively on an earlier version of this manuscript.

References

Bourke, W., McAvaney, B., Puri, K. & Thurling, R. 1977. Global modeling of atmospheric flow by spectral methods, Methods in Computational Physics, General Circulation

Models of the Atmosphere, Vol. 17, (ed. J. Chang), Academic Press, 267–324.

Charney, J. G. 1975. Dynamics of deserts and drought in the Sahel. Quart. J. Roy. Meteor. Soc. 101: 193–202.

Charney, J. G., Quirk, W. J., Chew, S. M. & Kornfield, J. 1977. Dynamics of deserts and droughts in the Sahel. J. Atmos. Sci. 34: 1366–138.

Dickinson, R. E. & Henderson-Sellers, A. 1988. Modelling tropical deforestation: a study of GCM land-surface parameterizations. Quart. J. Roy. Meteor. Soc. 114(B): 439–462.

Dickinson, R. E., Henderson-Sellers, A., Kennedy, P. J. & Wilson, M.F. 1986. Biosphere-atmosphere transfer scheme (BATS) for the NCAR Community Climate Model, NCAR Tech. Note, NCAR/TN-275 + STR, Boulder, CO, 69 pp.

Emanuel, W. R., Shugart, H. H. & Stevenson, M. P. 1985a. Climatic change and the broad-scale distribution of terrestrial ecosystem complexes. Clim. Change 7: 29–44.

Emanuel, W. R., Shugart, H. H. & Stevenson, M. P. 1985b. Response to Rowntree. Clim. Change 7: 457–460.

Ehhalt, D. H. 1988. How has the atmosphere concentration of CH_4 changed? In: F. S. Rowland & I. S. A. Isaksen, eds., The Changing Atmosphere, John Wiley & Sons, Chichester, 25–32.

Giorgi, F. & Bates, G. T. 1989. On the climatological skill of a regional model over complex terrain. Mon. Wea. Rev. 2325–2347.

Haurwitz, B. & Austin, I. M. 1944. Climatology, McGraw Hill, 410 pp.

Holdridge, L. R. 1947. Determination of world plant formations from simple climatic data. Science 105: 367–368

Holdridge, L. R. 1964. Life Zone Ecology, Tropical Science Center, San José, Costa Rica.

Jenne, R. L. 1989. Data from climate models; the CO_2 warming, Environmental Protection Agency Report on Climatic Change (unpubl).

Kiehl, J. T., Wolski, R. J., Briegleb, B. P. & Ramanathan, V. 1987. Documentation of radiation and cloud routines in the NCAR Community Climate Model (CCM1), NCAR Tech. Note, NCAR/TN-288 + IA, Boulder, CO.

Linacre, E. & Hobbs, J. 1977. The Australian Climatic Environment, John Wiley, Chichester, 354 pp.

Manabe, S. & Wetherald, R. T. 1987. Large-scale changes of soil wetness induced by an increase in atmospheric carbon dioxide. J. Atmos. Sci. 44: 1211–1235.

Marshall, S. E. & Washington, W. M. 1989. Classification of continental climates using the NCAR CCM with surface hydrology and a simple mixed-layer ocean model for $1 \times CO_2$ and $2 \times CO_2$, NCAR Tech. Note (unpubl).

Matthews, E. 1984. Prescription of land-surface boundary conditions in GISS GCMII and Vegetation, land-use and seasonal albedo data sets: Documentation of archived data tape, NASA Technical Memos 86096 and 86107, NASA, Goddard Institute for Space Studies, New York, NY, 20 pp and 9 pp.

Meehl, G. A. & Washington, W. M. 1988. A comparison of soil-moisture sensitivity in two global climate models. J. Atmos. Sci. 45: 1476–1492.

Pitman, A. J. 1988. The development and implementation of a new land surface parameterization for use in GCMs, unpubl. Ph.D. thesis, Liverpool University, 481 pp.

Ramanathan, V., Pitcher, E. J., Malone, R. C. & Blackmon, M. L. 1983. The response of a spectral general circulation model to refinements in radiative processes. J. Atmos. Sci. 40: 605–630.

Rowntree, P. R. 1985. Comment on 'Climatic change and the broad-scale distribution of terrestrial ecosystem complexes Clim. Change 7: 455–456.

Schlesinger, M. E. & Mitchell, J. F. B. 1985. Model projections of the equilibrium climatic response to increased carbon dioxide, in Projecting the climatic effects of increasing carbon dioxide. United States Department of Energy, DOE/ER 0237, Washington DC, pp. 81–148.

Sellers, P. J., Mintz, Y., Sud, Y. C. & Dalcher, A. 1986. A simple biosphere model (SiB) for use within general circulation models. J. Atmos. Sci. 43: 505–531.

Shukla, J. & Mintz, Y. 1982. Influence of land-surface evapo-transpiration on the Earth's climate. Science 215: 1498–1501.

Tucker, C. J., Fung, I. Y., Keeling, C. D. & Gammon, R. H. 1986. Relationship between atmospheric CO_2 variations and a satellite-derived vegetation index. Nature 319: 195–199.

Washington, W. M. & Meehl, G. A. 1984. Seasonal cycle experiment on the climate sensitivity due to a doubling of CO_2 with an atmospheric general circulation model coupled to a simple mixed-layer ocean model. J. Geophys. Res. 89: 9475–9503.

Wilson, M. F. & Henderson-Sellers, A. 1985. A global archive of land cover and soils data for use in general circulation climate models. J. Climatol. 5: 119–143.

Wilson, M. F., Henderson-Sellers, A., Dickinson, R. E. & Kennedy, P. J. 1987a. Investigation of the sensitivity of the land-surface parameterization of the NCAR Community Climate Model in regions of tundra vegetation. J. Climatol. 7: 319–343.

Wilson, M. F., Henderson-Sellers, A., Dickinson, R. E. & Kennedy, P. J. 1987b. Sensitivity of the biosphere-atmosphere transfer scheme (BATS) to the inclusion of variable soil characteristics. J. Clim. Appl. Meteor. 26: 341–362.

Woodward, F. I. 1987. Climate and Plant Distribution. Cambridge Studies in Ecology, Cambridge University Press, Cambridge, 174 pp.

Management

Introduction

The crucial role of land management in semi-arid regions, with particular examples drawn from New South Wales and throughout Australia, is the focus of this final section. Mitchell discusses vegetation and soil changes in New South Wales and argues that the accepted view of the effects of European settlement on western New South Wales is somewhat flawed. Thomas and Squires review assessment methods for semi-arid agricultural and recreational land development. Pickard discusses land management in semi-arid areas of New South Wales. He argues that while the technical aspects of land degradation, and the theoretical solutions to the problem are reasonably well understood, the social aspects of management continue to be neglected and thus many attempts to manage semi-arid areas are severely limited. Norris, Mitchell and Hart re-evaluate the evidence covering vegetation changes in the Pilliga forests of northern New South Wales. They conclude that the established view of the consequences of European settlement is too simplistic and that a more complex scenario involving more factors is required. Finally, Heathcote describes the historical development of public perception of drought in Australia. In particular, the evolution of government and public responses to drought are discussed in the context of the expected future modification in planning policy.

Vegetatio **91**: 169–182, 1991.
A. Henderson-Sellers and A. J. Pitman (eds).
Vegetation and climate interactions in semi-arid regions.
© 1991 *Kluwer Academic Publishers. Printed in Belgium.*

Historical perspectives on some vegetation and soil changes in semi-arid New South Wales

P.B. Mitchell
School of Earth Sciences, Macquarie University, North Ryde NSW, 2109 Australia

Accepted 24.8.1990

Abstract

The history of settlement of the semi-arid rangelands of western New South Wales is reviewed with respect to changes in the vegetation and soil which occurred under a regime of european land management. Simple dynamics of the vegetation response to grazing are illustrated and primary archival data is explored to verify the status of traditional wisdom about three examples of perceived change; the extent of the pioneers knowledge of land degradation, the timing and causes of nineteenth century *Callitris* pine regrowth events, and the importance of soil compaction. In each case it is shown that the traditional wisdom surrounding these issues is partly erroneous and that folklore is in danger of becoming accepted fact. Such errors must be avoided if we are to improve range dynamics models and management.

Introduction

Land degradation is a widespread and serious problem evident all over the world which is perhaps worst in the semi-arid zones of marginal agriculture. It takes many forms; serious wind and water erosion, depletion of native pastures, invasion by weeds, expansion of woody shrubs, loss of tree cover, deterioration in soil structure, and increases in salinity or acidity etc. In the semi-arid zone of New South Wales Graham *et al.* (1987/8) have documented the extent of such changes and it is undeniable that many of them were induced by european land use over the past century. It is necessary to ask however, are the generally accepted views that all such changes have anthropogenic causes, and that the early settlers 'did not know any better', correct? Or are we repeating folklore and creating myths which may be hindering the management of some of these problems?

It is the purpose of this paper to develop a theme opened by Ludwig (1987) and to show that our beliefs about some vegetation and soil changes in the semi-arid zone of New South Wales have become stereotyped as 'traditional wisdom'. These beliefs need to be reviewed if we are to seek remedies to the problems. It will be shown that one of the routes past the restrictions of traditional wisdom involves a careful analysis of primary historic sources which enables the researcher to view the landscape through the experience of others over a longer time span.

The background to settlement

The semi-arid zone nominally lies between the 250 and 500 mm (10 to 20 inches) rainfall isohyets and covers most of the western plains and the drier half of the western slopes of New South Wales from the Darling River east to a line through Wagga Wagga, Dubbo and Narrabri. Rainfall variability is typically between 25 and

30% of the mean, but can exceed 100%, and severe droughts have a frequency of recurrence of about 1 : 8 to 1 : 10 (Jeans 1972).

The slopes and plains were explored by Oxley, Sturt, and Mitchell between 1817 and 1845. All three explorers recorded detailed descriptions of the landscape and many of their observation points can be relocated. Pastoral settlement followed along the rivers in the 1830s and 1840s, with some of the first squatters and their stock moving out in front of the later explorers. Pastoralists were searching the north-western corner of the state in 1860 even as Burke and Wills were dying on Coopers Creek, and all of New South Wales was claimed by the mid-1880s. Three or four generations later, some of that land has been abandoned and a number of former stations have become National Parks.

Cattle were almost universally the pioneer stock but as the country was settled they were replaced by sheep which dominated most parts of the range by the 1860s. High wool prices in the 1870s encouraged increased stocking and gave large profits. This was a period of above average rainfall and changes were occurring in the vegetation with grasses replacing shrubs. These coincident events confirmed optimism and lead to the belief that 'sheep pushed back the margins of the desert'.

Wire fencing was started in the period when labour was abundant and cheap (Buxton 1965) after the peak of the gold rushes. Ringbarking, clearing and cropping began in the mid-1870s and accelerated during the late 1880s with closer settlement and expansion of the railways. The dry backblocks away from the rivers were all taken up (but not necessarily stocked) when groundwater supplies and surface tanks were developed and the last phase of land allocation was the establishment of State Forests in the early twentieth century, often on lands not previously selected because of poor soils, or thick scrub and timber.

New South Wales faced the first rural crisis in 1841/3. Most of the better country in eastern and central parts of the state was stocked, nobody wanted more breeders, meat was oversupplied and the London price of wool fell. For the next ten years most production went into candles and soap. The gold rushes of the 1850s re-established a meat market and the Riverina became a vital fattening area for the goldfields. The large population of ex-miners wanted to own some land and politicians began to reduce the state into smaller holdings regardless of the ability of the land to sustain a family. Such subdivision into sub-marginal units was a basic mistake repeated several times even into the mid-twentieth century (for later examples see Lake 1987).

More crises and problems followed, drought was an irregular but certain event, even though average rainfalls were higher in the thirty years before the mid-1890s (Foley 1957). Prices fell in the 1880s but loans had to be serviced, land tenure was uncertain and lessees faced a trebling of rents as a result of the Crown Lands Act 1884. All this stimulated an increase in production to offset the lower returns and higher costs. Australian graziers were to become very familiar with this cost/price squeeze over the next century and the end result was not more profit for the pastoralist, but short term economic survival at the expense of the ecological capital of the range. Into this scene in the 1890s came the rabbits.

At the end of the century the financial effects of an urban land-boom and depression extended to the rural sector and in 1895-1902 drought struck. In western New South Wales, stock figures crashed from a peak of 15.4 million sheep equivalent in 1891 to a low of 3.5 million in 1902. More than two million hectares of lease was abandoned. Stock numbers have stayed down ever since, never exceeding about 6 million and all of the rangeland degradation problems have traditionally been attributed to that drought. Figures from some properties show that they were carrying 1 sheep/ha in 1890 where later they needed between 3 and 8 ha/sheep (Beadle 1948).

That is the usual story; as it came from the evidence given to the 1901 Royal Commission (Anon. 1901). It was not, however, that particular drought which wrecked the pastoral industry, there had been at least seven earlier ones. A series of lesser changes had started 30 years earlier which were more important and these had passed almost unnoticed, nevertheless it is clear from original accounts that much of the ecological degrade of the western vegetation had occurred before the arrival of rabbits and before the mid-1890s shift in rainfall patterns (Jeans 1972).

Clues to unravelling this story can be found by reading early descriptions of the country such as that made by Anthony Trollope who described the saltbush country of the Riverina in 1873:

> 'The secret of the wealth of the country for pastoral purposes lies in the salt which the soil possesses. A great proportion of the Riverina used to produce salt-bush, – a shrub about three feet high, pale in colour, and ugly to look at when it covers a whole plain, on which the sheep feed willingly and which can stand great heat and great drought. I was told that the salt-bush was disappearing on runs which had carried sheep for many years, and that it certainly receded as the squatters advanced. But, though the salt-bush may go the salt remains. Australian squatters... are all agreed that a salt country is the best for sheep. In a salt country, though it seems to be as bare as a board, sheep will keep their condition',... (Trollope 1873, p 330)

In this quotation there are three interesting pointers to some of today's problems. In the 1870s the settlers knew that the Riverina soils contained salt; that salt, is now a serious problem in the irrigation areas. It is also clear that the composition of the native vegetation had shifted away from saltbush in only 25 years of grazing, and probably the country which Trollope described as being; '... as bare as a board...' was subject to some degree of scald erosion. This is a reliable report obtained by Trollope from eye witnesses which reveals that even in the early 1870s the country was changing and that the squatters knew it. Other changes took place over the turn of the century and they continue today as evidenced by the bladder saltbush 'dieback' problem between 1977 and 1983 (Clift *et al.* 1987).

The nature of the changes – vegetation dynamics on the range

One of the unique things about Australian rangelands in comparison with the rest of the world is that native herbivores were only a minor feature of the ecosystem and when they were replaced by domestic stock it was an invasion of a system which had little resistance to heavy grazing. For the most part the natural vegetation lacks adaptations like subterranean buds and protected vegetative parts except those species which happen to have them as fire or drought adaptations. No wonder then, that degrade of the whole system occurred within the life cycles of the dominant plant species.

Plants in the semi-arid zone have to cope with stresses induced by water shortage, extreme desiccation, low nutrients, moderate to high salinity and occasional fire (see Pate & McComb 1982 for a detailed account). Onto this was imposed a grazing regime by five dominant herbivores (cattle, sheep, horses, rabbits and goats). The plants were tolerant of the natural stresses but respond to the additional load of selective grazing in three ways.

By decreasing in abundance

Plant species which are more palatable and nutritious, are consumed in greater quantity and are rapidly depleted in a pasture. These are generally softer herbs and forbs, some grasses and some saltbushes, many of which are susceptible to grazing because their growth tissue is exposed and vulnerable.

Although opinions vary about the palatability of oldman saltbush (*Atriplex nummularia*), the species Trollope probably referred to (see above), it is a 'decreaser' species. This perennial shrub has the reputation of being valuable sheep fodder when in fact it is not. Like most chenopods it is high in digestible protein (15 to 20%), and it is eaten, but only after more palatable species have been taken from the range. Sheep feeding almost exclusively on saltbush require much more water than usual and most studies show that they only just maintain condition (Lynch 1988). As a less preferred species oldman saltbush tends to be eaten in dry times under heavy stocking rates and in these conditions the plant is also stressed and does not recover from defoliation. Trollope's description is evidence that the Riverina had been heavily overgrazed even in 1873, an observation supported by Dixon (1880), Turner (1891), and Beadle (1948).

Moore (1953) attributed the demise of oldman saltbush in the eastern Riverina to sheep grazing and the drought of 1875-77, but Williams (n.d.) suggested that it was part of an earlier vegetation change in the 1860s which he argued was so great that it forced the change from cattle to sheep.

Another typical decreaser species on which there is more data than for most plants is bladder saltbush (*Atriplex vesicaria*). Total defoliation usually kills it because new growth comes from the terminal bud on young stems and these are preferentially removed by stock. However it will stand periodic lighter grazing in which case the form of the plant changes because of increased leaf growth resulting in a lower, and more compact plant (Graetz 1973).

When the saltbushes were eliminated annual grasses tended to take over on the Riverina (Moore 1953) but in South Australia bare ground was the more common end point (Lange *et al.* 1984).

By increasing in abundance

The second response comes from the plants which are not generally eaten, or which respond positively to grazing by tillering and/or producing more seed. These are the 'increasers' and typically they are less nutritious, less palatable, possess damaging seeds or prickles and are often poisonous. Good examples are the bluebushes (*Maireana* sp.), numerous spiny bassias and burrs, annual wire grasses, and unpalatable woody shrubs.

Being native plants the shrubs are also encouraged by a reduction in the frequency of wildfire

which accompanies heavy grazing, they survive drought and when climatic conditions are favourable can rapidly take over the entire range. The consequences of shrub invasion are; reduced grazing capacity, more difficult musters, higher stock losses especially from fly strike, lower lambing rates, and ultimately, abandonment of the property or sale to a neighbour because grazing is no longer viable with a reduced flock (Booth n.d.).

Invasion by exotics

The third reaction of the pasture to grazing is the inevitable introduction of exotic weed species, most of which are increaser types. They commonly come from similar climatic environments overseas, arrive without their natural range of consumers and diseases and grow unhindered. As colonisers of bare ground they have been important in revegetating many of the scald eroded areas that were very widespread in the 1940s but their fodder value is limited. The first 'invaders' arrived with the earliest settlers, Mitchell (1848) for example, records the common occurrence of horehound (*Marrubium vulgare*) on Bogan River runs only ten years after occupation. Similar weeds are still colonising the range and marginal agricultural land at a rate of between two and five species per year (Parsons 1973, and Robards & Michalk 1979).

There are many variations on these themes so far as individual species are concerned, but despite extensive general knowledge and a number of clipping/grazing experiments we know very little about the responses of most native plants, although the Soil Conservation Service of New South Wales is well advanced with an important series of literature surveys which are being published as Technical Reports with National Soil Conservation Program funding (see for example; Scriven 1988 and Dalton 1989).

This simple pasture response model of decreasers, increasers and invaders was applied to the Riverina by Moore (1953) who showed that much of the eastern part was originally an open woodland and shrubland of boree (*Acacia pendula*), oldman saltbush and numerous annual chenopods and grasses. It supported small numbers of kangaroo and a range of small marsupials such as; rat-kangaroo, bilbies and hopping mice. The first plants to decrease were the succulent annual chenopods and the more nutritious grasses. As pressure increased and droughts came, the oldman saltbush vanished and the only surviving species were perennial grasses and the prickly/poisonous increasers. The proportion of bare ground increased, especially after drought, and weeds moved in.

The rapid change from a chenopod shrubland to a disclimax grassland favoured the larger kangaroos and led the smaller animals to extinction. Today the boree/saltbush country has almost vanished as a vegetation community, it is not well represented in any National Park and there are few large areas available which could be included in one (Leigh & Noble 1972, Benson 1987).

Underlying models

The simple model of vegetation dynamics used in the previous discussion is based on the observed response of a few species and invokes traditional ideas of climax, succession and stability. Many practical range management models are based on it and these aim for the maintenance of quasi-stable disclimax communities. If stock were to be removed from the disclimax range most models assume that there would be a reversion to the climax (Slatyer 1973). This is not necessarily so and although there are very few long term stock exclusure plots being monitored in Australia studies such as those at the T.G. Osborn Reserve (Koonamore) in South Australia (Noble 1977) show three things:

(i) That revegetation of heavily grazed land after destocking is very slow (the inertia effect of Noble 1986) and depends on the conjunction of favourable low frequency events especially exceptional rainfalls, available seed and reduced stocking levels, usually including rabbits.

(ii) The resulting communities are not necessarily stable.

(iii) The new communities do not have the same composition as the original supposed climax.

Climax and succession models assume a degree of constancy in environmental conditions which is clearly not true in the time span of decades which must be considered in the management of the semi-arid zone. Whilst such models have been useful in defining trends in range condition and identifying indicator species, it is important to ask; are these models adequate for 'stop-go' ecosystems in which the apparent dominance of rainfall may be masking other important disturbance factors? Such systems might be better understood by applying some of the ideas of patch dynamics (Pickett & White 1985). In practice this question has been answered in the negative and some management strategies now impose major disturbances on seemingly stable communities in order to force a favourable change. For example; the use of fire in lands infested with woody shrubs, or water ponding on scalds.

Several alternative models have been presented in recent years (Westoby 1979/80, Noble & Crisp 1979/80) but all are more complex than those based on observation of indicator species and they are likely to fall into the communications gap which exists between academic researchers and graziers. This gap seriously affects the rate of adoption of new management ideas and although current trends for model presentation are to incorporate them in PC based 'expert systems' (for example, Ludwig 1988), which should make them available to all, it has not yet been shown that these are either acceptable or useful to the average grazier.

The value of archival sources in improving understanding

Sound knowledge of the composition and dynamics of the vegetation in the semi-arid zone is so limited that it is inevitable that simplistic models and half truths become enshrined in the literature as traditional wisdom. To better manage these landscapes in the future it is essential that this wisdom and the models based on it be questioned. One of the means of doing this is to review early landscape descriptions found in archival files.

Throughout the period of settlement on the western slopes there is a good historical record, both published and unpublished. Several explorer's journals, numerous land surveys, travellers descriptions, railway surveys, sketches, and early photographs exist. State archives hold material concerning land ownership and stock numbers, and Parliamentary debates, Royal Commissions etc., are legion. Private archives and records are also held by individuals, large pastoral companies and in the Business Archives Department of the Australian National University. Little of this material has been used by researchers but interest is increasing as evident in the works of Heathcote (1965), Jeans (1978), Harrington *et al.* (1979), Griffin & Friedel (1984), and Oxley (1987 a,b).

There are some difficulties. Primary historic source material needs to be examined with care. Most works have an inherent bias which must be identified, early descriptions must be read with a nineteenth century view of natural history and landscape description in mind, and it is often difficult to locate specific places to make comparative studies (see Finlayson 1984, for an illustration of these problems). Carefully used however, the historic record can illuminate the history of development and early perceptions of land degradation and assist in differentiating the 'natural' from the 'human' impacts as the following examples will show. Only when we are reasonably certain that we have a good appreciation of the nature of the changes will we be able to assemble better predictive models for management.

Did the pioneers recognise landscape change?

A typical item of traditional wisdom which is often used to excuse pioneer settlers of any blame for land degradation is; 'that they came from a different environment, did not know any better and did not see that the landscape was changing adversely'.

Even a cursory examination of this claim will show that it is often not true and is probably best regarded as a piece of Australian folklore or mythology. For example; as early as 1828 Charles Sturt observed that on some of the land in the Macquarie valley north of Wellington;

The cattle had consumed all the food, and the ground on both sides of the river looked bare and arid (Sturt 1833 p 9).

Sturt made no critical comment on this observation.

A few years later (December 1845), after the first wave of settlement had advanced and retreated from the Bogan valley Mitchell was more specific and more thoughtful;

… poor country… but everywhere it was taken up for sheep; and these looked fat; yet not a blade of grass could be seen; and, but for the late timely supply of rain, it had been in contemplation to withdraw these flocks to the Macquarie. (Mitchell 1848, p 9).

We had encamped near those very springs mentioned as seen on my former journey (1835), but instead of being limpid and surrounded by verdant grass, as they had been then, they were trodden by cattle into muddy holes, where the poor natives had been endeavouring to protect a small portion from the cattle's feet, and keep it pure, by laying over it trees they had cut down for the purpose. The change produced in the aspect of this formerly happy secluded valley, by the intrusion of cattle and the white man, was by no means favourable, and I could easily conceive how I, had I been an aboriginal native, should have felt and regretted that change (Mitchell 1848, p 14).

After the 1870s there are many descriptions of changing land condition, most of which refer to the vegetation; for example Trollope (1873) as quoted above and in other parts of the same work, Anon. (1881), Hamilton (1892), Millen (1899), and the large body of anecdotal evidence presented to the 1901 Royal Commission (Anon. 1901). An example of how this material can be used may be seen in the work of Oxley (1987a and b) who presented a comprehensive account of vegetation change on south eastern Queensland stations using archival material from several sources dating from the late 1870s.

All the authors listed above write on a similar theme and leave the modern reader with little doubt that observant travellers and settlers were very aware of the damage that was being done to the rangelands well before the invasion by rabbits and the drought at the turn of the century. A contemporary assessment of the basic cause was presented by Millen (1899) when writing of conditions in western New South wales under extreme drought:

> The fundamental error has been in regarding the nominal condition as one of fair seasons punctuated with occasional drought. Those who settled this country held this error as an article of faith; those who succeeded them clung to it with the tenacity of despair. (Millen 1899, p 4)

The consequences of optimism and overgrazing were well summarised by Dixon (1892) and can be reasonably applied to all areas of the semi-arid and arid range:

> ... after 30 years of settlement... in the Riverina and northern South Australia, the injury to the original vegetation by overstocking has assumed such a great magnitude as to entail a national loss. (Dixon 1892, p 202)

Woody shrubs in the poplar box and cypress pine country

The second topic reviewed here concerns the remarkable expansion and increase in density of white cypress pine (*Callitris glaucophylla*, Thompson & Johnson (1986), formerly *C. columellaris, syn., C. glauca*) in the late nineteenth century and in the early 1950s. That there have only been two periods of pine regeneration and that extensive areas of formerly open, grassy, woodlands have become choked by pine is another piece of traditional wisdom that is now firmly established as a consequence of the popularity of an historical work by Rolls (1981), and despite some evidence to the contrary.

The region of interest is the semi-arid shrub woodland where poplar box (*Eucalyptus populnea*) and white cypress pine form an overstory to shrubs, herbs and grasses on harsh texture contrast soils and areas of deeper sands or clays of riverine depositional facies in the north and mid-west of New South Wales and extending into Queensland. Rainfall is between 400 and 750 mm and these communities are the driest (sub-economic) forest zones in the state. In many places this land has been extensively cleared for grazing and grain cropping.

There is no argument that woody shrubs (typical increaser species) are a serious management problem in many parts of the rangelands as described by Booth (n.d.) and that cypress pine is one of the problem species on lighter soils on the higher rainfall margin. What this paper takes issue with are the beliefs that there were only two main periods of pine regeneration and that dense shrub cover was unknown in the west at the time of first settlement.

Rolls (1981) claims that the Pilliga region north of Coonabarabran was an open, grassy, forest of pine and ironbark in which settlement began in the late 1830s and was more or less complete by the 1870s. At that time he claims there had been no regular burning for about 25 years and that domestic stock had displaced native herbivores. Some pine had moved off the ridges onto the clear valley floors and wire and spear grass had replaced more palatable fodder species. The graziers began to burn to try and control grass seeds and pine scrub encroachment. Weather patterns also had an important influence. Rolls claims that the period 1875–1878 was droughty and that 1879 was very wet, the cattle market was depressed and additional sheep were put on the grass which was burnt to clear seed and give the stock a green pick, with the result that white cypress pine germinated at 'wheatfield density' and the land was subsequently abandoned for grazing. Two quotations summarise this story:

> And where the fires ran years of pine seed came to life. (Rolls 1981, p 183-184).

> The four or five good years between 1879 and 1887 were the only years in which it was possi-

ble for the new forest to come away. By the next good rains in the 1890s there were sufficient rabbits, as enthusiastic eaters of seedlings as the disappearing rat-kangaroos, to stop most new tree growth. The extent of the country to be abandoned was determined by the 1890s. Except for a thickening of the undergrowth in places in the several wet years following the breaking of the 1902 drought, there was little more growth of pine or scrub until 1951, when a huge fire germinated seedlings (sic) on land soaked by heavy rain in 1950. At the same time myxomatosis destroyed the rabbits. And the lovely tangle which is the modern forest came to life. (Rolls 1981, p 205).

Rolls extended this story to virtually all accounts of scrub encroachment on woodlands and forests everywhere in New South Wales, (Rolls, 1981, Chapter 14), but at least in the case of the Pilliga, his arguments need to be questioned (Norris *et al.* in press) with respect to the extent, timing and possibly the causes of this pine regrowth event. His model was uncritically accepted by Austin & Williams (1988) who also cited Chiswell (1982) in support, without apparently realising that his (Chiswell's) article was also derived from Rolls (1981). Such a use of uncertain data to establish new models is a classic example of the persistence of error described by James (1967).

By examining primary sources; explorer's accounts, railway surveyor's letters, Land Commissioner's reports on the scrub problem, Forestry Department and later Forestry Commission Annual Reports, a very different story emerges which shows that pine scrubs have always been present in these parts of the state and that Rolls was probably wrong in attributing the Pilliga regrowth event to the 1880s.

There are three questions to consider under this topic:
– Were pine scrubs present at the time of european exploration?
– When did the regeneration events take place?
– What were the causes?

The previous existence of dense scrubs

The journals of Oxley (1820), Sturt (1833), & Mitchell (1839, 1848) describe journeys down the Lachlan, Macquarie and Bogan rivers respectively, and all specifically mention, on many occasions, that the lighter textured red soils away from the rivers often supported dense scrubs of *Eucalyptus* sp., (sometimes mallee), *Acacia* sp., (sometimes mulga), and *Callitris* sp. It is notable that there are more references to the difficulties of traversing such 'dreadful scrub' country when the explorers made side trips away from the river, or were using pack animals rather than wheeled vehicles which confined them to the clear country on the floodplains.

Some examples:
Oxley (1820) on the Lachlan; June 6, 1817 '… after going through about eight miles of very thick cypress scrub;… (p 58).

June 7 '…ten miles… it was a continued scrub, and where there was timber it chiefly consisted of small cypress: …' (p 59).

June 26 '… ten miles… occasional clear spaces, but for the greater part thick cypress bushes, acacia and other low shrubs, rendered it difficult for the horses to pass.' (p 84).

Sturt (1833) on the Macquarie; December 1826 '… through coppices of cypresses and *Acacia pendula*' (p 14).

Mitchell (1848) on the Bogan; January 1846 'Our route was rather circuitous, chiefly to avoid a thick scrub of *Callitris* and other trees, which having been recently burnt, presented spikes so thickly set together, that anyway round them seemed preferable to going through.' (p 25).

In southern Queensland; April 17 'We then got forward on foot as fast as the men could walk, or rather as fast as they could clear a way for the cart. We passed through much scrub, but none was of the very worst sort.' (p 126).

September 11 'A very dense forest of young *Callitris* trees next impeded us, and were more formidable than even the vines.' (p 303).

The date of nineteenth century regrowth event(s) and theories on causes

In 1878 railway surveyors working on the eastern side of the Pilliga forests commented on 'the marvellously dense scrubs on the poor sandy soils' in a way which implied that they had long been there (Carver 1878).

In 1880 the Surveyor General called for reports from all his Land Commissioners and surveyors on; the extent, age and significance of the pine scrub on leased lands in the west, after about 40 Lachlan and Murrumbidgee tenants had complained to him about the rapid spread of scrub and the difficulties they faced making a living from affected land without security of tenure (Anon. 1881).

In summary the replies of the Commissioners were:

In the Lachlan district a real problem existed. It consisted of young pine of recent date which it was claimed, had germinated because of a reduction in the number of fires. One Commissioner suggested that fire would control the scrub but that the tenants argued they could not afford to loose the grass. Another added that pine scrub regrowth was evident in 1875 and that these areas contained some good timber which needed thinning. Another suggested it had been spreading over the previous 15 years, that is, it had started about 1865/6.

In the Murrumbidgee District the problem was also real and it was said that the pine regrowth had set in during the 1866 drought when the soil surface was bare and pine seeds lodged in the dust. It was evident in 1868 and was caused by stock treading seed into the soil. The scrub was considered to have considerable future timber value if it was thinned.

Around Narromine and Dubbo, and on the upper Macquarie and Castlereagh rivers the problem was limited. The area of pine was not extending but scrub patches were believed to be increasing in density. The respondents noted that this scrub country contained the best timber and suggested that it should be preserved. They added that most settlers avoided the pine country because it was very poor.

On the Namoi, through the Pilliga and north to Warialda, the country contained very extensive indigenous scrubs on poor sandy soils which were of no value for grazing or agriculture. The scrubs were not considered to be recent but had been present since before the first settlement.

Other suggested causes of the spread of scrub north of the Murrumbidgee were; seed dropping into the soil after ringbarking, or that it was due to hardening of the ground and more runoff bringing seed from trees on the ridges down to the flats, or that it was the germination of dormant seed in areas that had been protected from fire and were heavily grazed (Fosbery 1913).

It is apparent from these accounts that some pine scrub existed in the Pilliga region in 1878/1880 and that pine regrowth in the Lachlan and Murrumbidgee districts preceded the regrowth event around the Pilliga by at least 4 to 13 years if Rolls is correct but if the Pilliga event was actually after 1889 as argued by Norris *et al.* (in press), this period increases to 14 to 23 years. In either case, the events in the northern and southern parts of the state were not coeval. It is also apparent that most contemporary observers believed that fire, drought and long term seed storage, all played some role in regrowth events but strangely, none of them mentioned periods of above average rainfall which is now recognised as the critical factor (Lacey 1973).

In the case of the 1950s regrowth event, this was first recognised over most of the natural range of white cypress pine in 1953/4 when the seedlings from the 1952 seed year (Forestry Commission 1953/4) were overtopping the grasses. 1950 was an exceptionally wet year and had been preceded by two or three wet years. Another event in this period was the rapid spread of the *Myxomatosis* virus in the rabbit population of eastern Australia in the summer of 1950/51 with an initial mortality rate higher than 99% (Myers & Calaby 1988). The conjunction of these events was probably important in the successful dramatic establishment of pine seedlings, the survival of which was probably enhanced by subsequent wet years in 1955 and 1956. It is also significant that in the period 1893 to 1946 there were only three sets of consecutive wet years, none of which were as extreme as the 1890s or 1950 and that although these allowed some pine germination, rabbits probably eliminated most of the seedlings as they were recorded doing at Gilgandra in 1917/18 (Forestry Commission 1918).

The role of fire in pine regrowth is less clear. It is known to be an important thinning mechanism in young pine because seedlings have a high mortality (Wilson & Mulham 1979), but no certain relationship has been established between the extensive fires in the Pilliga in the summer of 1951/2 and the germination of 1952 seed. Many mature cypress pines survived these fires even when defoliated (Forestry Commission 1951/2) and pine regeneration was just as successful elsewhere in the state in areas which were not burnt.

Soil compaction

Soil erosion by wind and running water is an enormous problem in the semi- arid zone and is one obvious consequence of the reduction in ground cover caused by heavy stocking (see Noble & Tongway 1986a,b, for reviews, and Graham *et al.* 1987/8 for location and extent of this problem in New South Wales). Heavy stocking is also blamed for another perceived problem that has some of the same elements of folklore as the pine scrub story. This is the question of topsoil compaction 'by the hard hooves' of introduced stock and the consequent reduction in soil porosity which is said to cause an increase in runoff and erosion.

This is a simple concept which 'common sense' suggests is true and is thus readily accepted, but as Lee (1977) and Noble & Tongway (1986b) pointed out, there are very few adequate studies demonstrating the importance of this widely accepted 'truth'.

Soil compaction by stock is known to be important on grazing lands in higher rainfall areas especially on clayey soils (Gradwell 1968; Willett & Pullar 1983). Here the combination of high soil moisture content, plastic and well graded (in the engineering sense) soil materials and high compactive effort is easily understood. In the semi-arid zone however these factors combine less fre-

quently and are not uniformly distributed in space and time. In poorly graded soils, sands for example, surface disturbance by stock is just as likely as compaction, and in other soils compaction achieved during a wet season under heavy stocking may well be reversed by soil dwelling invertebrates and plant root penetration in the following years with reduced or normal stocking.

Trampling, compaction and erosion, is certainly present around water points and in holding paddocks, but on the open range it has yet to be convincingly demonstrated and seems difficult to accept as a universal problem especially in the light of soil moisture studies by Johns (1983) in poplar box lands near Byrock. The soils of this area are reputed to have low infiltration and high runoff rates even on very low angle slopes as a result of compaction and sheet erosion. Johns found however, that significant quantities of runoff were only generated from parts of the slope by intense summer storms. One implication of this finding is that perhaps soil compaction is much less significant than is generally accepted.

Other stock trampling effects also exist which are likely to be at least as important as soil compaction:

– Trampling damages cryptogramic crusts of algae and lichen which possibly play an important role in the nitrogen cycle, probably provide an important protective surface to the forces of erosion, and certainly improve water infiltration and (Rogers & Lange 1971, Rogers 1972).
– Trampling generally reduces ground cover by damaging vegetation and may favour sod forming grasses (Klemmedson 1977).
– Surface aggregates may be crushed and sandy surfaces disturbed producing fines that are more readily eroded.
– Stock movement on bare dry soils raises dust which probably contributes to dust storms.
– Stock create incised trackways which often channel runoff and initiate gullies. Such trackways have been blamed for the breakdown of mulga groves in Western Australia by diverting natural run-on flow patterns (Mabbutt & Fanning 1989).

None of these effects have been studied in any detail but all of them warrant further attention because they may well be more important than the accepted but largely unconfirmed soil compaction story. Such a study should take full account of the relative compactive effort of the herbivores involved, stocking rates, and animal behaviour. They should provide full details of the nature of the soil materials, their moisture contents and any biocrusts present, and should attempt to differentiate erosive effects caused by soil compaction from those caused by ground cover reduction. The point of such work should be to identify a stocking level above which soil damage occurs which is not reversed by natural soil bioturbation processes. In other words, is there a soil carrying capacity as distinct from a vegetation carrying capacity on the range?

Our perception of soil compaction today is that it is 'a bad thing'; reducing infiltration and increasing erosion on slopes, shifting seed to run-on areas, and hindering vegetation establishment. Interestingly, this view was not shared by all nineteenth century observers many of whom describe the difficulties of travelling with heavy vehicles across loose soils in both dry and wet conditions, for example; Stephens (1878) who lamented the absence of stock trampling on red earths between Cobar and Bourke.

Conclusions

By researching original historic accounts it becomes apparent that most of the damage to the arid and semi-arid rangelands of New South Wales was done within the occupation span of the first generation of european settlers. Archival records also show that many of these people knew what was happening to their land resource, but as with todays farmer and grazier, they were often locked into an economic situation where the choices were either to abuse the country in order to make a living, or abandon their investment. Favourable climatic conditions in the late nineteenth century encouraged optimism and this too is a scenario which is being repeated at the present time. The grainbelt is still moving west into lighter soils and more marginal climates, pressures exist to clear and develop the poplar box woodlands, and the only indication that these 'developments' may not result in further land degradation comes from the early modelling of the greenhouse effect which suggests that the present pattern of higher rainfall might be a continuing trend, at least in the northern regions (see Pearman 1988, for reviews).

The careful analysis of archival data can provide many insights into the nature and dynamics of some of the present problems. Not only does it put these problems in perspective as in the case of the exaggerated pine scrub story, but it may also suggest some alternative management strategies when combined with well structured research programs such as the development of an index of soil carrying capacity involving assessment of the importance of soil compaction. The full potential of archival data for constructing better range dynamics models is only just being realised.

It is probably inevitable that this review should conclude by asking: What is the future of the arid and semi-arid rangeland in New South Wales?

Ten years ago some graziers said that in parts of the west, there were only 20 years left in the industry unless some cheap means of controlling woody shrubs was developed (Stanley 1981). The use of prescribed fire has become more common since then, but this is not always successful and may only be a short term solution wherever fire tolerant shrubs comprise even a small proportion of the problem species.

Economic forces in the past 20 years have made many properties in western New South Wales submarginal (Hassall and Assoc. 1982). Since that time there has been some improvement with high wool prices but these have fallen substantially in 1989/90 and the spectre of failure and further property amalgamation looms again. Amalgamation may be an option for some, but there is a flaw in the basic argument about the value of bigger runs; 'economies of scale' do not apply on the range to the same extent that they do in farming, more sheep mean; more country, more fences, more water points, and greater distances. Getting bigger on country with a reduced carrying may well be another road to ruin.

Can a balance be achieved in the animal production / range condition / climate equation? Does it involve conversion to simplified and more(?) productive seral grasslands? Can animal behaviour traits be exploited in management and lead to stability through the better manipulation of stock numbers? Should alternative species of herbivore and new pasture species be considered? Or is this whole scenario an example of one of a Hardin's (1968) dilemmas to which there are no technical solutions and perhaps rangelands utilisation should be phased out or handled in an entirely different way?

Beneath the green veneer of invader plant species, the western rangelands of New South Wales today are actually in better condition so far as ground cover is concerned, than they have probably been at any time since the 1860s. However large numbers of stock carried into a single extended drought could reverse this picture and reveal the truth of 150 years of land degradation. Existing problems are immense and to make matters worse, the land values and productivity of most of the range are so low that it must be asked whether the landholders can afford to fix it and whether the technology is available?

The knowledge exists to manage the range, and some managers have succeeded where others failed (Lange *et al.* 1984). In New South Wales essential basic data also exists in the form of Land System maps, Western Lands Lease Management Plans and Rangeland Literature Reviews all prepared by the Soil Conservation Service over recent decades. Practical technical solutions to soil erosion and the scald problem also exist. Some answers to the woody shrub problem have been devised but although some solutions are economic (Burgess 1988) others seem little better than financial suicide (Penman 1987) and it is unreasonable to expect the individual grazier to apply them unless some financial incentive is available.

The western rangelands have always faced some sort of crisis and the situation today is little different. So far the range has always come through and production has been maintained by individual graziers adapting to the circumstances with or without the assistance of government. History suggests that this process will continue but the scale of today's degradation problem is such that the graziers are going to need assistance. Who then is to pay?

The clear answer to this vital question is that there is only one sufficient source of funds and this is the Government, or in other words, the community. Australians should consider these costs as returning a small part of the capital borrowed from the arid zone over the past 100-150 years and if the 1990s 'Decade of Land Care' is to achieve anything, the authorities and the experts must convince the urban population of Australia that this bill must be met.

References

Anon. 1881, Report on the spread of pine scrubs. N.S.W. Legislative Assembly. Votes and Proceedings 3; 253–266.

Anon. 1901. Royal commission to enquire into the condition of the Crown tenants. Western Division of New South Wales, Gov. Printer Sydney.

Austin, M. P. & Williams, O. B. 1988. Influence of climate and community composition on the population demography of pasture species in semi-arid Australia. Vegetatio 77: 43–49.

Beadle, N. C. W. 1948. The vegetation and pastures of western New South Wales: with special reference to soil erosion, Gov. Printer, Sydney. 281 p.

Benson, J. S. 1987. The effect of 200 years of european settlement on the vegetation of New South Wales, Australia: An overview, XIVth International Botanical Congress, West Berlin, Reprint from; National Parks and Wildlife Service N.S.W. 44 p.

Booth, C. A. n.d. (circa 1987) Woody weeds, Their ecology and control, Soil Conservation Service of N.S.W. 24 p.

Burgess, D. M. N. 1988. The economics of prescribed burning for shrub control in the semi-arid woodlands of north-west New South Wales. Aust. Rangel. J. 10: 48–59.

Buxton, G. L. 1965. The Riverina; 1861–1891 an Australian regional study, Melbourne Uni. Press, Melbourne. 338 p.

Carver, N. P. 1878. Letter to engineer in charge of railway trial surveys 30/6/1878. N.S.W. Legislative Assembly, Votes and Proceedings 1881 4, 310.

Clift, D. K., Semple, W. S. & Prior, J. C. 1987. A survey of bladder saltbush (Atriplex vesicaria Howard ex Benth.) dieback on the riverine plain of south- eastern Australia from the late 1970s to 1983. Aust. Rangel. J., 9: 39–48.

Chiswell, E. L. 1982. The Pilliga story, Forest & Timber 18: 23–24.

Dalton, K. L. 1989. A review of information relevant to the riverine woodland and forest rangelands of south-western New South Wales. Tech. Rep. No. 15. Soil Conservation Service of N.S.W. 305 p.

Dixon, W. A. 1880. On salt-bush and native fodder plants of New South Wales. J. and Proc. Roy. Soc. N.S.W. 14: 133–143.

Dixon, S. 1892. The effects of settlement and pastoral occupation in Australia upon the indigenous vegetation. Trans. Roy. Soc. South Aust 15: 195–206.

Finlayson, B. 1984. Sir Thomas Mitchell and Lake Salvatore: An essay in ecological history. Hist. Studies 21: 212–228.

Foley, J. C. 1957. Droughts in Australia; review of records from the earliest years of settlement to 1955. Bureau of Met. Bull. 43: 281 p.

Forestry Commission. 1918. Annual Report of the Forestry Commission of New South Wales.

Forestry Commission. 1951/2. Annual Report of the Forestry Commission of New South Wales.

Forestry Commission. 1953/4. Annual Report of the Forestry Commission of New South Wales.

Fosbery, L. A. 1913. Climatic influence of forests. Appendix to the Annual Report of the Forestry Department, Bull. 4.

Graetz, R. D. 1973. Biological characteristics of Australian Acacia and Chenopodiaceous shrublands relevant to their pastoral use. In: D. N. Hyder (Ed.) 1973. Arid shrublands – Proceedings of the third workshop of the United States/Australia rangelands panel Tucson, Arizona. March 26 – April 5, 1973, Society for Range Management, Denver, Col. p 33–39.

Gradwell, M. W. 1968. Compaction of pasture topsoils under winter grazing. Trans. 9th Int. Soil Sci. Congr. Adelaide 3: 429–435.

Graham, O. P., Emery, K. A., Abraham, N. A., Johnson, D., Pattemore, V. J. & Cunningham, G. M. 1987/8. Land Degradation Survey. New South Wales 1987–1988. Soil Conservation Service of N.S.W. 32 p.

Griffin, G. F. & Friedel, M. H. 1985. Discontinuous change in central Australia: some implications of major ecological events for land management. J. Arid Envir. 9: 63–80.

Hamilton, A. G. 1892. On the effect which settlement in Australia has produced upon indigenous vegetation. J. and Proc. of the Royal Soc. of New South Wales 26: 178–240.

Hardin G. 1968. The tragedy of the commons. Science 162: 1243–1248.

Harrington, G. N., Oxley, R. E. & Tongway D. J. 1979. The effects of european settlement and domestic livestock on the biological system in poplar box (Eucalyptus populnea) lands. Aust. Rangel. J. 1: 271–279.

Hassall and Associates P/L. 1982. An economic study of the Western Division of New South Wales, Vol. 1. N.S.W. Gov. Printer, 185 p.

Heathcote, R. L. 1965. Back of Bourke: A study of land appraisal and settlement in semi-arid Australia, Melb. Univ. Press. 244 p.

James, P. E. 1967. On the origin and persistence of error in geography. Annals of the Assoc. of Amer. Geographers 57: 1–24.

Jeans, D. N. 1972. An historical geography of New South Wales to 1901, Reed Education, Sydney. 328 p.

Jeans, D. N. 1978. Use of historical evidence for vegetation mapping in New South Wales. Aust. Geog. 14: 93–97.

Johns, G. G. 1983. Runoff and soil loss in a semi-arid shrub invaded poplar box (Eucalyptus populnea) woodland, Aust. Rangel. J. 5: 3–12.

Klemmedson, J. O. 1977. Physical effects of herbivores on arid and semiarid ecosystems. In: Anon. 1977. The impact of herbivores on arid and semi-arid rangelands. Proc. of the 2nd United States/Australian Rangeland Panel, Adelaide 1972, Aust. Rangel. Soc. Perth, pp. 187-210.

Lacey, C. J. 1973. Silvicultural characteristics of white cypress pine. For. Comm. N.S.W. Res. Note 26. 51 p.

Lake, M. 1987. The limits of hope: Soldier settlement in Victoria 1915–1938, Oxford Univ. Press. Melbourne. 317 p.

Lange, R. T., Nicholson, A. D. & Nicholson, D. A. 1984. Vegetation management of chenopod rangelands in South Australia, Aust. Rangl. J. 6: 46–54.

Lee, K. E. 1977. Physical effects of herbivores on arid and semi-arid rangeland ecosystems. In: Anon. 1977, The impact of herbivores on arid and semi-arid rangelands. Proc. of the 2nd United States/Australian Rangeland Panel, Adelaide 1972, Aust. Rangel. Soc. Perth, p 17–186.

Leigh, J. H. & Noble, J. C. 1972. Riverine Plain of New South Wales – Its pastoral and irrigation development, CSIRO Div. Plant Industry, Canberra.

Ludwig, J. A. 1987. Primary productivity in arid lands: myths and realities. J. Arid Env. 13: 1–7.

Ludwig, J. A. 1988. Expert advice for shrub control. Aust. Rangel. J. 10: 100–105.

Lynch, I. F. 1988. Review of the literature pertaining to Atriplex nummularia (oldman saltbush) with special reference to soil conservation and rangeland management. Review of chenopod species working document No. 2. Soil Conservation Service of N.S.W. 109 p.

Mabbutt, J. A. & Fanning, P. C. 1989. Vegetation banding in arid Western Australia. J. Arid Env. 12: 41–59.

Millen, E. D. 1899. Our western lands. Sydney Morning Herald. 18 Nov. 1899. p 4.

Mitchell, T. L. 1839. Three expeditions into the interior of eastern Australia. T. & W. Boone London. Public Library of S A. Facsimile edn. 1965. 2 Vols.

Mitchell, T. L. Sir. 1848. Journal of an expedition into the interior of tropical Australia, in search of a route from Sydney to the Gulf of Carpentaria. Greenwood Press, Publishers, New York Reprint 1969. 437 p.

Moore, C. W. E. 1953. The vegetation of the south-eastern Riverina, New South Wales. 2. The disclimax communities. Aust. J. Bot. 1: 548–567.

Myers, K. & Calaby, J. H. 1988. Rabbit. In: The Australian Encyclopaedia 5th ed. p 2440–2443.

Noble, I. R. 1977. Long-term biomass dynamics in an arid

chenopod shrub community at Koonamore, South Australia. Aust. J. Bot. 25: 639–653.

Noble, I. R. 1986. The dynamics of range ecosystems. In: P. J. Joss, P. W. Lynch & O. B. Williams (Eds.) Rangelands: A resource under siege. Cambridge Univ. Press, Cambridge. p 3–5.

Noble, I. R. & Crisp, M. D. 1979/80. Germination and growth models of short-lived grass and forb populations based on long term photo-point data at Koonamore, South Australia. Israel J. Bot. 28: 195–210.

Noble, J. C. & Tongway, D. J. 1986(a). Pastoral settlement in arid and semi-arid rangelands. In: J. S. Russell & R. F. Isbell (Eds.) Australian soils. The human impact. Univ. of Qld. Press, St Lucia, p 215–242.

Noble, J. C. & Tongway, D. J. 1986(b). Herbivores in arid and semi-arid rangelands. In: J. S. Russell & R. F. Isbell (Eds.) Australian soils. The human impact. Univ. of Qld. Press, St Lucia, p 243–270.

Norris, E. H., Mitchell, P. B. & Hart, D. M. (in press). Vegetation changes in the Pilliga forests: a preliminary evaluation of the evidence. Vegetation.

Oxley, J. 1820. Journals of two expeditions into the interior of New South Wales, undertaken by order of the British Government in the years 1817–1819, John Murray, London. Libraries Board of South Australia. Australian Facsimile editions No 6. 1964. 408 p.

Oxley, R. E. 1987(a). Analysis of historical records of a grazing property in south-eastern Queensland. 1. Aspects of the patterns of development and productivity. Aust. Rangel. J. 9: 21–29.

Oxley, R. E. 1987(b). Analysis of historical records of a grazing property in south-eastern Queensland. 2. Vegetation Changes. Aust. Rangel. J. 9: 30–38.

Parsons, W. T. 1973. Noxious weeds of Victoria. Inkata Press, Melbourne. 300 p.

Pate, J. S. & McComb, A. J. (Eds.) 1982. The biology of Australian plants. Univ. of West Aust. Press, Nedlands W.A. 412 p.

Pearman, G. I. (Ed.) 1988. Greenhouse: Planning for climate change. CSIRO. Australia. 752 p.

Penman, P. 1987. An economic evaluation of waterponding. J. Soil Conservation N.S.W. 43: 68–72.

Pickett, S. T. A. & White, P. S. (Eds.) 1985. The ecology of natural disturbance and patch dynamics. Academic Press, Orlando, Florida. 472 p.

Robards, G. E. & Michalk, D. L. 1979. Appearance of new species in pastures at Trangie in central-western New South Wales. Aust. Rangel. J. 1: 369–373.

Rogers, R. W. 1972. Soil surface lichens in arid and subarid south-eastern Australia. III. The relationship between distribution and environment. Aust. J. Bot. 20: 301–316.

Rogers, R. W. & Lange, R. T. 1971. Lichen populations on arid soil crusts around sheep watering places in South Australia. Oikos 22: 93–100.

Rolls, E. 1981. A Million Wild Acres. 200 years of man and an Australian forest. Thos. Nelson, Melbourne. 465 p.

Scriven, R. N. 1988. A review of information relevant to the belah and bluebush rangelands of western New South Wales. Tech. Rep. No. 10. Soil Conservation Service of N.S.W. 287 p.

Slatyer, R. O. 1973. Structure and function of Australian arid shrublands. In: D. N. Hyder (Ed.) 1973. Arid shrublands – Proceedings of the third workshop of the United States/Australia rangelands panel Tucson, Arizona March 26–April 5, 1973. Society for Range Management, Denver, Col.

Stanley, R.J. 1981. Report of a seminar on practical scrub control organized by the Broken Hill branch of the Australian Rangeland Society. Aust. Rangeland J. 3: 171–172.

Stephens, E. P. 1878. Letter to engineer in charge of railway trial surveys. N.S.W. Legislative Assembly. Votes and Proceedings 1881. 4.

Sturt, C. 1833. Two expeditions into the interior of Australia, during the years 1828, 1829, 1830, and 1831: with observations on the soil, climate, and general resources of the colony of New South Wales. Smith, Elder & Co., London. Vol 1. Facsimile 1982, Doubleday Aust. Ltd.

Thompson, J. & Johnson, L. A. S. 1986. Callitris glaucophylla, Australia's 'White cypress pine' – a new name for an old species. Telopea 2: 731–736

Trollope, A. 1873. Australia and New Zealand. Vol. 1. Dawsons of Pall Mall, London. 533 p.

Turner, F. 1891. The forage plants of Australia. Gov. Print., Sydney. 94 p.

Westoby, M. 1979/80. Elements of a theory of vegetation dynamics in arid rangelands. Israel J. of Botany 28: 169–194.

Williams, O. B. n.d, Atriplex nummularia Lindl. Oldman saltbush, Review and bibliography. Unpublished. Cited in: I. F. Lynch 1988. Review of literature pertaining to Atriplex nummularia (oldman saltbush) with special reference to soil conservation and rangeland management. Review of chenopod species working document No. 2. Soil Conservation Service of N.S.W. p 70-71.

Willett, S. T. & Pullar, D. M. 1983. Changes in soil physical properties under grazed pastures. Aust. J. Soil Res. 22: 343–348.

Wilson, A. D. & Mulham W. E. 1979. A survey of the regeneration of some problem shrubs and trees after wildfire in western New South Wales. Aust. Rangel. J. 1: 363–368.

Vegetatio **91**: 183–189, 1991.
A. Henderson-Sellers and A. J. Pitman (eds).
Vegetation and climate interactions in semi-arid regions.
© 1991 *Kluwer Academic Publishers. Printed in Belgium.*

Available soil moisture as a basis for land capability assessment in semi arid regions

D. A. Thomas & V. R. Squires
National Key Centre for Dryland Agriculture and Land Use Systems, Roseworthy Campus, University of Adelaide, Roseworthy, Australia, 5371

Accepted 7.9.1990

Abstract

Rangelands are extensive areas in arid or semi arid regions. They have many uses (e.g. grazing, dry land farming, wildlife habitat, recreation and mining). Generally vegetation is sparse and the principal plant species are adapted to erratic rainfall events. Rainfall use efficiency (RUE) is the quotient of annual primary production divided by annual rainfall i.e. the number of kilograms aerial dry matter produced by 1 ha in 1 year per millimetre of rain. It decreases with increasing aridity. Reasonably well managed arid and semi arid grazing lands are usually in the 3.0-6.0 value range. Available soil moisture is the principal determinant of productivity. The role and significance of the major parameters (rainfall, soil depth, slope and salinity, texture, cover, erosion etc) are considered. Available moisture can be predicted if the appropriate relative productivity indices (RPI) are used in a parametric way. The predictions can be used to classify land according to its capability to support plant growth. Equations have been derived which enable land to be classified on the basis of its potential productivity. Links between land capability and vegetation cover are given.

Introduction

On a global basis rangelands represent about one third of the land surface. In many regions, especially the semi-arid climatic zones they are now being severely overused. Because they are vast and vary widely in space and time rangelands are hard to manage. Many rangelands are degrading rapidly into deserts. Dryland degradation is a widespread phenomena which affects grazing lands around the world (Joss *et al.* 1986).

To help prevent further over utilisation and degradation the vegetation and land must be used within its capability to support grazing and dryland agriculture. Most existing systems of biophysical and ecological land classification have a major short coming. They have a limited ability to predict the relative productivity of land units for purposes of development planning and environmental management. There is a need to focus more clearly on the biophysical processes which operate within the active land component and which have relevance to respond within a time-frame of current land development practices (Moss 1983). A basis of land capability assessment which uses determinants of available soil moisture is described.

Production rainfall relations

The main determinant of annual primary production in arid and semi-arid lands throughout the world is the annual rainfall. An extensive review of rain use efficiency, (RUE) kg ha^{-1} mm^{-1} rainfall as a unifying concept in arid-land ecology has been given by Le Houerou (1984). World wide the RUE tends to decrease with aridity, low intensity of useful rain and increases in the potential evapotranspiration. It depends strongly on soil condition and very strongly on the vegetation state and its dynamics. It gives a good indicator of ecosystem productivity allowing comparisons between ecosystems for various climatic zones having totally different botanical and structural characteristics. The RUE varies from less than 0.5 in depleted sub-desert ecosystems to over 100.0 in highly productive well managed steppes, prairies and savannas.

Many studies show that dryland/rangeland dry matter productivity can be related to rainfall by linear to curvilinear relations for dry regions throughout the world.

The basic linear equation used by many workers is

$$D_M = a + b(r_f) \qquad (1)$$

where D_M is dry matter (kg ha^{-1}yr^{-1}), r_f is the annual rainfall (mm yr^{-1}) and a, b are constants defined for different regions and forage types in Table 1. Linear regression relationships between rainfall and annual dry matter production have been given for semi-arid Kenya, (Wijngaarden 1985); the Sudano-Zambazion region, (Rutherford 1978); East and South Africa (Deshmukh and Baig 1983); North China (Ting Cheng et al. 1983). Data from semi arid Australia are shown in Table 2.

Estimates of forage production in semi-arid regions of Australia have been made from the calculated seasonal soil moisture status. Using canonical correlation, highly statistically significant predictive relations were produced to predict drought conditions, seasonal carrying capacity, reproduction rates, wool production and stock deaths (Thomas & Morris 1973, Reid & Thomas 1973).

Slightly curvilinear relationships, where production was related to a power of the rainfall, have been given for the Mediterranean region (Le Houerou & Hoste 1977) and for the stock carrying capacity in semi-arid Australia for both the winter and summer rainfall zones (Wilson and Harrington 1984). Seasonal soil moisture status over extensive regions has been extended using rainfall and simple climatic data in the WATBAL model (Keig & McAlpine 1974).

More elaborate models include other environmental factors to obtain water balances (e.g.

Table 1. Mean annual rainfall and the values of constants used in relation to arid and semi arid zones by various workers.

Region	Type Herbage only	Forage Herbage & Browse	Rainfall range mm yr^{-1}	Constants		Source
				a	b	
Mediterranean*		+	20–900	− 200	4.4	Le Houerou & Hoste (1977)
Sahelo-sudanian*	+		200–1400	100	2.6	Le Houerou & Hoste (1977)
Semiarid Kenya	+		50–400	− 180	6.3	Wijngaarden (1985)
Semiarid Kenya		+	50–400	− 400	10.0	Wijngaarden (1985)
Sudano-Zambazian	+		200–800	0	2.0	Rutherford (1978)
Karroo-Namib	+		50–500	− 100	4.8	Walter (1973)
East & South Africa	+		500–800	− 200	8.5	Deshmukh & Baig (1983)
Northern China	+		>85	− 530	6.4	Ting Cheng (1983)

* The correlation and determination coefficients for the Mediterranean Basin and Sahel were 0.9 and 0.89 respectively.

Table 2. Total dry matter production – rainfall relations determined in Australian arid and semi-arid regions*

Location type	Production* (kg D_M/ha/mm)	Plant association/or Plant community	Reference
Deniliquin NSW	4.8–8.3	Danthonia & Stipa Grassland	Williams (1974)
Central Australia	1.3	Grassland	Williams (1974)
Winton QLD	5.2	Astrebla Grassland	Davis *et al.* (1938)
Cunamulla QLD	3.7–6.6	Astrebla Grassland	Roe & Allen (1945)
South Western QLD	6.1	Digitaria-Aristida Paspalum-Egragrostis Community	Ebersohn (1979)
Charleville QLD	2.0	Aristida-Digitaria Penotis-Rhyncheletrium Community	Ebersohn (1979)
Charleville QLD	2.0	Dichanthium Chrysopogon Community	Ebersohn (1979)

* Referred to as Rain Use Efficiency (RUE) by Le Houerou (1984)

Cornet 1983), the availability of water and nutrients (Penning de Vries & Djiteye, 1982; Lok & Keulen 1986; Keulen *et al.* 1986), actual evapotranspiration, or mean annual precipitation and temperature (Lieth & Whittaker 1975). The most detailed and elaborate is the Simulation of Production and Utilisation of Rangelands (SPUR) model developed by United States Department of Agriculture Research Service. Predictions of dryland farming production using rainfall and water balance models have shown yields to be a linear function of cumulative water use (Greacen & Hignett 1976; Hamblin, Tennant & Perry 1987).

The problem with these models is they are very data demanding. In most arid and semi-arid areas it is impossible to get adequate long term rainfall and temperature data. They are not simple models and simplicity is essential to the present level of management available for most to the world's rangelands.

Available moisture – its use as the key determinant of land capability

The available soil moisture (A_M) of a given homogenous land can be parametrically related to rainfall and the other parameters which together determine the water available to support plant growth. (Steeley *et al.* 1986; Thomas *et al.* 1986)

$$A_M = (k_1 \times k_2 \times k_3 \ldots)R \qquad (2)$$

where A_M is available moisture (mm) and k_1, k_2, $k_3 \ldots \ldots k_n$ are relative production indices (RPI) ratings ranging from 1 to 0 which give weightings for the way in which a given parameter affects the A_M.

As the relation is multiplicative, it will be realised that the greater the number of RPI considered the smaller the product, hence the same number of quantities are required in all A_M's that are to be compared. If some quantities are not needed in a particular consideration, i.e. in a particular land unit compared to another, they are included as 1. An absolute constraint to plant growth is given the value 0.

Steeley et al. (1986) give RPIs to account for;

1. Soil depth, the primary determinant of the volume of water a soil can hold.
2. Slope, which determines the time available for water to infiltrate the soil.
3. Salinity, as this determines the amount of water extractable from the soil by plants.

From data collected on shrub steppe communities given in Le Houerou and Hoste (1977), Thomas et al. (1986) and Steeley et al. (1986) the annual dry matter production at ecological potential (Y_{ep})

$$Y_{ep} = 2.33A_M^{1.09} \qquad (3)$$

where 2.33 is the water use efficiency (Kg D_M mm^{-1})

Rainfall considerations

The mean annual rainfall is taken and not the effective rainfall. This is because in semi arid regions there is often a lack of reliable long term rainfall data from the few meteorological stations in semi arid regions. In these circumstances a regression analysis of altitude, latitude and longitude can be used to estimate annual rainfall (Yevjevich 1972). However if the amount of effective rainfall can be determined the accuracy of production estimates can be increased. The effective rainfall is that which promotes plant growth. For example, in Mediterranean regions of the northern hemisphere the effective rainfall falls in November to April while in the southern hemisphere it is in the period May to October. Any rain falling outside these periods is quickly lost by evaporation and hence not available to support plant growth.

By taking the total effective rainfall of late autumn, winter and early spring period, production of the forage available over the crucial late spring, summer and autumn period can be estimated. In this way stocking rate adjustments can be made before this crucial period when there is little effective or no growth which would prevent over stocking and utilisation, the major factors in rangeland degradation.

Relations using the rainfall falling in the effective growth period to predict stocking rate adjustment for the dry season have been developed and tested in the pastoral areas of Queensland and New South Wales (Reid & Thomas 1973; Thomas & Morris 1973; Easter 1975).

Other relative production indices

Soil texture

Soil texture should be considered particularly if there is a large range in textures – e.g. sands to clays. This is because texture has an effect on the amount of water held in soils to support plant growth. Table 3 and 4 give the amount of water available between field capacity and permanent wilting point (– 10 to – 1500 kPa) as it is controlled by texture together with the RPI that can be used in relation 2.

The RPI are only relative to the region in which land production capabilities are being compared. Hence the RPI's need only be calculated relative to the soils found in the region, i.e. not all texture classes have to be considered.

Table 3. Soil texture, water retained between – 10 and – 1500 kPa, and RPI on the basis of clay loam = 1 for Australian soils. Source Williams (1983).

Field texture Class	Water retained between – 10 & – 1500 kP (cm m^{-1})	Relative Productivity Index (RPI) (Clayloam = 1)
Sands	13.5	0.80
Sandy loams	15.5	0.97
Loams	15.8	0.94
Clay loams	16.8	1.00
Light clays	13.8	0.82
Medium/heavy clays	11.5	0.68
Self mulching clays	21.4	1.27

Table 4. Soil texture, water holding capacity between − 1/3 bar (Field capacity) and − 15 bar (Permanent wilting point). Water content expressed in percent by volume and the RPI based on silt = 1. Based on Lane and Stone (1983) for American soils.

Texture Class	Available Water % (− 1/3 bar to − 15 bar)	RPI (silt = 1)
Sand	6	0.32
Loamy sand	6	0.32
Sandy loam	11	0.58
Loam	14	0.74
Silt loam	18	0.95
Silt	19	1.00
Sandy clay loam	15	0.79
Clay loam	14	0.74
Silty clay loam	15	0.79
Sandy clay	10	0.53
Silty clay	13	0.68
Clay	13	0.68

Evaporation

Evaporation should be considered however, over the extensive semi-arid and arid areas advection of hot dry air tends to make the evaporation high and uniform. In all cases evaporation exceeds precipitation by over three fold.

Compared to available water holding capacity the evaporation from soils of different textures is relatively uniform (Table 5).

Table 5. Evaporation from soils of different textures.

Soil texture	Evaporation (mm day^{-1})
Sand	3.3
Loamy Sand	3.3
Sandy Loam	3.5
Loam	4.5
Silt Loam	4.5
Silt	4.0
Sandy Clay loam	3.8
Clay Loam	3.8
Silty Clay Loam	3.8
Sand Clay	3.4
Silty Clay	3.5
Clay	3.4

Aspect

Where the topography is undulating to rolling and hills are high, aspect may have to be considered in assessing land capability and the vegetation they can support. This is particularly so in higher and lower latitudes where sun angles are lower.

In the northern hemisphere, north- and in the southern, south-facing aspects will receive less radiation. This will have two effects.
1. There will be more Photosynthetically Active Radiation (PAR) on one aspect than the other.
2. However the evapotranspiration on the aspect receiving the greater radiation will be higher hence the A_M and productivity could be lower. The incidence, intensity and the prevailing direction of rainfall may favour one aspect more than another. This would result in one aspect having a higher rainfall.

The vegetation associations on the various aspects can be different. Their productivities could be the same, due to adaptation, or different as a result of lower or higher PAR and evapotranspiration.

Under these conflicting conditions the effect of aspect on moisture availability and land capability assessment must be determined in the region under study.

Effect of vegetative cover on rainfall retention

Vegetation cover has a marked effect on rainfall acceptance – hence A_M. This is because the cover maintains the topsoil and provides litter both of which increase soil permeability. As many dryland areas throughout the world are badly degraded as a result of vegetation loss the effect on A_M should be considered. From the work of Berman et al. (1983) a relation between foliage cover and % rainfall acceptance can be obtained.

$$R_a = 66.8 + 0.83F_c \qquad (4)$$

where R_a is the rainfall acceptance and F_c is the % foliage cover. From this RPI can be estimated (Table 6).

Table 6. Relation between % vegetative ground cover and the Relative Productive Index (RPI).

Ground Cover %	RPI
20	1.00
15	0.87
10	0.70
5	0.45
3	0.28

Erosional soil loss related to vegetative ground cover

From the work of Lang & McCaffrey (1984) at Gunnedah New South Wales on a gravelly fine sandy clay loam, a relation between percentage ground cover and the annual accelerated soil loss is obtained.

$$S_1 = 9.68\,e^{-0.041 G_c} \tag{5}$$

where S_1 is the soil loss (tonnes $ha^{-1} yr^{-1}$) and G_c is the % Ground cover.

Estimating soil loss is an important consideration as it means the loss of the most friable, permeable fertile surface layers. As well there is the loss of seed, organic matter and nutrients and a gradual loss in soil depth which can be translated finally to a decrease in A_M and productivity. A loss of 1 cm of top soil could result in 1.6 mm of A_M i.e. a loss in potential productivity of 4.8 to 9.6 kg D_M ha^{-1} from values given by Le Houerou (1984). For low ground covers soil loss is around 1 mm/ha/year. i.e. a loss of about 0.5 to 1.0 kg D_M yr^{-1} due to the reduction in A_M alone. These rates of loss seem very small. However if continued over long periods there is a steady attrition in productivity due to the decrease in available soil moisture. Shorter term decreases in production due to loss in seed, nutrients, organic matter and top soil structure would be much higher.

Estimates of the loss in productivity due to erosion are important in calculating the cost benefits of regenerative treatments (Thomas *et al.* 1986).

Ground cover and productivity

There is often insufficient data to obtain a direct relationship between ground cover and productivity but data is available from *Artemisia herba alba* associations growing on silty soils in Algeria and *Rhanthericum suaveatens*, *Stipa lagascae* steppe on sandy soils of Southern Tunisia (Le Houerou 1984).

The relation is:

$$P = P_{max}(1 - e^{-0.09 G_c}) \tag{6}$$

where P is the dry matter production, $Kg\,D_M\,ha^{-1}\,yr^{-1}$, P_{max} is the potential maximum production and G_c is the Ground cover.

This enables stocking rates to be calculated directly from estimates of ground cover if the nutritive value of the vegetation is known. The RPIs can be calculated for the effect cover has on productivity (Table 6). Using Equation (2) and RPIs the potential product (Y_{ep}) for a given land unit can be calculated. These can be used in resource assessment and land use planning.

Climate change and land capability

Rangelands that are already being over-utilised will suffer still further degradation if it is assumed that future climates will be warmer and drier. The vulnerable regions are the marginal rangelands. Land utilisation in these regions will have to change as land capability decreases. Former rangelands will become deserts and dry land farming areas, grazing lands.

What is required under these conditions is a system of land capability which can taken into account climatic change. The methodology described enables this to be considered as it focuses on available soil moisture as the key biophysical process determining land capability attributes. These attributes can be used in a Geographic Information System (GIS) to monitor and predict the effects of climate change on land capability. This gives planners a method of estimating land support capacities and adjustments that have to be made to maintain stability.

The socio-economic disruption will be greatest in regions where use of marginal lands comprises a relatively large segment of the economy of a state, province, or country, and where other means of livelihood are few.

References

Cornet, A. 1983. Utilisation de modeles simples de bilan hydrique et de production de biomass pour determiner les potentialites de production de parcours en zone sahelienne Senegalese. In: Van pract L. (ed) Actes du Colloque tenu a Dakar, les 16–18 Novembre.

Davis, J. G., Scott, A. E. & Kennedy, J. F. 1938. The yield and composition of a Mitchell grass pasture for a period of twelve months. J. Coun. Scient. Ind. Res. Melbourne 11: 127–139.

Easter, C., 1975, Some agronomic factors underlying production on the New South Wales grazing industry. Q. Rev. Agric. Econ. 28: 177–199.

Ebersohn, J. 1979. Herbage production from native grasses and sown pastures in southwest Queensland. Trop. Grassld. 4: 37.

Greacen, E. L. & Hignett, C. T. 1976. A water balance model and supply index for wheat in South Australia. CSIRO Aust. Div. Soils. Tech. Pap. No. 27: 1–33.

Joss, P., Lynch, P. & Williams, O. B. (eds). 1986. Rangelands – A Resource Under Seige. Australian Academy of Science, Canberra.

Keig, G. & McAlpine, J. R. 1974. WATBAL – A computer system for the estimation and analysis of soil moisture regimes from simple climatic data (2nd Ed). Tech Mem Div Land Res. CSIRO No 74/4 45 pp.

Keulen, H. van & Wolf, J. 1986. Modelling of agricultural production: weather, soils and crops. Pudoc, Wageningen.

Hamblin, A., Tennant, D. & Perry, M. W. 1987. Management of soil water for wheat production in Western Australia. Soil Use and Management 3: 63–69.

Lane, L. & Stone, J. 1983. Water balance calculations, water use efficiency and above ground net production. Hydrology and Water Resources in Arizona and the South West 13: 27–43.

Le Houerou, H. N. 1984. Rain use efficiency: a unifying concept in arid land ecology. J. Arid Environments 7: 213–247.

Le Houerou, H. N. & Hoste, C. H. 1977. Rangeland production and annual rainfall relations in the Mediterranean Basin and in the African sahelian and Sudanian zones. J. Range man. 30: 181–189.

Lieth, H. G. & Whittaker, R. H. 1975. (eds) Primary productivity of the biosphere. Springer-Verlag New York pp. 202–215.

Lok, J. J. and Keulen, H. van 1986. Calculation method for regional annual forage availability of natural pastures in the Sahel. CABO Wageningen, 86 pp.

Moss, M. R. 1983. Land processes and land classification. J. Envir. Management 20: 295–319.

Penning de Vries, T. W. T. & Dijiteye, M. A. 1982. La productivite des paturages saheliens. Pudoc, Wageningen.

Reid, G. K. R. & Thomas, D. A. 1973. Pastoral production, stocking rate and seasonal conditions. Quarterly Review of Agricultural Economics 26: 217–27.

Roe, R. & Allen, G. H. 1945. Studies on the Mitchell grass association in south-western Queensland II. The effect of grazing on Mitchell grass pasture. Counc. Sci. Indus. Res. Aust. Bull. No 185.

Steeley, C., Thomas, D. A., Squires, V. R. & Buddee, W. 1986. Methodology of a range resource survey for steppe regions of the Mediterranean Basin. In: Rangelands – a resource under siege, Proc. 2nd International Rangeland Congress Adelaide 1984. Aust Academy of Science, Canberra pp 538–539.

Thomas, D. A., Squires, V. R., Buddee, W. & Turner, J. 1986. Rangeland regeneration in steppic regions of the mediterranean basin. In: Rangelands-a resource under siege. Proc 2nd International Rangeland Congress, Adelaide, 1984 Australian Academy of Science Canberra pp 280–287.

Thomas, D. A. & Morris, J. G. 1973. Soil moisture storage and stock carrying capacity in Pastoral Queensland. Research Report, Bureau of Agricultural Economics, Canberra Australia.

Ting-Cheng, Z., Jiang-Dong, L. & Dian-Chen, Y. 1983. A study on the ecology of Yangcao (Leymus chinensis) Grassland in Northern China. Proc 14th Int. Grassl. Congress. Lexington, Kentucky 1981: 429–431.

Walter, H. 1973. Ecology of tropical and subtropical vegetation. Oliver and Boyd. Edinburgh.

Wijngaarden, W. van 1985. Elephants – Trees – Grass – Grazers. Relationship between climate, soils, vegetation and large herbivores in a semi-arid savanna ecosystem (Tsavo, Kenya). ITC publ, 4 Enschede. pp 159

Williams, O. B. 1974. Vegetation improvement and grazing management. In: Studies of the Australian Arid Zone. II Animal Production (ed A.D. Wilson) CSIRO: Melbourne.

Williams, J. 1983. Soil Hydrology. In: Soils: an Australian Viewpoint, Division of Soils CSIRO (CSIRO Melbourne/Academic Press, London) pp 507–530.

Wilson, A. D. & Harrington G. N. 1984. Grazing ecology and animal production. In: Management of Australian Rangelands. G.N. Harrington, A.D. Wilson and M.D. Young (eds) CSIRO Division of Wildlife and Rangelands Research.

Yevjevich, V. 1972. Probability and Statistics in Hydrology. Water Resource Publications.

Vegetatio **91**: 191–208, 1991.
A. Henderson-Sellers and A. J. Pitman (eds).
Vegetation and climate interactions in semi-arid regions.
© 1991 *Kluwer Academic Publishers. Printed in Belgium.*

Land management in semi-arid environments of New South Wales

John Pickard
Graduate School of the Environment, Macquarie University, NSW Australia 2109

Accepted 24.8.1990

Abstract

Land of the semi-arid zone of Australia is generally managed to produce wool or beef. Past management has caused many changes in the land. These changes may be difficult to detect and assess. Much of the available information is at too coarse a scale to be really useful in assessing change. Graziers' perceptions of change are unknown but survey results from the agricultural zone suggest that their perceptions are probably incorrect. Apportioning the causes of change is very difficult as the main agents (climate, stocking rates, bushfires, legislation and economics) are not independent. Three different approaches to separating cause are described: use of historical information, integrating all information and using unpalatable plants as proxies for key economic species. Some difficulties with the historical approach are outlined. The major issues in semi-arid land management are social rather than technical. However, key aspects such as perceptions, motivation, and sources of information used by graziers are neglected research subjects. Recent research into an objective basis for assessing stocking rates from forage biomass production will replace traditional estimates based on extrapolating from similar country. This will significantly assist graziers in determining appropriate stocking rates to maximise their incomes. Other research by graziers has demonstrated the benefits of low stocking rates leading to increased incomes on both an animal and area basis. Such advances by graziers provide keys for future extension programs to achieve the desired goals of a stable grazing industry with good financial rewards, and improved land management.

Abbreviations: dse: dry sheep equivalent. A standard animal unit equivalent to a medium sized wether or non-lactating ewe used to assess total grazing pressure on an area. All other vertebrates can be converted to dse using laboratory data. For this paper, I adopt the conversions factors of 1 beast (cattle, horse) = 10 dse, 1.6 kangaroos = 1 dse, 16 rabbits = 1 dse.

Introduction

Land degradation is recognised by the Australian Government as the major environmental problem facing Australia in the 1990s. Elsewhere around the world, other nations are facing similar problems locally, regionally, and transnationally. Degradation occur in all biomes from polar to tropical, coastal to alpine and humid to arid. Probably no biome is completely free of degradation in some form or degree. The arid and semi-arid areas of the world are particularly prone to degradation, frequently as a consequence of inappropriate land management.

Land management may be defined in many ways, but all involve or specify some purpose: production, conservation, recreation, extraction etc. Thus a general definition would be the human intervention in and manipulation of the land to achieve some desired end (Nix 1985). The bulk of this paper deals with managing semi-arid lands in Australia, and specifically New South Wales for extensive grazing usually to produce wool or beef. The land is also used for mining, timber production, irrigated agriculture, dryland agriculture, conservation, recreation and urban areas. However, the rangelands which make up most of the area are generally leased specifically for grazing and are likely to remain so for the foreseeable future.

Given the paramount position of grazing in semi-arid lands, then the first of Wilson's (1986) fifteen 'Principles of grazing management systems' is highly relevant: '...total stocking intensity (the total number of animals grazed on an area of land for the full grazing season) is the most important factor affecting rangeland productivity and stability.' The task of the land manager is more complex as he strives to '...relate livestock numbers to the conditions of fluctuating herbage supply and still maintain stability of the soil and vegetation resource and cash flow!' (Christie 1984).

This is an important concept and one which is frequently not grasped by those bemoaning the extent of degradation in semi-arid lands. Very few, if any, graziers degrade land deliberately. However, while struggling to maintain a reasonable standard of living (which is often well below standards considered acceptable in cities), graziers need cash flow. The land frequently suffers in this struggle. Thus degradation or landscape change is not just a technical problem of geomorphology or ecology. If this were the case, then the solutions would be almost trivial. Landscape change is basically a social problem and because it has this human side, solutions which neglect the aspirations and needs of the land holders are doomed to failure.

In this paper, I consider several aspects of changes that have occurred in both the land and the vegetation of semi-arid New South Wales (Fig. 1) as a consequence of land management. I consider alternatives to assessing the causes of the changes and examine some of the difficulties in each approach. The most important lessons from this are that the major problems are in fact social and not technical. I discuss the implications of this, and finish by looking at recent research undertaken by graziers themselves. The encouraging results from these studies will become the main approach to extension in management of semi-arid lands in the 1990s.

Changes in landscape and vegetation

All land management involves change in some attributes of the land. The changes may or may not be due to the direct action of the managers, but they occur nonetheless. Thus, gradual shifts in populations of animals or plants caused by a sequence of very wet (or dry) years are beyond the control of the manager and may be independent of any stocking levels imposed. In the absence of human management, the cause may be identifiable after both field and laboratory study. Such instances are infrequent in semi-arid zones because of the now ubiquitous presence of grazing domestic animals. Thus while rainfall patterns are both varying seasonally and annually, and changing over decades (Fig. 2), so are other factors: numbers of domestic animals (Fig. 3), fire regimes, numbers of predators, etc. The problem of apportioning cause is important for future land management and is discussed in the next section. Before examining the nature and scale of changes, it is worth pausing to reflect on our value judgments of change.

'Judgment of the importance of these changes depends on each user's land management objectives. It is presumed that the maintenance of soil condition is an inalienable prerequisite for all land uses, but the evaluation of the desirability of a change in plant populations should be based on a consideration of all land use objec-

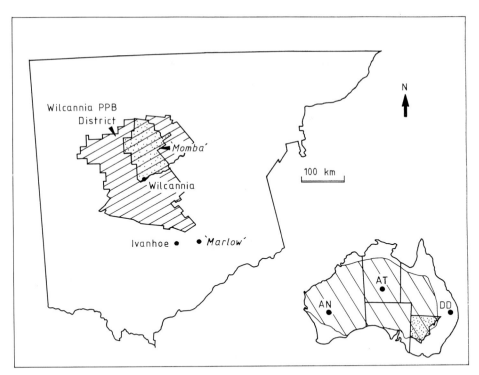

Fig. 1. Map of Western Division of New South Wales showing location of various places and features mentioned in text. 'Momba Station' is stippled, Wilcannia Pastures Protection Board District is hatched. Inset map of Australia shows the arid zone (hatched), the Western Division (stippled), Darling Downs (DD), 'Annean Station' (AN) and 'Atartinga Station' (AT).

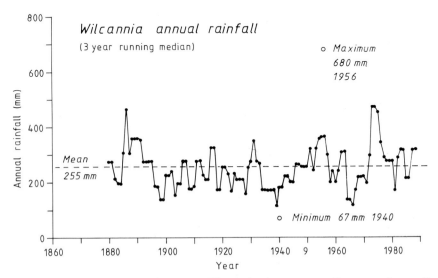

Fig. 2. Annual rainfall at Wilcannia from 1880s to the present. Line and points represent 3 year running medians. Source: Bureau of Meteorology records.

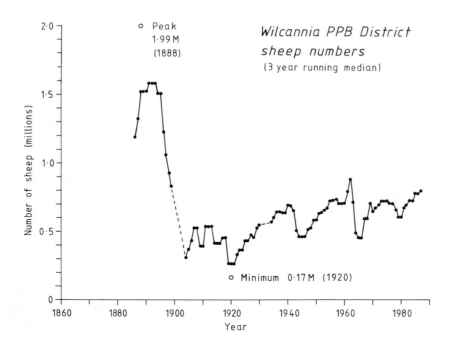

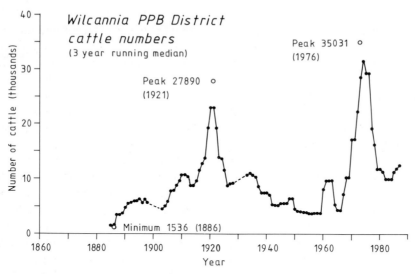

Fig. 3. Numbers of sheep (upper) and cattle (lower) recorded in the Wilcannia Pastures Protection Board District from 1880s to the present. Line and points represent 3 year running medians. Source: Western Lands Commission Annual Reports, Wilcannia Pastures Protection Board.

tives... The pre-European plant community was not adapted to grazing by domestic stock. The changes wrought by such grazing are not intrinsically deleterious... The significance of a vegetation change must be assessed by the direct measurement of animal production or indirectly by measure of the palatability, productivity and quality of the species. Such assessment must take a long term view'. (Wilson & Harrington 1984, p. 68)

Nature and perception of changes

In principle, change is easy to detect and assess, but in practice, it may be extremely difficult. Not the least problem is knowing the original state or even preferred state of the land. Once these have been determined or (more likely) assumed, and the change has been estimated, then management action can follow. Whether the change is deleterious ('degradation' by definition) or advantageous ('development' by definition) depends on the viewpoint and the time-scale. For the remainder of this paper, I will use 'change' to indicate some alteration from an original or a desired state of the landscape. 'Degradation' has now acquired a pejorative meaning in Australia and is cause for considerable resentment from land managers (pastoralists and farmers). As I show later, there is much to be gained at this stage in history from acknowledging change but not trying to assign blame as though seeking to punish the alleged offenders.

The first point to consider is scale. Australia is a large continent and the semi-arid and arid zones occupy 70% of the area (Fig. 1). There is a common assumption that this vast area of some 3.5 M km^2 is uniform. This is not the case and can be demonstrated by considering droughts which are frequently continental in scale (e.g maps of droughts in Working Party... 1988). Even during the severe droughts of 1940 and 1944, approximately 30% of the semi-arid zone (especially in the north) received average or greater rainfall.

If the continent is the wrong scale to consider, then is the region an appropriate scale? The Western Division is the administrative region that includes all the semi-arid lands of New South Wales. It covers several major environments and climatic areas. The north east corner is dominated by summer rainfall, while the south west is winter rainfall. Although common factors may be responsible for the changes, the climate differs and changes are best treated by examining individual properties, or small groups of properties. This has the advantage of considering the localised climate while still having replicates to avoid the effect of a single better or worse manager.

The major example that I use in this paper is what is now a group of some 27 properties north of Wilcannia. In 1884 the area was a single property, 'Momba' (Fig. 1), and it was the largest property to ever exist in New South Wales. I am studying the area to answer the question 'why is the landscape in its present condition?'. The answer must necessarily involve integrating information on climate, landforms, vegetation, stocking rates, land management, economics, legislation and social expectations. The project is incomplete and the data reported here are only preliminary.

Even this type of study does not provide a complete answer to the question of appropriate scale. In part, it depends on the question and I suggest that the only question worth asking in this context is 'how can the land manager cope with change, maintain stable rangelands and stable income?' Note that the benefits or otherwise of the change are not included in the question. The reason for suggesting that this is the only correct question is that the individual land manager, operating on his block of land, actually decides what management regime is implemented. It is not government agencies, be they federal, state or local, soil conservation services or pastoral boards; it is the pastoralist. Current land tenure ethics give the individual almost unfettered freedom to do what he/she likes with their land. The fact that most of the land in the semi-arid and arid zones is leasehold makes little difference in practice. From this, it follows that the property is the optimum scale to consider change.

I should point out here that I believe that regional, state and national surveys of land change (or degradation) are essential. Such surveys may be instrumental in changing regulations, subsidies or taxation measures which benefit land owners and assist in managing land change. But it is still the individual land owner who must incorporate these administrative changes into his/her management strategies.

Perception of change to the land is a difficult area and one which has been generally neglected by social scientists. Many city people seeing rangelands for the first time conclude that because it has not been cleared, it is 'natural'. Spectacular flowering displays of wild hops (*Acetosa vesicaria*) or Patersons curse (*Echium plantagineum*) in the Flinders Ranges of South Australia are so widespread and dense that they are not recognised as exotic weeds. Stands of chenopod shrubs are not recognised as regeneration after severe scalds (Condon 1986). Dense patches of budda (*Eremophila mitchellii*) with its beautiful flowers are not seen as woody weeds that have grown in the last decade. Conversely, many graziers seem reluctant to admit that their properties and districts have altered. This is especially the case if the changes have been detrimental and gradual. The usual explanation is poor seasons, but fenceline effects indicate quite clearly that management frequently dominates over season rainfall variations.

Rickson *et al.* (1987) reported research on perception of degradation on farmed lands of the Darling Downs of southern Queensland (Fig. 1). Although not from grazing lands in the semi-arid zone, their general conclusions on perception can probably be extrapolated. One result was that 88% of farmers perceived soil erosion to be a major problem on the Darling Downs, 47% perceived major erosion on nearby farms, but only 11% perceived erosion to be a major problem on their own farms. Curiously, while they underestimate the extent of erosion as a major problem on their own land, they overestimate the long-term decrease of crop yield caused by soil loss.

Many farmers underestimate the risk of degradation on their properties. There is considerable publicity of dramatic erosion in the normal media and the conservation and extension literature. In marked contrast there is insufficient publicity given to the early signs of degradation (Vanclay 1989). Thus many farmers do not recognise the warning signs of degradation on their own properties. This may help explain the findings of Rickson *et al.* described above.

It would be interesting to determine the accuracy of grazier perceptions of the impact of changes to the land. Removing the top 5 cm of soil in mulga lands reduces above ground production of buffel grass (*Cenchrus ciliaris*) by over 90% (Pressland *et al.* 1988). These results prompt the question 'are graziers aware of the magnitude of the reduction, and its implications for their enterprise?' This in turn leads back to vital questions about the perception of land and land management. I return to this issue of social research later in the paper.

Before leaving the subject of perception of change, it is worth considering perceptions of drought in the semi-arid country. Drought may be defined in various ways but generally as reduced rainfall for a longer than seasonal period. One of the most obvious features of semi-arid lands is the high variability of the low rainfall. Put another way: expect droughts! An objective observer would expect that anyone running an enterprise in a risky environment (be it a natural, social, political, financial or business environment) should recognise the risks and learn to cope with them. This does not appear to be the case with recurring phenomena of the Australian environment such as droughts, floods and bushfires. Despite nearly 150 years of experience in the arid and semi-arid zones of Australia, perceptions of drought are polarised into those of graziers and those of scientists (Table 1). Heathcote (1988) considered this perception problem in some depth and found confusion and bitter debate over drought relief schemes funded by the Federal and State Governments. The same debate continues in 1990 (Table 1) with as much acrimony.

The views of the protagonists have not materially altered in nearly a century. There seems to be no middle ground of opinion, thus one side must be incorrect unless both are voicing value judgments. Given the available data, it is difficult to agree with the strident views of the President of the NSW Farmers' Association. It seems fairly clear that the perception of drought by some graziers is linked to the expectation of government handout as a substitute for good decisions about land management (CSIRO National Rangelands Program 1990). If it is so difficult to acknowledge the reality of droughts, then it is no wonder that perceptions of land change are so weak.

Table 1. Differing perceptions of drought in western New South Wales and Australia, 1901 to 1990.

Royal Commission 1901 (Volume 1, p. vi)

That the story of our western country makes such a gloomy page in the history of the pastoral industry of the State is probably mainly due to the general failure in the past of those interested – under the seductive influence of a short run of good seasons – to recognise that drought is the predominant characteristic of the west, and not merely an enemy to be occasionally encountered.

Evidence by individual graziers in the report of the Royal Commission all tends to support this view.

Holmes 1938, p. 34, p. 37

In a remote land like the Western Division climatic normality is often a matter of opinion or of memory. But human memory is short and selects for remembrance only that which is pleasantly or unpleasantly outstanding; so that in the matter of weather memory has been credited with a surety it does not possess.' '...widespread droughts are frequently common...

Working Party on the Effects of Drought Assistance Measures and Policies on Land Degradation 1988, p. 2

Drought is a natural process, a normal extreme event of climatic variability and a recurring feature of the Australian environment.'

Drought Policy Review Task Force 1989, p.5

In most circumstances, drought is a normal commercial risk that should be included in the management decisions of Australian rural enterprises.

NSW Farmers' Association 1989

Association President, Peter Taylor, said that the [Drought Policy Review] Task Force had not been able to comprehend the reality of drought. 'The Task Force belives (sic) drought should be regarded as a natural and recurring condition. This is an absurd statement,' he said. 'While it is highly likely a drought is occurring somewhere in Australia at any time, by definition a drought is an extreme and unusual experience for a particular area, and should not be confused with a poor season.'

CSIRO National Rangelands Program 1990, p.14.

Drought is a time of crisis. A crisis for the land, the animals and its people. The standard of management at these times will often determine the survival of the enterprise and the land on which it depends. Yet **drought is a natural feature of rangelands**. [emphasis in original]

Apportioning cause for change

The historical approach

While there is ample evidence for change, and many would assert degradation, determining the cause of that change is far from simple. It is usual for several factors to vary simultaneously, some in parallel, others in opposition. How do we partition causation between these major elements? This is an important question because until we have determined the major causes for change,

then we have little idea of how to manage land to either achieve or avoid similar changes in future. I will use two studies from south-western USA, because they are among the most complete examples of their kind. Both investigations integrate historical changes in climate, vegetation, and land use much as one should expect. There are similar historical studies in Australia (e.g. Oxley 1987a, b) but none as detailed as the American examples.

The first study documents change in the vegetation of the American south-west by comparing old photographs with modern photos from the same site and describing the changes qualitatively (Hastings & Turner 1965). While there may be difficulties in interpreting old photos, this technique provides incontrovertible evidence of change, and it can be adapted to give semi-quantitative information on the changes (Noble 1977). Hastings & Turner examine existing causal hypotheses invoking rabbits, fire, cattle and climate; and conclude (p. 289) that a combination of climate and cattle is implicated:

'About cause, then, the best answer seems to be that the new vegetation – if one may call it that – has not arisen from climatic variation alone, but in response to the unique combination of climatic and cultural stress imposed by the events of the past eighty years; that climate and cattle have united to produce it'.

The second example considers a more conventional and dramatic form of degradation – development of gullies or arroyos in ephemeral and permanent streams in southern Arizona and coastal California in the American south-west. Cooke & Reeves (1976) used detailed surveyed plans and cross-sections of various streams, climate records and contemporary descriptions of the valleys. There was ample incontrovertible evidence of considerable change in the hydrologic regimes of streams over a wide geographic area. Like Hastings & Turner (1965), they assessed a range of evidence to determine the causes. The model they erected is complex as befits a complex process like gully initiation. Major contributing factors are secular changes in climate, especially droughts; secular changes in vegetation, probably induced by overgrazing; changes in valley-floor vegetation; and most importantly, creation of drainage-concentration features such as irrigation canals, roads etc. along valley floors.

'The final conclusion from this brief comparison [of gullying in Arizona and California] is perhaps the simplest and most obvious: apparently similar arroyos can be formed in different areas as a result of different combinations of initial conditions and environmental changes'. (Cooke & Reeves 1976, p. 189)

One interesting aspect of these two studies is the coincidence of the 1880s as a period of marked environmental change and the start of certain environmental problems in the American south-west. For example, unpalatable native shrubs such as mesquite (*Prosopis juliflora* var. *velutina*) became a pest in Arizona about the 1880s. Herds of Spanish cattle grazed in Arizona from the late 1600s but there are no reports of problems with either mesquite or initiation of gully erosion until the 1880s (Hastings & Turner 1965, p. 30). Both forms of degradation appeared with the coincidence of climatic change and increase in cattle numbers by Anglo-Americans in the late 1800s. Unpalatable native shrubs were first recognised as 'woody weeds' in semi-arid Australia at about the same time (Royal Commission 1901). Sheep numbers were increasing dramatically during this period, and several severe droughts also occurred (Figs 3 and 2). This may be coincidence, or it may suggest that both regions, with markedly different environments and plant species have previously unrecognised common critical features.

Extrapolating halfway around the globe on such slender evidence borders on the foolhardy, but nevertheless, I suggest that such coincidences are worth examining in more detail. Adamson *et al.* (1987) have demonstrated links between floods of the Nile (Africa), Darling (Australia) and Krishna (India) Rivers and the El Niño Southern Oscillation. Such global connections in climate may help explain contemporaneous initiation of problems in America and Australia. Similarly, the development of the cattle (Arizona) and sheep (Australia) grazing industries may have reached equivalent stages of development and passed some threshold together. Whatever the cause, the coincidence may tell us a lot about the initiation of woody weeds and thus help to find an ecological solution which may be applied through land management.

Integrated approach

One problem with both the American studies quoted above is that they are regional and thus do not deal with individual properties. As land is managed on a property basis, the property is the ideal unit to examine to determine the historical changes. By doing this, we can document changes in management by successive owners, the impact of legislative and economic changes and try to integrate these into a coherent understanding of the condition of the land.

In the 'Momba' project, I am addressing the question 'Why is the country like it is?' by examining a range of landscape elements, property management, legislation and economics. The last two are major driving forces behind many changes (Pickard 1988), but they are usually and inexplicably omitted from similar studies.

The project has raised several problems of access to information and interpretation of apparently good information. I will discuss several of these as an indication of the type of problem that arises in such studies.

Access to information

Information is not always easy to gather. The personal credibility and acceptability of the investigator are crucial to the success of any research program. If graziers do not trust the investigator and question his or her motives, then little information will be forthcoming. Official records are available, but accessing many of these requires permission from the individual grazier. Again the problem of trust. Freedom of Information legislation may be used to bypass such problems, but at the cost of alienating all the graziers. Such a solution is short-term and could effectively destroy a research program. Media attention to degradation in recent years has exacerbated the difficulties. The majority of land owners who are good managers feel victimised and are now very defensive. They see any scientist asking questions as yet another attempt to brand them as 'land rapers'. This reaction can effectively stop a well-designed and funded program. Virtually any question can be interpreted as having some bearing on how the grazier or his parents or grandparents managed (abused?) the property. Such problems can only be overcome by individual perseverance and by being prepared to consider all the evidence.

Difficulties in interpreting survey plans

Survey plans are primary source material and generally regarded as reliable. However, various information is often omitted because it was not relevant to the purpose of the survey, or the plan. Consider the plans from 1886 to 1977 of the paddock containing Well A on Property X at 'Momba' (Fig. 4). I do not reveal names and exact locations to preserve anonymity of the present grazier.

Plans can be regarded as most reliable when information cross-checks. Thus the 1886 and 1977 plans both show Well A and nearby Tank B in the same relative positions. However the fences are markedly different suggesting that successive graziers have implemented different subdivision patterns over the years.

The 1926 sale plan (Fig. 4) illustrates a more complex problem. It was prepared to illustrate a poster advertising the auction of a very large and therefore important property. We would expect that the plan was derived from accurate plans kept by the owners of the property and thus the poster would provide reliable information. Taken at face value and in isolation, there is no reason to question the information on the plan. However, when compared with the other plans it is clear that the plan is oblique, and for no apparent reason. Accurate plans were already available in 1913, and from the then Western Lands Board (i.e. 1926 subdivision plan). The positions of several natural features, e.g. the lake and the creeks differ from other plans. Even Well A is confused with Tank B. However, the 1926 sale plan does in fact show that two watering points were present. Both the other 1926 plans omit Tank B altogether.

The 1926 survey plan was prepared from a boundary survey immediately after the auction, Thus it shows the boundary fences accurately but almost no internal detail except where subdivision fences joined the boundary. Even the watering points are omitted. The plan compiled by the Western Lands Board in 1926 shows the subdivision accurately but omits Tank B. Thus, no single plan from 1926 provides all the available information, all three plans are necessary.

200

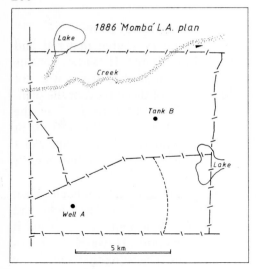

1886 'Momba' L.A. plan

Lake

Creek

Tank B

Lake

Well A

5 km

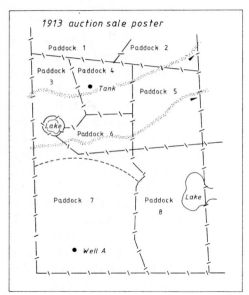

1913 auction sale poster

Paddock 1   Paddock 2

Paddock 3   Paddock 4

Tank

Lake

Paddock 5

Paddock 6

Paddock 7   Paddock 8

Lake

Well A

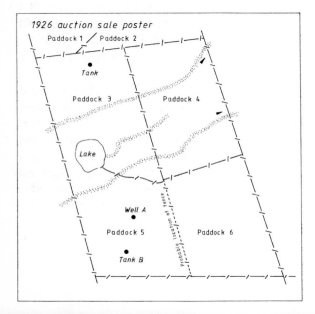

1926 auction sale poster

Paddock 1   Paddock 2

Tank

Paddock 3   Paddock 4

Lake

probable location of fence

Well A

Paddock 5   Paddock 6

Tank B

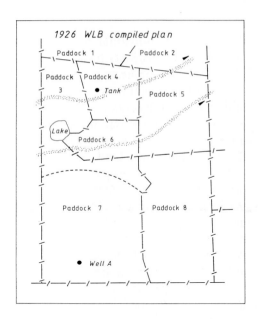

1926 WLB compiled plan

Paddock 1   Paddock 2

Paddock 3   Paddock 4

Tank

Paddock 5

Lake

Paddock 6

Paddock 7   Paddock 8

Well A

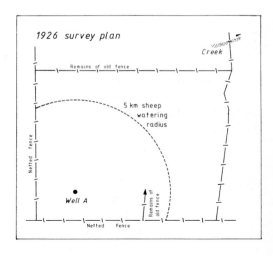

1926 survey plan

Creek

Remains of old fence

5 km sheep watering radius

Netted fence

Well A

Remains of old fence

Netted fence

1977 SCS property resource plan

Lake

Creek

Tank B

Well A

Estimating carrying capacities

Interpreting the data is more critical for information on carrying capacity of the same Well A on Property X. Estimating safe carrying capacities is critical in managing semi-arid lands as stock numbers are often the primary determinant of land condition. Well A was sunk in the late 1870s. Various reports and plans provide information on the numbers of sheep that the well could support (Table 2). Taken at face value, these numbers indicate stocking rates well below 1 ha/dse. Such high rates of stocking have never been achieved 'Momba' (Pickard 1990) and are unbelievable by modern standards. How then we do interpret this vital primary source material?

Perhaps the basis for these 'carrying capacity' estimates can be found in the quantity of water available at the well. Rankin (1887) estimated that the well could produce 109-136,000 L day^{-1}. In this type of semi-arid vegetation (shrublands and open woodlands) merinos require 2–4 L day^{-1} (Squires 1981), thus the well could water up to 34,000 sheep. While this estimate is even higher than those in the primary sources (Table 2), it is of the correct magnitude. The discrepancy could be due to different estimates of the water requirements of merinos. Alternatively, it could be due to inefficiencies and difficulties is raising the water

from the well. I conclude that the early estimates of carrying capacity of wells and waterholes were based solely on the amount of water available, and the question of feed was ignored. So far, I have been unable to confirm that such high numbers of stock were actually carried at the well.

Subsequently, (i.e. since about 1900) carrying capacity has been estimated by comparing the rangeland with another area of similar vegetation and known stocking history. Initially, this was done by rough rules of thumb known only to experienced graziers. More recently, the Soil Conservation Service has formalised this approach by considering the carrying capacity of each land class on a property. Land classes are areas of similar topography, soils and vegetation and (by inference and observation) capacity for and response to grazing. Each land class has a different safe carrying capacity and by combining the area of each land class in the paddock and the safe stocking rate, the safe carrying capacity of the paddock is estimated. Stanley & Lawrie (1977) mapped Property X into five land classes (Fig. 4) providing the most accurate estimate of carrying capacity (Table 3).

There is essentially no relationship between estimated carrying capacity on Property X (Fig. 5) and the number of sheep reported in the

Table 2. Estimated carrying capacity of Well A, Property X, 1886 to 1977.

Date	Carrying capacity (sheep)	Equivalent stocking rate (ha/dse)[a]	Source
1886	15,000	0.1	1886 plan
1891	20,000	0.1[b]	Wright 1891
1896	15,000	0.1[b]	Woodbine 1896
1926	7-10,000	0.3–0.4	1926 sale plan
1977	633[c]	5.1	Stanley and Lawrie

[a] Calculated assuming that the sheep would only use that area of the paddock within 5 km of the well.
[b] Assuming the same paddocks as shown on the 1886 plan.
[c] Estimated carrying capacity for paddock (see Table 3).

Fig. 4. Plans redrawn to the same scale (using southern boundary) showing paddocks and natural features around Well A on Property X in 1886, 1913, 1926 (3 plans) and 1977. Numerals on the 1977 plan show land types (see Table 3 for explanation). To preserve anonymity of the present grazier, the exact location is not given. Sources: 1886 plan, 1913 sale plan, 1926 subdivision plan, 1926 sale plan, 1926 survey plan, Stanley & Lawrie 1977.

Table 3. Estimated carrying capacity of land classes in paddock containing Well A on Property X.

Class	Country	Assessed grazing rate (ha/dse)	Area ha	Sheep
1.	Sandplain with swamps and flats	6.6	1178	179
2.	Calcareous sandplains	5.8	676	116
3.	Creek flood-outs	3.1	863	275
4.	Heavily timbered creek channels	7.4	106	14
5.	Swamps and sub-terminal lakes	8.4	401	48
	Mean 5.1		Total 3224	633

Source: Stanley and Lawrie (1977)

Wilcannia Pastures Protection Board District over the last century (Fig. 3a). This is quite disturbing because it demonstrates the lack of any link between estimated carrying capacity and actual numbers. If the estimates were of any value, than the numbers should show some relationship.

In fact, there is a link between some of the data. Every decade the Western Lands Commission determines or reassesses a 'rental carrying capaci-

ty' for every grazing lease in the Western Division. This figure approximates what the land should be able to carry year in, year out, over the long term. It is used to estimate the income-generating potential of the lease, and thus forms the basis for the lease rental. Many of the rental carrying capacities were determined by averaging available data from the same and adjacent properties. Typically, information was obtained from the returns made every year to the Pastures Protection Board. Thus, the previous decade's returns determine the next decade's rental carrying capacity. This gives a direct connection between sheep carried and estimates of carrying capacity but this connection is not evident when the data are compared (Fig. 5). The estimate of rental carrying capacity has not changed in 40 years (Fig. 5). This is curious given that both rainfall (Fig. 2) and stock numbers (Fig. 3) have varied considerably in the Wilcannia area over the same period. It would appear that the original 1949 estimate is simply reused at each reassessment period.

Also of note is the marked increase in estimated carrying capacity from the 1870s to the 1880s. The earlier figure is among the lowest capacities ever estimated, the later is one of the highest. The

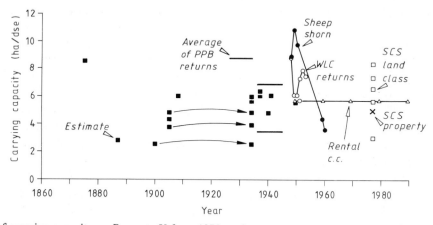

Fig. 5. Estimates of carrying capacity on Property X from 1875 to the present.
Different symbols show estimates by: stock inspectors, land inspectors and surveyors (solid squares); returns of numbers of sheep shorn (solid circles); returns of numbers of sheep to Western Lands Commission (open circles); rental carrying capacity determined by the Western Lands Commission every decade from 1949 to 1989 (open triangles); Soil Conservation Service for each land class (open squares), and the entire property (cross). Curved lines with arrows join estimates from 1900–1905 which were quoted and used in the 1930s. Thick horizontal bars show time period over which stock returns to the Pastures Protection Board were averaged to estimate carrying capacity. Sources: unpublished data in files of Western Lands Commission, and Stanley & Lawrie 1977.

only change in this period was a vast increase in the number of stock carried.

Using unpalatable species to monitor change

Perhaps the main problem in apportioning cause for changes is that the very objects of interest (plants and vegetation) are acted on simultaneously by climate, grazing stock and bushfires. If stock numbers responded only to climate, then there would be little difficulty, but they do not (cf Figs 2 and 3). One quite different approach is to chose an unpalatable species in a grazed situation and follow its population trends. If the plant is readily available to all the herbivores (sheep, cattle, goats, horses, kangaroos, rabbits, pigs), then side effects of grazing such as trampling are present. However, the critical element of continued and continual defoliation is absent. Such a plant is then free to respond primarily (but not solely) to seasonal and climatic change. Measurements on the response of this plant may then be useful as a proxy of the effect of climatic change on the entire ecosystem.

Kippistia suaedifolia is one such plant. This perennial shrub is restricted to gypsiferous soils in the winter rainfall area of semi-arid Australia, and is found in New South Wales on a single 400 ha site on 'Marlow' Station, some 40 km east of Ivanhoe (Fig. 1). Although quite restricted in New South Wales, its geographic distribution elsewhere in Australia is broadly similar to that of Bluebushes (*Maireana* spp.) and Saltbushes (*Atriplex* spp.), both important forage plants in these rangelands. *K. suaedifolia* is smaller than the most common Bluebushes, but is similar to the most widespread Saltbush (*A. vesicaria*).

On 'Marlow', the *Kippistia* community is open to grazing at all times. Sheep, goats, kangaroos, cattle and rabbits all move freely through the community but never eat the shrub, presumably because of the high content of volatile oils (Brophy *et al.* 1982). Field observations during several droughts indicate that even when rabbits are starving they will not eat the shrub. As a consequence of its distribution, unpalatability and sim-

ilar size, *K. suaedifolia* can be used as a proxy for the economically more important Bluebushes and Saltbushes.

In 1972 I established permanent transects in the essentially monotypic population of *K. suaedifolia*. The objective at the time was to examine population changes in the shrub over both time and space. The study was expanded in 1976 by permanently tagging plants to follow individuals through time. The data available (Fig. 6) have not previously been analysed because for the first 10 years there were remarkably few changes in the populations. However, numerous changes have occurred since then and the full data sets are now being analysed because they show the response of the community and the species to weather and climate patterns. Given the present interest in changes consequent on the Greenhouse Effect, the *Kippistia* data are an invaluable starting point for an examination of Greenhouse changes on semi-arid rangelands.

Several points can be made from results so far. Firstly, the small and large plants behave as different populations with little similarity (Fig. 6). Secondly, there was no overall increase in small individuals from 1976 to 1984. During this time, numerous seedlings sprouted and it is likely that as small plants died, they were replaced by growing seedlings and the overall numbers maintained. However, in 1984 the population of small plants increased 10-fold. This followed January rainfall of 210 mm. However, January rainfall is not the only threshold as January 1974 was similar but there was no subsequent flush of small plants. Finally, it is difficult to detect any significant impact of droughts, including the severe 1982 drought.

The study is incomplete but it bears promise of providing key information to apportion cause for change in a major section of the semi-arid lands.

Lessons

Landscape change in the semi-arid zone is a complex involving more than just the land. I offer the following as a starting point for discussion on

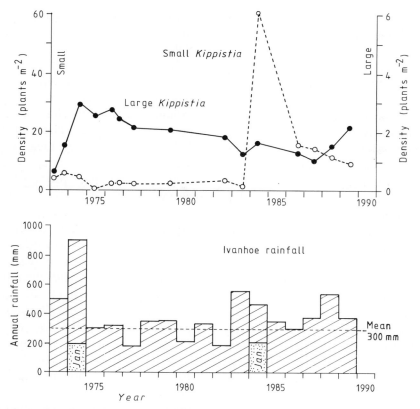

Fig. 6. Upper: Changes in density of small and large individuals of *Kippistia suaedifolia* on abandoned gypsum mine on 'Marlow' Station, western New South Wales. Each point represents mean of 128 measurements, symbols fully enclose standard errors of means. Note that the density scales differ by a factor of 10.
Lower: Annual rainfall at Ivanhoe, 40 km west of 'Marlow' showing extremely high January rainfall recorded in 1974 and 1984. Sources: *Kippistia* density: Pickard unpublished data, Rainfall: Bureau of Meteorology records

approaches to recognising and managing the changes.

i) Vegetation change is caused by complex interactions of the rangeland vegetation and its species with secular changes in climate and management decisions. This is self-evident and should not need stating, but it must be repeated because it is so frequently misunderstood. Also misunderstood or, at least under-rated, is the part played by economic and legislative changes in affecting management decisions.

ii) Management is usually (invariably) short-term oriented. Most of the critical inputs to decisions are short-term: price fluctuations, showers of rain, need to meet interest payments etc. There is little doubt that long-term

conservation is a luxury that only the well-financed property can afford. The critical task is to determine why some managers can achieve this position.

iii) Long-term changes to rangeland are barely recognised by most graziers. Most changes are regarded as seasonal with the invariable view that 'she's good country and she always comes back after a good fall of rain'. There is virtually no recognition that rangelands are currently changing. Nor is there acknowledgement of the long-term implications of these current changes. Further, today's changes are a consequence of management practices that are currently regarded as acceptable.

These are all basically social rather than technical issues. I think that this is the most neglected aspect of rangelands research in Australia. The motivation of most graziers is only poorly understood and there is virtually no information on where they obtain information and how they make decisions. Without such information, extension programs may well fail abysmally. Perhaps a model can be found by looking at the development of land-care groups addressing salinity problems in Victoria. Until each land holder acknowledged that he had a problem on his farm, and until others stopped shouting that it was someone's fault, little progress was achieved. The land-care groups of farmers are cooperatives working with the common goal of tackling a community problem. Several land care groups have been recently established in New South Wales with similar objectives.

In the rangelands of New South Wales there is still insufficient recognition of change and the early signs of degradation. We are also still at the stage of blaming graziers for the condition of the land. Perhaps we should adopt a neutral attitude, acknowledge that the land has changed (perhaps even degraded) and try as a community to reverse the changes where necessary.

The future

Lest I sound too pessimistic I offer what I believe to be examples of how we should address one technical problem (estimating carrying capacity) and one social problem (increasing financial status). Both are essential for good land management, however we define 'good'!

Traditional methods of estimating carrying capacity are based on what similar country has carried in the past (Fig. 5). In many instances this has not worked because the criterion was often health or production from the animals rather than overall income. Maximum income is generally achieved at well below maximum possible stocking rates. Typically, it is reached at about 30% utilisation of the available feed (Beale *et al.* 1986). The grazier needs some method of being able to estimate the amount of feed, and convert this to stock numbers equivalent to 30% utilisation. Christie (1984) describes a simple method of doing this provided some initial research has defined photographic or other field standards so the grazier can estimate amount of feed at the start of the season. Alternatively, the amount of feed can be estimated by modelling net primary production from rainfall information. In non-seasonal environments, some modification is necessary, but the approach is both more reliable and predictable than the traditional methods.

In contrast, it is not agency researchers who are leading the way in addressing the question of increasing financial status. Considerable research (e.g. Beale *et al.* 1986) has shown that returns are maximised at lower stocking rates. However, with the odd notable exception (e.g. Lange *et al.* 1984) it appears that this has only recently been demonstrated at a property-scale. In all cases, it was by shown by graziers acting independently. Recognition and adoption of these minimal stocking rate systems is probably the single most important advance in rangeland management from the 1980s.

On 'Annean Station' in Western Australia (Fig. 1), the grazier, O'Connor, had two major objectives in his management:
'1. To provide sufficient income.
2. To increase plant populations and plant cover, particularly the population of perennial and facultative perennial pasture plants on the property' (Morrissey & O'Connor 1988)
After pursuing these objective for 28 years, 'Annean' has the lowest stocking rates in the district, but the highest returns on either land area or sheep basis (Fig. 7).

'The management system described in this paper has allowed a sheep station business to invest in new equipment and structures, replace and maintain the original infrastructure and it has allowed the manager and his family to significantly expand their equity in station businesses while maintaining a modest level of consumption expenditure. On limited survey evidence, the management system currently yields higher net income per hectare than other surveyed properties, however the relative con-

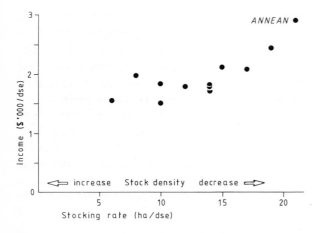

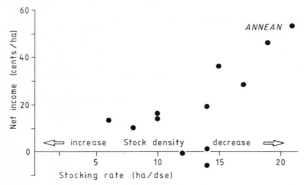

Fig. 7. Relationship between income per dse (a) and net income ha⁻¹ (b) and stocking rate from 12 properties in semi-arid Western Australia. The property 'Annean' is discussed in the text. Note that at a stocking rate of 20 ha/dse, sheep are less dense than at 5 ha/dse. Source: Morrissey & O'Connor 1988.

tributions from the long term management strategy as opposed to that from the short term tactical decisions of the manager, cannot be assessed from the information available'. (Morrissey & O'Connor 1988)

An essentially similar result has been achieved in Central Australia on 'Atartinga' (Fig. 1) a beef cattle property (Purvis 1986). Purvis took over the degraded property with a large debt in 1960 and as he says '...[I] decided to attack the problem as follows:

– I had to establish a productive breeding herd;
– I had to reclaim my degraded country;
– I had to manage my debt in the short term just to survive;
– I had to develop my own philosophy of management because the only advice that the industry or government could offer was the 'conventional wisdom' which had produced the crisis I was in!'

Since then, he has managed to reduce stock numbers, clear the debts, improve the station and the quality of the herd, reclaim the land and improve the condition of the rangeland. While this is a remarkable achievement in itself, it is all the more notable because Purvis achieved it while fighting local opinion and government regulations such as required minimum stocking numbers.

There are undoubtedly several other graziers who have been successful through similar attitudes and philosophies. A major task for the next decade is to recognise their achievements and the contribution they can make in the battle against degradation. Other graziers will listen to their peers whom they regard as successful, and will eventually adopt the methods which increase their incomes. As these minimal stocking rates both increase income and restore degraded county, then there is considerable hope for the future of Australia's semi-arid lands.

Acknowledgements

The data used in this paper were collected while I was employed at the National Herbarium of New South Wales, and the Western Lands Commission. I thank the many graziers who have cooperated with my research at 'Marlow' and 'Momba'. Comments by Dr Peter Mitchell (School of Earth Sciences, Macquarie University) improved the paper. Naturally, the opinions and conclusions are mine and must not be taken as a statement of official policy of any organisation.

References

Adamson, D. A., Williamson, M. A. J. & Baxter, J. T. 1987. Complex late Quaternary alluvial history in the Nile, Murray-Darling, and Ganges basins: three river systems presently linked to the Southern Oscillation. In: Gardiner, V. (ed) International geomorphology. Volume 2. pp. 875–887. John Wiley, London.

Beale, I. F., Orr, D. M., Holmes, W. E., Palmer, N., Evenson, C. J. & Bowly, P. S. 1986. The effect of forage utilization levels in sheep production in the semi arid south west of Queensland. In: Joss, P. J., Lynch, P. W. & Williams, O. B. (eds) Rangelands: a resource under siege. p. 30. Cambridge University Press, Cambridge.

Brophy, J. J., Flynn, T. M., Lassak, E. V. & Pickard, J. 1982. The volatile herb oil of *Kippistia suaedifolia*. Phytochemistry 21: 812–814.

Christie, E. K. 1984. Production and stability of semi-arid grassland. In: Parkes, D. (ed) Northern Australia. The arenas of life and ecosystems on half a continent. pp. 157–171. Academic Press, Sydney,

Condon, R. W. 1986. Recovery of catastrophic erosion in western New South Wales. In: Joss, P. J., Lynch, P. W. & Williams, O. B. (eds) Rangelands: a resource under siege. p. 39. Cambridge University Press, Cambridge.

Cooke, R. U. & Reeves, R. W. 1976. Arroyos and environmental change in the American south-west. Oxford University Press, Oxford.

CSIRO National Rangelands Program. 1990. A policy for the future of Australia's rangelands. CSIRO Division of Wildlife and Ecology, Canberra.

Drought Policy Review Task Force. 1989. Managing for drought. Interim report July 1989. Australian Government Publishing Service, Canberra.

Hastings, J. R. & Turner, R. M. 1965. The changing mile, University of Arizona Press, Tucson.

Heathcote, R. L. 1988. Drought in Australia: still a problem of perception? Geojournal 6: 387–397.

Holmes, M. 1938. The erosion-pastoral problem of the Western Division of New South Wales. University of Sydney Publications in Geography 2: 1–51.

Lange, R. T., Nicolson, A. D. & Nicolson, D. A. 1984. Vegetation management of chenopod rangelands in South Australia Australian Rangeland Journal 6: 46–54.

Morrissey, J. G. & O'Connor, R. E. Y. 1988. 28 years of station management. Paper presented to 5th Biennial Conference of the Australian Rangeland Society, Longreach, Queensland, June 1988.

Noble, I. R. 1977. Long-term biomass dynamics in an arid chenopod shrub community at Koonamore, South Australia. Australian Journal of Botany 25: 639–653.

Nix, H. 1985. What is environmental planning? In: Basinski, J. J. & Cocks, K. D. (eds) Environmental planning and management. pp. 31–36. CSIRO Division of Water and Land Resources, Canberra.

NSW Farmers' Association. 1989. Report is a cop out. Farmer October 1989, p. 5.

Oxley, R E. 1987a. Analysis of historical records of a grazing property in south-western Queensland. 1. Aspects of the patterns of development and productivity. Australian Rangeland Journal 9: 21–29.

Oxley, R. E. 1987b. Analysis of historical records of a grazing property in south-western Queensland. 2. Vegetation changes. Australian Rangeland Journal 9: 30–38.

Pickard, J. 1988. Impact of recent legislation on the Western Division of New South Wales. Working Papers, 5th Biennial Conference of Australian rangelands Society, Longreach Queensland, June 1988, pp. 103–106.

Pickard, J. 1990. Analysis of stocking records from 1884 to 1988 during the subdivision of Momba, the largest property in semi-arid New South Wales. Proceedings of the Ecological Society of Australia 16: 245–253.

Pressland, A. J., Mills, J. R. & Cummins, V. G. 1988. Landscape degradation in native pasture. In: Burrows, W. H., Scanlan, J. C. & Rutherford, M. T. (eds) Native pastures in Queensland. The resources and their management. Queensland Department of Primary Industries Information Series Q187023, 174–197.

Purvis, J. R. 1986. Nurture the land: my philosophies of pastoral management in Central Australia. Australian Rangeland Journal 8: 110–117.

Rickson, R., Saffigna, P., Vanclay, F. & McTainsh, G. 1987. Social bases of farmers' responses to land degradation. In: Chisholm, A. & Dumsday, R. (eds) Land degradation problems and policies. pp. 187–200. Cambridge University Press, Cambridge.

Royal Commission, 1901. Royal Commission to inquire into the condition of the crown tenants of the Western Division of New South Wales. Legislative Assembly of New South Wales, Sydney. 2 volumes.

Squires, V. 1981. Livestock management in the arid zone. Inkata Press, Melbourne.

Stanley, R. J. & Lawrie, J. W. 1977. Western lands lease management plan, property 'X'. Soil Conservation Service of New South Wales, Sydney.

Vanclay, F. 1989. Stewardship and conservationism in Australian farmers. Paper presented to Eighth Biennial Conference of Australian Sociological Association, La Trobe University, December 1989.

Western Lands Commission 1901–1980. Annual reports. Department of Lands, Sydney.

Wilson, A. D. 1986. Principles of grazing management systems. In: Joss, P. J., Lynch, P. W. & Williams, O. B. (eds) Rangelands: a resource under siege. pp. 221–225. Cambridge University Press, Cambridge.

Wilson, A. D. & Harrington, G. N. 1984. Grazing ecology and animal production. In: Harrington, G. N., Wilson, A. D. & Young, M. D. (eds) Management of Australia's rangelands. pp. 63–77. CSIRO, Melbourne.

Working Party on the Effects of Drought Assistance Measures and Policies on Land Degradation 1988, Report of the Working Party on the Effects of Drought Assistance Measures and Policies on Land Degradation. Australian Government Publishing Service, Canberra.

Unpublished files, archives, maps and plans

The following sources are all unpublished reports or plans. They are generally attached to either current or archived files of the Western Lands Commission. Access to the files is generally via the Commission.

d'Apice, L. 1905. Recommending areas to be offered under Part VII of the Western Lands Act within the Momba Resumed Area No. 55, Counties of Fitzgerald, Killara, Yungnulgra and Young, Western Division. [File WLC 51- 4060, Government Records Repository 258124]

Rankin, J.T.C. 1887. Appraisment report on runs. [File PL 55 Momba, Archives Office AO 10-43868]

Woodbine, F.W. 1896. Evidence at land board. [File PL 55 Momba, Archives Office AO 10-43868]

Wright, W.C. 1891. Evidence at land board. (record no. 91/1699 and 02/11998) [File PL 55 Momba, Archives Office AO 10-43868]

'1886 Plan' Momba Holding Leasehold Area, Counties of Fitzgerald, Killara and Yungnulgra. Scale 2 miles to 1 inch.

'1907 Plan' Plan shewing areas offered for lease under Part VII of the Western Lands Act, Counties of Fitzgerald, Killara, Yungnulgra and Young, Western Division. Department of Lands, Sydney [Lithograph 07-16M]

'1913 Sale Plan' Poster advertising subdivisional sale of 'Momba' 1913.

'1926 Subdivision Plan' Sketch Application for subdivision of Portion 'C', subdivided Momba WLL 32. County of Killara and Yungnulgra. [Western Lands Commission registered plan WLB 2488, copy available on file WLL 3234, Government Records Repository K257506]

'1926 Sale Plan' Poster advertising subdivisional sale, Mount Murchison Station, Dalgety and Company Limited, Broken Hill, Wednesday March 10, 1926. (Printed by J.H. Sherring & Co, Adelaide)

Vegetatio **91**: 209–218, 1991.
A. Henderson-Sellers and A. J. Pitman (eds).
Vegetation and climate interactions in semi-arid regions.
© 1991 *Kluwer Academic Publishers. Printed in Belgium.*

Vegetation changes in the Pilliga forests: a preliminary evaluation of the evidence

E. H. Norris,[1] P. B. Mitchell[2] & D. M. Hart[2]
[1] *National Herbarium, Royal Botanic Gardens, Sydney, Australia*
[2] *School of Earth Sciences, Macquarie University, Sydney, Australia*

Accepted 24.8.1990

Abstract

Changes in the vegetation of Australia since white settlement have been much discussed in recent times. In particular, the changes that have been reported to have occurred in the Pilliga forests in northern New South Wales have been used as a reference for other areas of the State. Two periods of pine regeneration are believed to have occurred in the Pilliga, but preliminary research concerning the history of these forests has uncovered various sources of information indicating that the story is a more complex. Climatic data, archival records and the biology and ecology of various flora and fauna are examined in this paper in a preliminary attempt to gain a more accurate picture of change or stability in the vegetation of this region.

Introduction

The Pilliga Forests of northern New South Wales are often referred to as the 'Pilliga Scrub' and these two terms, 'forest' and 'scrub', reflect long held conflicting perceptions about the nature of this environment.

The most widely accepted view of the origin of the forest country is presented by E. Rolls (1981) in: *A Million Wild Acres*. This book, hailed as an Australian classic by Murray (1984), offers the opinion that prior to European settlement the present forest country comprised a mosaic of open woodland and grassy plains that was maintained by regular Aboriginal fires. The initial exclusion of fire by early graziers, heavy stocking, above average rainfalls between 1879 and 1887, reduced stocking and the reintroduction of fire,

encouraged extensive pine regeneration and the eventual abandonment of many grazing runs. No similar pine regeneration event is believed to have occurred until after the wet year of 1950, the fire of 1951 and the reduction of rabbit numbers by *Myxomatosis*. This paper will question the details of this sequence of events because there is a scattered body of evidence which is in conflict with the general model. For example:

1. Although all of the forest area seems to have been claimed by pastoralists by the 1880s, there is little evidence that the core of the forest east of Baradine Creek was ever heavily grazed, of even entirely occupied.
2. Survey maps from the 1870s to the 1930s depict vegetation boundaries in this core area which are remarkably similar to those of today.
3. A number of primary sources describe thick

210

scrub and pine regeneration events at other times.

4. Although the Forestry Commission management plan (Forestry Commission 1986) generally accepts this model (but with a significant anomaly in the timing of the nineteenth century regeneration event), archival data and past management objectives indicate that earlier foresters had a different appreciation of the environment.

Resolution of these conflicts requires a detailed examination of all the evidence for change or stability in the vegetation and it is the purpose of this paper to begin this process.

Location

The State Forests of the Pilliga, (Figure 1), situated north of Coonabarabran in northern New

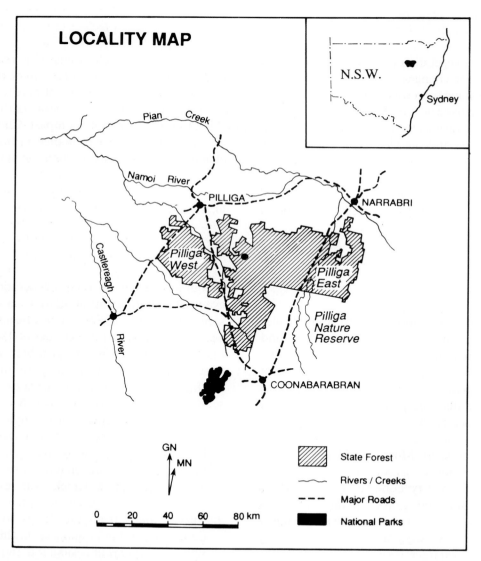

Fig. 1. Location of the Pilliga State Forests, New South Wales, Australia.

South Wales, cover an area of 400,000 ha and constitute the biggest single mass of dedicated native forest in the State.

The geology of this area is of non-marine, Jurassic, Pilliga sandstone which dips to the north-west and flanks part of the Great Artesian Basin (Brown *et al.* 1977). Outcropping sandstone is common in the southern sections of the forests whilst to the north it is covered by extensive sediments deposited by dendritic streams draining north and west. These sediments become finer towards the Namoi River (Mitchell *et al.* 1982). The climate is warm sub-humid with variable rainfall averaging 450–700 mm per annum and showing a slight summer maximum.

There are two parts to the forest; an area covering the main river valleys and Pilliga west which has been settled at various times, and an inner core which is predominantly covered in cypress pine and ironbark forests and woodlands and broom plain scrubs (mainly *Melaleuca uncinata*).

As part of an ongoing series of projects including detailed examination of the soil stratigraphy and vegetation, the history of pre-European and European settlement in the Pilliga Forests is being examined. In particular, we are looking at historical documents for evidence of changes in the vegetation over the past 172 years in an effort to distinguish between those changes which may have been driven by climate and those which may have been initiated by changes in land use.

The accepted view

As noted earlier, Rolls (1981) presents the most widely accepted view of the sequence of events which is assumed to have taken place in the forests of the Pilliga and which has been repeated by others including Austin and Williams (1988). Two periods of pine regeneration are recognised; between 1879 and 1887 and in 1950/51. Of the first period Rolls (1981) claims that by the 1870s settlement in the Pilliga area was more or less complete and at that time there had been no regular burning for about 25 years and that domestic stock had displaced native herbivores (rufous rat-

kangoroos). As a consequence of this, pine spread from the ridge country and invaded clear valley floors, and wire and spear grass replaced better species. The graziers began to burn in an effort to control unfavourable grasses and pine scrub encroachment. Weather patterns also had an important influence. Rolls claims that the period 1875–1878 was droughty and that 1879 was very wet, the cattle market was depressed and additional sheep were put on the grass which was burnt to clear seed and give the stock green pick with the consequence that '... where the fires ran years of pine seed came to life.' (Rolls 1981, p 183–184). Similar factors of a reduction in grazing pressure (from rabbits), wet seasons and a major fire are used to explain the second regeneration event in 1950/51. A quotation summarises his argument;

'The four or five good years between 1879 and 1887 were the only years in which it was possible for the new forest to come away. By the next good rains in the 1890s there were sufficient rabbits, as enthusiastic eaters of seedlings as the disappearing rat-kangaroos, to stop most new tree growth. The extent of the country to be abandoned was determined by the 1890s. Except for a thickening of the undergrowth in places in the several wet years following the breaking of the 1902 drought, there was little more growth of pine or scrub until 1951, when a huge fire germinated seedlings (sic) on land soaked by heavy rain in 1950. At the same time myxomatosis destroyed the rabbits. And the lovely tangle which is the modern forest came to life.' (Rolls 1981, p 205).

Evaluation of the model

The validity of this general model needs to be questioned on five important points:
1. The reality and significance of the stated climatic events.
2. The evidence for pre-existing pine scrubs, other periods of regeneration and the actual timing of the main events.
3. The significance of stored seed.

4. The significance of rat-kangoroos in reducing pine regrowth.
5. The significance of fire in relation to pine regeneration.

The weather patterns

Rolls argued that '... the four or five good years between 1879 and 1887...' (Rolls 1981, p 205) were the important years for pine regrowth in the Pilliga. This climatic pattern and the drought between 1875 and 1878 are difficult to confirm because the only official records starting that early are from Narrabri on the north eastern edge of the region. At this station the record shows that rainfall was 24% below average in 1875, average in 1876, and 26% below average in 1877; droughty perhaps but not extreme, although Nicholls (pers. comm.) has confirmed that 1877 was an El Niño year and that a large part of western New South Wales was in severe drought. Rainfall was 44% and 36% above average in 1878 and 1879 respectively and there were only three good years (above average rainfall), rather than 'four or five' within his critical period, these being 1879, 1885 and 1886 (Bureau of Meteorology 1989).

From 1881 rainfall records are also available for Baradine and Coonabarabran. All three stations show similar patterns and can be accepted as representing the Pilliga. At each station 1886 and 1887 were wet years, 1888 was dry, and the early 90s were very wet. This period finished in 1892 at Baradine, and 1894 at Coonabarabran and Narrabri. The years 1886, 1889 and 1890 rank in the ten highest rainfall records at all stations and at Narrabri 1890 is the wettest year on record with the total rainfall being 103% above average (Bureau of Meteorology 1989). If several consecutive wet years are significant in setting pine seed and allowing germination and establishment as Lacey (1973) indicates, then the period 1889 to 1892/94 seems likely to be more important climatically than the late 1870s as the actual regeneration period. This suggestion is supported by the acceptance of the 1890s as the period of regeneration by the Pilliga Management Plan (Forestry Commission 1986).

The circumstances of the 1950/51 regeneration event also supports this conclusion because the rainfall at all three stations was well above average between 1947 and 1950 (with the exception of 1948 at Coonabarabran) and 1950 was the wettest year on record at Coonabarabran and Baradine and the second wettest at Narrabri. Between 1892 and 1947 there were no other such extreme consecutive wet years.

These five decades of lower rainfall follow the patterns identified by Pittock (1975) and appear to have been generally unfavourable to cypress pine regeneration. There were some other periods of pine regeneration however, for example; at Gilgandra in 1917/18 when rabbits were recorded as attacking seedlings and destroying that crop (Forestry Commission 1918) and in 1932/34 in the east Pilliga forests where there had been no sheep grazing and very few cattle (Lindsay 1948).

The 1950/51 regeneration event first became apparent throughout most of the natural range of the white cypress pine in 1953/54 when the seedlings from the 1952 seed year (Forestry Commission 1953/54) were overtopping the grasses. This observation is consistent with the normal two year flowering and cone formation cycle (Lacey 1973) but also indicates that weather conditions for some years after 1953 must have been favourable for seedling survival. Soil moisture levels for a couple of years after the record wet of 1950 were probably high despite average or below average rainfall and the years 1954, 1955 and 1956 were again much wetter than average. We suspect that this coincidence of a subsequent wet period was important in consolidating the regeneration and that this also has a parallel in the 1890s rainfall sequence but not in the 1880s.

So far our review of weather patterns has only examined rainfall, but it is also believed that temperature is important in that mild summers are necessary for seedling establishment (Forestry Commission 1986).

Pre-existing scrubs and other regeneration events

No research has been done on the occupation of the forest area by the Kamilaroi aborigines but the few sites that have been recorded are only short distances from main creeks and it seems likely that they rarely visited the forest core.

The journals of Oxley (1820), Sturt (1833) and Mitchell (1839, 1848) describe journeys down the Lachlan, Bogan and Macquarie rivers and all specifically mention on many occasions that the lighter red soils away from the rivers often supported dense scrubs including *Callitris* sp. It is notable that there are more references to the difficulties of traversing such 'dreadful scrub' country when the explorers made side trips away from the river, or were using pack animals rather than wheeled vehicles which confined them to clearer country.

It is apparent that at the time of first exploration pine scrubs did exist in parts of northern and mid-western New South Wales and southern Queensland but unfortunately there are few records which relate directly to the Pilliga. Oxley crossed the area of the present Pilliga Nature Reserve in 1818 and commented on the density of ironbark saplings but apparently had no trouble in the open valleys and it was along these main valleys that settlement started in the 1840s. All of the forest areas seems to have been claimed by pastoralists by the 1880s but there is little evidence that the core of the forest east of Baradine was ever heavily grazed or even entirely occupied.

Our examination of records shows the majority of runs held were along the main watercourses; the Namoi River and Baradine and Bohena Creeks, as well as in the foothills of the Warrumbungles. These runs were maintained as cattle stations until the cattle market low of 1875 when sheep were introduced. As reported in Rolls, various runholders had access to other runs further into the core of the forest, but it is unknown at present how often they ventured in there and how many stock they ran. Rolls reports that one manager of a combined property never ventured into this area for fear of being lost (Rolls 1981 p 190).

As earlier settlers departed, runs were divided and/or combined and occupation licenses were issued, some of which cover our area of study. Stocking rates are as yet unknown.

It is reported that the Crown did not receive any rental from the Pilliga later than the year 1888 (Forestry Commission n.d.), a fairly good indication that its use was very little indeed. The settlement was along the rivers and on the flats – it was known that the country in the core was both poor and scrubby.

Other evidence of what the core area was like can be gained from reports and survey maps of the 1870s to the 1930s. In 1878 railway surveyors working on the eastern side of the Pilliga commented on 'the marvellously dense scrubs' of the poor sandy soils as if they had long been there (Carver 1878). In 1880 the Surveyor General called for reports from all his Land Commissioners and surveyors on the extent, age and significance of the pine scrub on leased lands all over the west and the response concerning the Pilliga region was that the country contained very extensive indigenous scrubs on poor sandy soils which were of no value for grazing or agriculture. The scrubs were not then believed to be recent but had been present since before first settlement (Anon 1881).

Survey maps depicting vegetation boundaries in the core of the forest show remarkable similarities to the present vegetation patterns.

Figures 2 (a) and 3 (a) are tracings from topographical surveys conducted in 1914 (Lands Dept. 1914a & b). Figures 2 (b & c) and 3 (b & c) are tracings from air photographs taken in 1938 and 1970. Figure 2 shows an area adjacent to Etoo and Rocky creeks (Central Mapping Authority 1974, Cubbo 1:50,000, grid ref. 603703). The area west of the creeks has been settled; the remainder has been little touched. Allowing for the nature of the 1914 survey it seems that the boundaries of the deep sands and the broom plains have not changed, at least in this century. In the 1914 survey, the surveyor recognised areas of thick forest ('thick forest of Pine, Ironbark, Oak, Box and Budda' [Lands Dept. 1914a]) to the north and south of the sands and

214

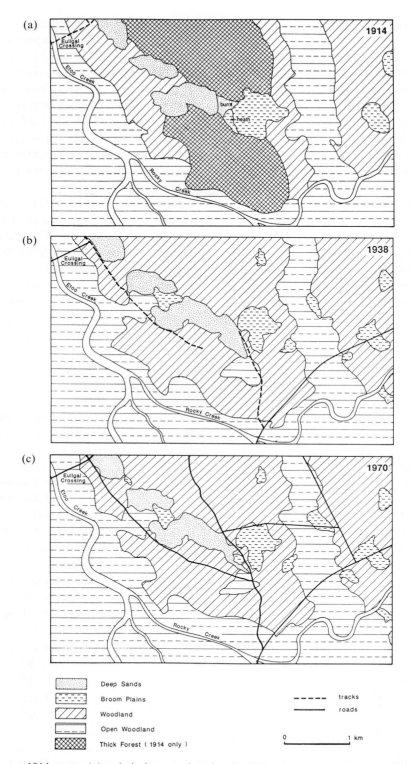

Fig. 2. Tracings from a 1914 survey (a) and airphotographs taken in 1938 (b) and 1970 (c), of the Etoo and Rocky Creek area in the central western portion of the Pilliga State Forests (Fig. 1).

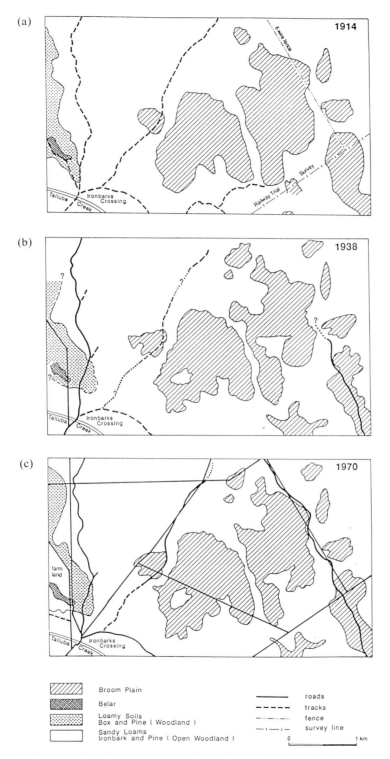

Fig. 3. Tracings from a 1914 survey (a) and airphotographs taken in 1938 (b) and 1970 (c), of the area north and east of Ironbarks Crossing slightly north east of the area in Figure 2.

the central broom plain. While problems arise here with the exact meaning of a 'thick forest', these denser areas are not obvious on the air photographs of later years; a reversal of the expected trend if the forest was becoming denser.

Figure 3 is of an area north and east of Ironbarks Crossing on Talluba Creek (Central Mapping Authority 1974, Cubbo 1 : 50,000, grid ref. 615714). Again, the vegetation patterns seem not to have changed since 1914 (Lands Dept. 1914b), and indications are that the density is also similar.

Nineteenth century maps covering the core of the forest are rare. McClean's 1847 map of squatting districts on the Liverpool Plains shows settlement along the Macquarie, Castlereagh and Namoi Rivers. Other maps include the 1874 map of the Coghill run which gives some information regarding the vegetation, and two portion plans dating from 1878 in the Parish of Dunwerian (Lands Dept. 1878) give similar information. Chatfield was the surveyor for these portions, and although his fieldbook numbers are noted on these plans, we have not yet been able to locate them. Portion plans only cover a small part of the forest, but the notes regarding vegetation detail in 1878 closely match the vegetation on 1970 air photographs and on the ground at the present time.

The importance of stored seed

Rolls (1981) suggests that large stores of pine seed were available on mature trees or in the soil prior to the regeneration years and he implies that this was another important factor in the successful events. This idea was widely held by early observers including Fosbery (1913), but is not supported by the studies of Lacey (1972, 1973) which showed that seed was normally shed in a period of about four weeks in summer and that viability under field conditions was as low as 1% several months after seedfall.

The evidence for the role of rat-kangoroos in reducing pine regrowth

Rolls (1981) suggested that grazing of young pine by rufous rat-kangoroos was an important factor limiting regrowth densities. He presented no evidence for this statement which was apparently drawn from a single comment on the prevalence of rat-kangoroos (species unidentified) made by Oxley (1820 p 270). We have so far been unable to find any other primary source confirming the identity of the rat-kangoroos in the Pilliga and no evidence that they were at all partial to young pine.

The significance of fire in relation to pine regeneration

Rolls (1981) and most nineteenth century observers believed that fire was an important factor in successful pine regeneration. There is, however, no clear relationship between the extensive fires in the Pilliga in November 1951 and the germination of 1952 seed. Many mature cypress pines survived these fires even after being defoliated (Forestry Commission 1951/52), pine regeneration was apparently just as successful in areas which were not burnt elsewhere in the Pilliga, and it was abundant over most of their natural range in the absence of fires elsewhere in the State.

Fire is known to be an important thinning mechanism in young pine stands (Lacey 1973) because seedlings have a high mortality (Wilson & Mulham 1979), but whether it is significant in other ways is not clear and this topic also requires further study.

Conclusions

We have no argument with the observation that increasing densities of woody shrubs are a very serious management problem in many parts of the rangelands in New South Wales as described by Booth (n.d.) and that white cypress pine is one of the problem species in lighter soils on the higher rainfall margin. What this paper takes issue with is the general belief that there were only two main periods of pine regeneration in the Pilliga area and dense shrub cover was virtually unknown at the time of first settlement when open woodlands and grassy plains were believed to be the norm.

Climatic records suggest that there were two main opportunities for extensive pine regeneration and that Rolls (1981) may have incorrectly identified the first of these by about a decade. We also have evidence that pine did regerate at other times in the twentieth century but may not have survived well because of rabbits or subsequent unfavourable weather.

To judge from the land settlement patterns around the Pilliga it seems to be important that we differentiate the central core of the forests where our evidence indicates that there has been little change in the vegetation, from areas to the west of Baradine Creek, especially Pilliga West State Forest where pine regeneration did close over former grazing lands in the 1890s. Even as recently as 1912 a soils map by Jensen (1912) labelled the central region 'almost unknown'. It is only since good road access was provided by the Forestry Commission after the 1930s that it has became accessible.

Most of the fieldwork that we have been doing in the past few years has centred on the Dunwerian area in the core of the forest. Here we are gathering evidence which seems to point to a remarkably stable vegetation pattern over the past century. The pattern is governed by a factoral complex dominated by soil characteristics, in particular moisture.

Work thus far has been of a preliminary nature only, but has opened up several interesting lines of evidence which we plan to follow up. These include:

1. Tree-ring studies which will help us to establish the pattern of pine regeneration over the past 100 or so years. Preliminary tree-ring counting from pine in the core area indicates a wide scatter of tree ages which would tend to support the hypothesis that regeneration is fairly well spread and not confined to two main events.

2. A closer examination of the available climatic records including temperature and the ENSO phenomenon.
3. A closer examination of the historical records which might exist in obscure places or in the memories of settlers descendants. We need to sort out what stock was in the forest, where, how many and when.

Acknowledgements

We wish to acknowledge the assistance of Macquarie University in supporting this research, and the Forestry Commission of N.S.W. for permission to conduct research in forests managed by them and to Dr. N. Nicholls for curtailing some of our excesses.

References

Aerial mosaics of the Pilliga scrub. 1938. N.S.W. Archives Office, Kingswood, AO18981–19040. 60 sheets.

Anon 1881. Reports by District and other Surveyors as to spread of pine and other scrubs. N.S.W. Legislative Assembly. Votes and Proceedings 3: 255–266.

Austin, M. P. & Williams, O. B. 1988. Influence of climate and community composition on the population demography of pasture species in semi-arid Australia. Vegetatio 77: 43–49.

Booth, C. A. n.d. (circa 1937). Woody weeds. Their ecology and control. Soil Conservation Service of New South Wales.

Brown, D. A., Campbell, K. S. W. & Crook, K. A. W. 1977. The geological evolution of Australia and New Zealand. Pergamon Press.

Bureau of Meteorology. 1989. Report of monthly and yearly rainfall, microfiche 053001 Baradine, 054120 Narrabri Bowling Club, and 064000 Conabarabran.

Central Mapping Authority of N.S.W. 1974. Cubbo 8736–N 1 : 50,000 map sheet, Bathurst.

Carver, N. P. 1878. Letter to engineer in charge of railway trial surveys 30/6/1878. N.S.W. Legislative Assembly. Votes and Proceedings 1881, 4, 310.

Forestry Commission. 1918. Annual Report of the Forestry Commission of New South Wales.

Forestry Commission. n.d. (circa 1939). Notes refering to the booklet The Pilliga National Forest by R. S. Vincent. Forestry Commission of New South Wales.

Forestry Commission. 1951/52. Annual Report of the Forestry Commission of New South Wales.

Forestry Commission. 1953/54. Annual Report of the Forestry Commission of New South Wales.

Forestry Commission. 1986. Management plan for Pilliga Management Area. Forestry Commission of New South Wales.

Fosbery, L. A. 1913. Climatic influence of forests. Appendix to the Report of the Forestry Department, Bull. 4.

Jenson, H. I. 1912. The agricultural prospects and soils of the Pilliga scrub. N.S.W. Farmers Bull. 54. Dept. of Agriculture.

Lands Department. 1878. Portion Plans, County of Baradine, Parish of Dunwerian. Portion 1, B110, Portion 2, B.111, N.S.W.

Lands Department. 1914a. Topographic Survey of part of the Pilliga Scrublands. County of Baradine, Parishes Cumbill, Coomore, Coomore South, Euligal. Land District of Narrabri. Ms. 1032. Me.

Lands Department. 1914b. Topographic Survey of part of the Pilliga Scrublands. County of Baradine, Parish of Bundill, Land District of Narrabri. Mis. 959. Me.

Lacey, C. J. 1972. Factors influencing occurrence of white cypress pine. Forestry Commission of New South Wales Research Note 20.

Lacey, C. J. 1973. Silvicultural characteristics of white cypress pine. Forestry Commission of New South Wales Research Note 26.

Lindsay, A. D. 1948. Notes on the cypress pine regeneration problem. N.S.W. Forestry Recorder 1(1): 5–16.

McClean. 1847. Plan showing the squatting districts of Gwyder and Liverpool Plains with parts of New England, Darling Downs, Bligh, Wellington and Lower Darling. Archives Authority of New South Wales, AO-5664.

Mitchell, P. B., Rundle, A. S. et al. 1982. Land systems of the Pilliga region, N.S.W. Unpubl. report to the New South Wales National Parks and Wildlife Service, Macquarie University.

Mitchell, T. L. 1839. Three expeditions into the interior of eastern Australia. 2 Vols. T. & W. Boone, London, Public Library of S.A. Facsimile edition, 1965.

Mitchell, T. L. 1848. Journal of an expedition into the interior of tropical Australia, in search of a route from Sydney to the Gulf of Carpentaria. Greenwood Press, Publishers, New York Reprint 1969.

Murray, L. A. 1984. Persistence in Folly. Sirius Books, Australia.

Nicholls, N. pers. comm. 1990. Bureau of Meteorology Research Centre, Melbourne.

New South Wales Lands Dept. 1970. Aerial Photographs, Baradine, Run 2, 1677: 5085; Run 4, 1677: 5177.

Oxley, J. 1820. Journals of two expeditions into the interior of New South Wales, undertaken by order of the British Government in the years 1817–1818. John Murray, London. Libraries Board of South Australia, Australian Facsimile editions No. 6 1964.

Pittock, A. B. 1975. Climatic change and the patterns of variation in Australian rainfall. Search 6: 498–504.

Rolls, E. C. 1981. A million wild acres. Nelson, Melbourne.

Sturt, C. 1833. Two expeditions into the interior of Australia, during the years 1828, 1829, 1830, and 1831, with observations on the soil, climate, and general resources of the colony of New South Wales. 2 vols, Smith, Elder and Co., London. Vol. 1. Facsimile 1982, Doubleday Aust. Ltd.

Wilson, A. D. & Mulham, W. E. 1979. A survey of the regeneration of some problem shrubs and trees after wildfire in western New South Wales. Aust. Rangel. J. 1: 363–368.

Vegetatio **91**: 219–230, 1991.
A. Henderson-Sellers and A. J. Pitman (eds).
Vegetation and climate interactions in semi-arid regions.
© 1991 *Kluwer Academic Publishers. Printed in Belgium.*

Managing the droughts? Perception of resource management in the face of the drought hazard in Australia

R. L. Heathcote
Flinders University of South Australia, GPO Box 2100, Adelaide 5001 SA, Australia

Accepted 24.8.1990

Abstract

The long history of drought occurrences in Australia is reflected not only in varying community perceptions of drought as a hazard, but also in the growing recognition of its role in Australian ecosystems. The history of drought occurrences and the evolution of private and governmental responses to drought are reviewed as the context for the expected modification of traditional official policies, due in 1990.

Introduction

The establishment of the Federal Drought Policy Review Task Force in May 1989 and the publication of its Interim Report in July (DPRTF 1989) is a convenient time to review past Australian resource management in the context of the long-standing drought hazard facing the Australian community. A recent review of the last 20 years of drought policies in Australia (Heathcote 1988) and a renewal of the drought subsidy debate (Smith 1989; Kraft & Piggott 1989) provides the basis for this update and forecast for future drought mitigation policies.

Contexts and definitions

Drought has been a long-time companion to resource management in Australia. Aboriginal legends encompass it and since 1788 its impacts, both positive and negative, have concerned Australian society. In 1813 it helped push settlers over the Blue Mountains into the interior; in 1865

Surveyor General Goyder delimited that year's drought in South Australia and initiated a regional resource management policy which is still relevant today (Heathcote 1981); in the Centennial Year of European settlement it was devastating southeastern Australia; in the 1930s it added to the miseries of the Depression Years; in February 1983 it helped to bring the Mallee soils to Melbourne as it had similarly in 1902; while in the Bicentennial Commonwealth Year Book it merited a six-page article (ABS 1988).

Defining drought, however, has been difficult and without reviewing all the arguments noted earlier (Heathcote 1969), it is necessary to remind ourselves that most definitions are relative, that is the shortage of moisture (whether rainfall, soil moisture or ground water) is assessed in relation to the demand for that moisture and drought is defined when that imbalance causes human hardship – usually in the Australian context an economic hardship. Given that context, the problems of definition arise when we realise that human management of the resources can either reduce or enhance the risk of drought occurring by reducing or increasing the demand for water – be it on a sheep/cattle property, on a wheat farm, or in an urban water supply system. Different thresholds of water need and different management systems create different potentials for drought occurrence.

Conventionally, however, research into the general effects of drought upon the Australian community has compared reports of drought occurrences with the most readily accessible climate statistics – rainfall data, and suggested that meteorological droughts (annual rainfalls in the lowest 10% of values on record) when occurring over at least 10% of the continent have in the past coincided with agricultural droughts (shortfalls of moisture causing significant economic losses of crops and livestock) (Gibbs & Maher, 1967; ABS, 1988). Following this convention therefore

I will focus on such meteorological droughts and their associated societal and environmental impacts.

Patterns of drought in space and time

Using the Commonwealth Bureau of Meteorology's annual maps of rainfalls by deciles which supplement the earlier study by Gibbs & Maher (1967), it is possible to survey the temporal and spatial patterns of the occurrence of meteorological drought in Australia (Fig. 1 and Table 1).

At the continental scale drought has occurred somewhere on the continent in 78% of the 103 years of record from 1885 to 1988. For 33% of

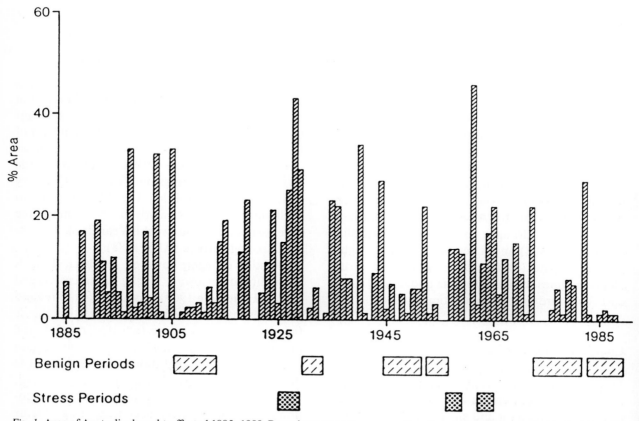

Fig. 1. Area of Australia drought-affected 1885–1989. Bars show percentage area of continent affected by meteorological drought (annual rainfall in the first decile). Benign periods are sequences of at least 4 years with less than 10% of the continental area drought-affected; stress periods are sequences of at least 3 years with more than 10% of the continental area drought-affected. (Source: Commonwealth Bureau of Meteorology *Monthly Rainfall Reviews*).

Table 1. Drought sequences in Australia.

Area sequence	Maximum area in drought in any one year		Maximum sequence of years with drought on 10% or more of area		Maximum of years with drought on less than 10% of area	
	%	year	N	period	N	period
Australia	46	1961	4	1926–29	9	1973–1981
					6	1983–1988
Victoria	100	1982	2	1907–08	20	1945–1964
				1914–15		
				1943–44		
Tasmania	90	1961	3	1919–21	11	1892–1902
					10	1951–1960
Northern Terr.	90	1961	3	1900–02	18	1971–1988
				1927–29		
				1963–65		
New South Wales	81	1940	3	1913–15	13	1889–1901
					11	1968–1978
Queensland	70	1902	4	1926–29	10	1936–1945
South Australia	70	1961	3	1927–29	9	1973–1981
		1982				
Western Aust.	66	1936	2	1894–95	9	1980–1988
				1928–29		
				1935–36		
				1958–59		

Note: Drought areas are areas with annual rainfall in the first decile.
Sources: Commonwealth Bureau of Meteorology.

that time over 10% of the continent has been affected and some significant societal impact likely. The maximum sequence of years (4) with over 10% of the continent drought-affected was 1926–29, a period of considerable stress for new farming settlements in both southeastern and Western Australia. Conversely, the maximum sequence of years relatively free of drought has occurred recently – since 1973; and with the exception of the bad drought of 1982, there have been 14 out of the 15 years with annual droughts affecting less than 10% of the continent. Such a long period relatively free of drought is unprecedented in our history.

At the regional scale of the States, the picture is different. While no more than 46% of the continent has experienced drought in any one year, the whole of Victoria was affected in 1982; 90% of Tasmania and the Northern Territory was affected in 1961; 81% of New South Wales was affected in 1940; 70% of Queensland and South Australia in 1902 and 1961 respectively, and 66% of Western Australia was affected in 1936. This varying spatial impact reflects the latitudinal contrasts between the tropical and temperate weather systems – since it has been rare for both the summer rains of northern tropical Australia to fail in sequence with the winter rains of southern tem-

222

perate Australia. This probably accounts for the low maximum area drought-affected in Western Australia, since it is the only state to span both temperate and tropical zones.

The spatial variation in drought occurrence is not merely latitudinal however, for there is evidence of longitudinal contrasts between droughts in eastern and western Australia (Fig. 2). Of the 100 years 1888–1987, in 17 years New South Wales had droughts when Western Australia was drought-free and Western Australia had droughts in 26 years when New South Wales was drought-free. The most recent example of this 'oscillation' of droughts occurred in 1982, when 58% of New

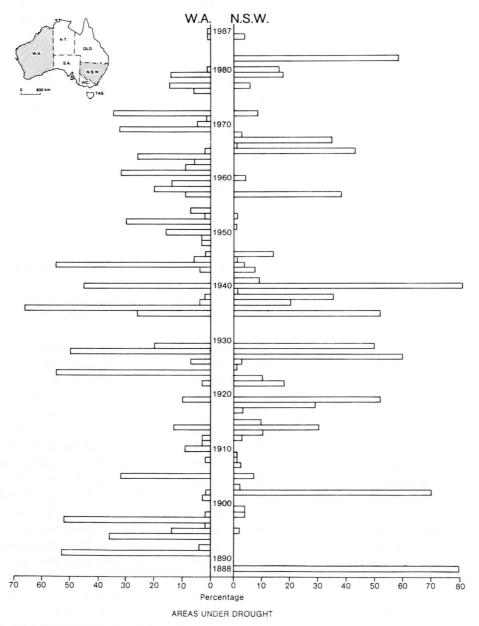

Fig. 2. Area drought-affected in New South Wales and Western Australia 1888–1987. Bars show percentage of each state area affected by meteorological drought in each year. (Source: Commonwealth Bureau of Meteorology *Monthly Rainfall Reviews*).

South Wales was in drought but Western Australia was completely drought-free.

What conclusions can we draw from this brief review of the historical record of drought in Australia? First is the obvious comment that meteorological drought is a frequent event, occurring somewhere in the continent every 3 out of 4 years and having the size to cause serious economic hardship on average every 1 in 3 years. Secondly, it is extremely unlikely that the whole of the continent will be affected in any one year and thus regional drought impacts in one part of this continental nation may be offset by normal seasons in other areas – as witness the good harvests in Western Australia in 1982 when the harvests in eastern Australia generally failed. For individual states, however, drought may affect the bulk of their area and result in major economic and environmental impacts, requiring major mitigation strategies.

Changing drought impacts

The historical background of meteorological droughts is paralleled by significant historical evidence of both positive and negative drought impacts on society and the environment. Summarising the earlier papers (Heathcote 1969 and 1988), the sizeable historical losses attributed to drought included shortfalls in agricultural and pastoral production leading to adverse national balance of payments and reduced Gross Domestic Product (GDP). Service costs were increased – for water provision both urban and rural, and farm and road maintenance costs were increased from drought-accelerated soil erosion. There was also some evidence of retreat of rural settlement from the riskier sites inland. Droughts did bring some benefits: drought-reduced domestic, feral and wild livestock numbers gave the hard-pressed grazing lands a breathing space to make at least a partial recovery; the private transport industries generally benefited from accelerated livestock and fodder movements in droughts; and official drought relief measures often channelled funds into areas which otherwise

would not have benefited and in many cases the money spent, for example on public road construction and maintenance, was of considerable long term benefit to the rural communities.

Bearing in mind the reservations on the complexities of the accounting noted in the original review – particularly, for example, the inclusion of losses from wheat not sown and livestock not conceived – there is continuing evidence of high economic losses (Dury 1983). The economic costs of the last major drought (1982) were claimed to multiply up to $7500 million (Allan & Heathcote 1987; Martin 1983 & Oliver 1986, 304), and this despite a declining share of Gross Domestic Product provided by the agricultural and pastoral sectors – from 20% in the 1950s to c.5% currently (Williams 1986).

The social and demographic effects of that drought have also been shown to resemble those of the droughts of the 1930s – particularly in rural communities already suffering from declining incomes and rising prices (Burnley 1986; Gregory 1984; Williams 1986). The onset of drought appeared to have speeded up the processes of rural change already present in those communities.

To the undisputed socio-economic impacts, however, have been added increasing concern for the environmental effects of droughts. National concern for resource conservation issues in Australia in the 1970s and 1980s has reflected the wider concerns of western societies. Thus the appearance in 1965 and subsequent work of the Australian Conservation Foundation (ACF), the first national survey of soil conservation needs (DEHCD 1978), the adoption of a National Conservation Strategy in 1986, and the joint ACF – National Farmers Federation proposal for a National Land Care Program in 1989 – subsequently adopted by the Commonwealth Government, were evidence of increasing awareness of land degradation or desertification, the loss of natural and cultural heritage items and of quality of life issues.

Earlier, at the turn of the century and in the 1930s and 1940s, there had been fears that severe droughts were responsible for 'desert outbreaks' – desertification as it would now be termed – on eroding arid rangelands in eastern Australia (Heathcote 1987). The national soil survey of 1978 showed that 29% of the land in agricultural or pastoral use needed soil conservation works or a change in land use (three-quarters of this being arid grazing land) and the United Nations World Map of Desertification of 1977, although criticised in detail, again raised the spectre of the spread of the desert lands. Further, the series of massive bushfires covering 500,000 hectares of the tinder-dry drought-stricken countryside in South Australia and Victoria in 1983 killed 72 people, over 300,000 domestic animals, an unknown number of wild and feral animals, and caused property damage of approximately $A400 million. The coincidence with drought was clear enough. Less clear was the link between drought and the soil erosion (up to 48 tonnes lost per hectare) from some of the burned areas in the drought-breaking rains of March 1983 (Allan & Heathcote 1987).

A complicating factor is that reports of drought occurrence may be being increased by increased pressure upon the land as a result perhaps of economic necessity, perhaps of ignorance, or perhaps from greed on the part of the resource managers. Historically in the pastoral industry in Australia overstocking the range initially paid off but made the subsequent operations more vulnerable to drought, since the drought reserves of edible scrub were destroyed. The history of the advance of the wheat farmers inland into areas where rainfall was not only lower but less reliable has effectively increased their vulnerability to drought, despite the efforts of the plant-breeders to produce more drought-resistant crop varieties. In Western Australia the official reports of drought occurrence certainly seem to favour the drier edge of the wheat country (Fig. 3 in Heathcote, 1988). In New South Wales the disparity between drought occurrences as defined by soil moisture criteria and as declared at the request of local rural producers has been noted by Smith & Callahan (1988) and increases in inverse proportion to the annual rainfalls (Fig. 3). It may be that drought, not rain, has 'followed the plough' in Australia (de Kantzow & Sutton 1981; Heathcote 1990).

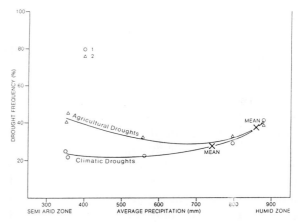

Fig. 3. Comparative cross-section of climatic versus agricultural droughts in New South Wales, 1976–82.
Key: 1-Percentage of the period when climatic droughts affected each location. 2-Percentage of the period when agricultural droughts affected each location. The two types of data are shown for five locations, sited on the horizontal axis according to their average annual rainfall, with a trend curve and mean value for both types of drought indicated. Climatic droughts are months with Palmer Index values below -2. Agricultural droughts are months officially declared as drought periods, usually officially verified after a request from primary producers in the location. (Source: Smith & Callahan 1988, 87).

Responding to droughts

Australian responses to the stress of drought may be usefully classified into two groups: private (the strategies of the settlers themselves) and official (the policies of governments).

Private strategies

Whether the Australians have ever fully accepted drought as endemic to the continent is debatable (Heathcote 1969) but there is no doubt that as settlement spread down the rainfall gradient into the continental interior, pragmatic, often technologically-based, strategies to cope with agricultural drought were evolved by the settlers independent of any official inputs.

For the pastoralists with a good sense of geography and the finances to back it, an interregional spread of their properties from summer to winter rainfall areas was a favoured strategy, helping to create millionaires such as Sir Sidney Kidman (1857–1935) (Bowen 1987) and James Tyson (1819–1898). Technological responses were also favoured, from fencing to control access to the range, through tank sinking to store the fleeting surface run off, to deep drilling and more recently drought-resistant livestock breeds. Storage of surplus feed, however, has only proved economic for valued breeding stock (Dillon & Lloyd 1962; Powell 1963). Evacuation of starving stock to purchased agistment beyond the drought area or exploitation of the public 'long paddock' – the travelling stock routes – were also possibilities.

For farmers the march out onto the droughty plains has been hard-fought, with human ingenuity stretched to its limits to cope with the variable seasons. The optimal time for seeding was learned by trial and error; fallowing of land to save moisture from one season to the next began privately in South Australia in the 1880s before the technique received official blessing following official contacts with the Campbell Dry Farming System of the Great Plains in 1906. Machinery to harvest drought-stunted crops – Ridley's 'Stripper' of 1843 – similarly was developed in South Australia before diffusing rapidly through Victoria and New South Wales. South Australian farmers also were experimenting with drought-resistant wheat varieties (Williams 1974) before William J. Farrer began to cross drought-resistant Indian wheats with good quality Canadian baking wheats and with the well established, high yielding but vulnerable, Purple Straw wheats. The results, especially 'Federation' produced during the last years of the 1895–1902 drought, were to push the wheat fields further down the rainfall gradient into the semi-arid lands over the next 25 years.

The apparent success of highly mechanised broad acre grain farming using controlled fertilisers and associated pastures in the semi-arid portions of South Australia led to the formation of a specialised government agency in the 1970s (South Australian Agricultural Research Investment Company – SAGRIC) which has been invited to develop dry farming systems in north Africa and south west Asia (SAGRIC n.d.). The emphasis is upon adapting successful technologies developed in South Australia to areas of similar climate around the world.

An earlier private strategy for water management on farm, by which an attempt was made to conserve all the rainfall on farm and spread it evenly over the land surface, had been developed in New South Wales in the late 1940s. Despite strong support from the professor of geography at Sydney (Macdonald Holmes 1960) it appears to have been forgotten, although there have been some recent private attempts (with official support, see *Journal of Soil Conservation New South Wales*, 43(2), 1987) to pond runoff on the paddocks as an aid in soil conservation.

On the Australian farms, however, when the crop starts to wilt there is nothing really to be done if irrigation is impossible. And when the crop has died and the soil begins to blow there is only emergency tillage to stave off disaster, and that only as long as the subsoil is moist enough to hold the clods together.

Official policies towards drought

The appointment of the Commonwealth Drought Policy Review Task Force in 1989 marked an acceleration of an already established trend towards a revision of long standing official policies towards drought. The 1970s and early 1980s had witnessed a series of natural disasters with severe impacts upon Australian society: the destruction of Darwin by Cyclone Tracy and massive flooding in eastern Australia in 1974, serious bush fires in 1981 and 1983 and the drought of 1982-83. Various conferences and symposia brought together academics, civil defence personnel and disaster victims, to share experiences and hopefully mitigate any future impacts (Heathcote & Thom 1979; Minor & Pickup 1980; Oliver 1980 & 1986; Pickup 1978; Reid 1979; Smith *et al.*, 1979). Among these responses was some evidence of rethinking, particularly on drought.

Traditionally drought occurrence had been seen in effect as an Act of God requiring commu-

226

nity support for the disaster victims. For drought, as with all other 'natural disasters', the States had to bear the initial costs with the Commonwealth providing further support (currently on a $3 for $1 basis) for costs above a threshold tied to the States' population. In effect drought relief has been for many years the largest component of official disaster relief – from 1962/63 to 1987/88 it absorbed 57.6% of Commonwealth disaster payments (Smith & Callahan, 1988), and for South Australia from 1977 to 1983 drought relief absorbed 82% of the State's disaster payments.

Since the occurrences of drought, as we have seen, tended to be regional and the *claims* of the occurrences tended to be more frequent in the climatically marginal semi-arid country (Fig. 3), over the years the provision of public funds via the States and Commonwealth probably has provided a fairly regular subsidy to resource users in those areas. Further this subsidy has in the main been provided irrespective of whether their management systems adequately reflected the constraints of their environment. The irrational character of this traditional policy was exacerbated by significant variation in the procedures for drought declaration between the States – responsibilities ranging from local government authorities and pasture protection boards to rural producer groups – and the potential to cry 'wolf' seems to have been exploited locally (Heathcote 1969; Smith & Callahan 1988).

The 1980s, however, have brought some revision of traditional thinking. First was the national review of water resources to the year 2000. This recognised the inevitability of drought in Australia and the political relevance of the associated problems:

Water supply restrictions in periods of severe rainfall deficiencies must be expected in Australia with its high rainfall variability, and the resultant inconvenience and losses must be expected by all classes of users.... The acceptable degree of restrictions is essentially a matter for political decision, as is the extent of public investment in storage to reduce the impact of drought. (DRE 1983, 61)

Significantly, the review also suggested that ir-

rigation not only did not protect agriculture against drought, but might exacerbate the effects by encouraging maximum use of available water resources without adequate reserves. This, coming after the final abandonment of cloud-seeding trials by the CSIRO in 1981 and other evidence, seemed to mark the end of a purely technological approach to the problem of drought (Heathcote 1986).

In 1984 two separate activities to promote a national policy on drought were initiated, one of which was to fade away, the other to result in a policy statement. A Joint Committee of the Australian Academies of Science, Social Sciences and Technological Sciences met in June 1984 'to consider the need for, and the feasibility of, establishing a concerted national effort to minimise the economic, social and environmental impact of drought' (pers. comm.). Its report was considered by the Academies in October but the initiative lapsed, probably because its function had been pre-empted by the National Drought Consultative Committee (NDCC) which had also been formed in 1984. This committee, comprising officials from each State and Territory and the Commonwealth Departments of Primary Industry and Finance, along with three private representatives of primary producers, was formed to advise the Commonwealth Minister of Agriculture on drought matters and its first report appeared in 1985 (NDCC 1985). The objectives of drought policy were spelled out, as in general, '– to encourage the efficient allocation of national, regional farm resources; and – to minimise the economic hardship to individuals resulting from drought'. These general objectives were to be fleshed-out by the specific national objectives:

(i) maintenance of land resource;
(ii) maintenance of breeding herds, at a level which would enable post-drought recovery of livestock industries within a reasonable period;
(iii) enable post-drought sowing of crops and pastures;
(iv) maintenance of minimal living standards for farmers assessed as viable in the long term;

(v) short term income support for farmers assessed as non-viable in the longer term;
(vi) minimisation of stock distress. (NDCC 1985, 2).

To achieve this policy a series of measures were proposed for the various levels of government. Joint action by the Commonwealth and States was to take the form of the traditional concessional loans and freight rebates for transport of livestock, feed and water, along with soil conservation programs and other possible measures to be determined. The Commonwealth alone, through its basic control of income tax raising, was to apply taxation measures, an Income Equalisation Deposits Scheme (by which the highly variable annual incomes are smoothed for income tax purposes and which replaced an earlier unsuccessful Drought Bonds scheme) and 'household support and rehabilitation' under the Rural Assistance Scheme. The States, in addition, were to develop and provide information on the policies to the rural managers. Finally, a proposal for crop and rainfall insurance was referred to the Commonwealth Industries Assistance Commission for investigation, and a working party of the NDCC was set up to review the principles relating to the declaration and revocation of drought which, as we have seen, varied considerably between the States.

Much of this was a continuation of traditional policies but there were some innovations. Concern for the links between drought and land degradation was evident in the effort to maintain the land resource and associated soil conservation measures. For the first time also drought relief measures were linked with broader policies on rural restructuring – with the implication that farmers non-viable economically in the longer term would only get short-term income support, i.e. would be expected to leave the industry eventually. Earlier drought relief policies had attempted to maintain the rural population through the crisis, now the poorer managers were to be encouraged to leave.

Concern for the distress to livestock was re-emphasised and it was proposed that crop and/or rainfall insurance should be considered. A national Natural Disaster Insurance Scheme had been considered (and rejected as too costly) as part of the response to the heavy insurance losses from the cyclone and flood disasters of 1974. Drought in fact was not included in the hazards (DNDE 1980). The specific new proposal for insurance was considered but initial reactions have been negative, arguing that this would be difficult to implement and would be economic only if it were to replace the existing drought relief policies (BAE 1986; RMSAB 1986).

Just to confirm the impression that officialdom was beginning to look more carefully at the principles of drought relief, a subsequent statement by the NDCC in May 1986, noted what had been evident throughout Australia's history, namely that drought vulnerability is a function of resource management:

Production systems fail at different rainfall depending on the nature of the components and the intensity at which it (sic) is operating. This feature makes the distinction between normal production risk and 'drought, worthy of public assistance' difficult to define. Declaration processes should not encourage individuals to adopt production strategies which result in inappropriate use of particular land systems and declarations made at a greater frequency than one might otherwise expect. (NDCC 1986, 2).

The gamblers on the environmental margins were not to be encouraged.

In this context therefore it was not really surprising that evidence of rorts (devious practices) in the application of drought relief measures in Queensland in April 1989 was used as the excuse by the Commonwealth Minister for Finance to remove drought assistance from the national Natural Disaster Relief Arrangements, and led to the setting up of the Drought Policy Review Task Force a month later.

So far, the Interim Report has suggested interim drought assistance under the existing Rural Assistance Scheme – basically for carry-on finance, which is the traditional response, but for the future has signalled that all primary producers will be encouraged (required?) 'to manage their enterprises across the full range of climatic conditions' (DPRTF 1989, 27). The final report, expected in March 1990 will no doubt spell out the incentives to place the aims for sustainable revenue management upon the primary producers themselves.

Future management for drought, however, may be easier for both officials and private resource managers if the current hopes for the prediction of meteorological drought are realised. What are the implications?

Drought forecasts and drought management

Drought forecasts have appeared as part of the current high-profile scientific concern for climate change, receiving a significant stimulus from the El Niño-Southern Oscillation (ENSO) of 1982–83 (Glantz *et al.* 1987). Nicholls has claimed that the 1982 drought could have been predicted from the Darwin mean sea level air pressure positive anomalies of a few months previously (Nicholls 1983). He subsequently went on to suggest that major Australian droughts (particularly the failure of the winter-spring rains, June to November, which are essential for the cereal grain crops), might be predicted by cooler sea surface temperatures and higher air pressure anomalies off northern Australia near the start of the calendar year (Nicholls 1985a, 1986) and recent studies have confirmed this (Whetton 1989).

Despite some reservations about the stability of these teleconnections (Pittock 1984), support has come from global modelling exercises (Voice & Hunt 1984) and Nicholls has suggested that Australian crop production in general is related to ENSO phenomenon:

Observing the SST [sea surface temperature off northern Australia] anomaly change from September-November to March-May seems to provide a means for predicting yields of some Australian crops [wheat, barley, oats and sugar cane] and also the gross value of Australian crops. (Nicholls 1985b, 559)

The key is the dependence upon rainfall for most of Australia's crop production.

But what are the real possibilities for, and implications of drought forecasting? In 1985 a gathering of drought researchers concluded that three avenues for drought forecasting existed. These were probabilistic forecasts – thought useful for small scale and short periods of drought; statistical forecasts – useful for regional and severe droughts; and deterministic forecasts based upon climate modelling for periods up to 2 to 3 months ahead (Hunt 1985, 3).

It would seem that the chances of *regional* drought forecasts sufficiently in advance of crucial management decision-making, e.g. crop planting times (or possibly animal mating times?) are good and seasonal weather forecasts have already begun to be issued by the Commonwealth National Climate Centre. Scientific probability forecasts for those able to interpret them have existed for some time (Verhagen & Hirst 1961) and I suspect individual farmers and pastoralists make use of their own 'probability forecasts' based upon their personal assessment of the rainfall records which they all keep anyhow.

If improved forecasts, of whatever type, are becoming available, what effect will they have on the community? Will the forecast of a drought mitigate its potential impact? The answer is not as simple as it might appear, and depends first upon the accuracy of the forecast, second upon the form it takes, and finally upon the response of the decision-makers, given the commercial context of their activities.

Resource managers may or may not accept the forecasts – and in either case they could be in error (Pittock 1986)! The resource managers, however, face not only the problems of the accuracy of the forecast and whether or not he/she believes it, but also what might happen if it *was* accurate, if he/she *did* believe and act upon it and find that the future market was glutted by produce from other resource managers who had also believed and also acted upon the forecast? The forecasters attempt to predict only the weather which might affect production but not the value of that production in the free market!

Conclusion

Historically drought has played a significant role in the development of Australian resources. Traditionally it has been viewed as a national disaster, the victims deserving of public disaster relief, the response being crisis management. Technology has been harnessed to try to buffer society against drought, but the recognition that the real threat to the society is the continental aridity and its associated rainfall variability has been learned only at substantial cost – both economic and social. Public funds have maintained resource use on lands both economically and climatically marginal and at the cost of land degradation. Unscrupulous operators have exploited official drought relief policies, while the better managers have coped with the drought stresses, neither asking for relief nor receiving any official recognition for their skills.

What of the future? The meteorological droughts will continue. Indeed the current unprecedented drought-free spell does not augur well for the next few years – we are due for a major drought.

When that drought comes, what will eventuate? We can be certain that the agricultural production will decline and in the process economic processes will have forced some primary producers into bankruptcy, others into gambling unsuccessfully on the seasons or pushing cropping or grazing into unsuitable locations. We can also be certain that some urban areas will experience water shortages as water demand continues to escalate, existing storages prove inadequate and popular opposition to further big dam construction grows. Further, if the Greenhouse Effect is as forecast, the temperate cereal-growing areas will be more vulnerable to the failure of the winter rains.

The impacts, however, may be reduced by the continued reduction of the GDP share provided by agriculture; by more off-farm income sources for primary producers; by better management strategies on the rural properties (helped by the removal of the failed managers); by soil conservation measures undertaken as part of the National Land Care Program; by advance warnings through drought forecasting; and perhaps most of all by the removal of traditional official drought relief policies. This latter, of course, depends upon what the final report of the Drought Policy Review Task Force has to say in mid-1990.

Acknowledgment

Research for this paper was funded by Flinders University Research Committees.

References

ABS. 1988. Year Book Australia 1988. Aust. Bur. Statistics, Canberra.

Allan, R. & Heathcote, R. L. (eds.) 1987. 1982–83 Drought in Australia, in Glantz *et al.* 19–23.

BAE. 1986, Crop and rainfall insurance: a BAE submission to the IAC, Commonwealth Bureau of Agricultural Economics, Australian Government Publication Service, Canberra.

Bowen, J. 1987. Kidman, the forgotten King, Angus & Robertson, North Ryde, Sydney, Australia.

Burnley, I. 1986. What we can learn from drought and depression. Inside Australia 2(3): 7–10.

de Kantzow, D. R. & Sutton, B. G. (eds.) 1981. Cropping at the Margins: potential for overuse of semi-arid lands, Australian Institute of Agricultural Science and Water Research Foundation of Australia, Sydney.

DEHCD. 1978. A Basis for Soil Conservation Policy in Australia, Report No. 1, Department Environment, Housing and Community Development, AGPS, Canberra.

Dillon J. L. & Lloyd A. G. 1962. Inventory analysis of drought reserves for Queensland graziers: some empirical evidence, Australian J. Agric. Econ 6: 50–67.

DNDE. 1980. Policies and practices for adjustment to natural hazard risk, Report of the (Natural Hazards) Mitigation Committee, Commonwealth Department of National Development and Energy, Canberra, Unpublished draft.

DPRTF. 1989. Managing for Drought, Drought Policy Review Task Force, Interim Report, AGPS, Canberra.

DRE. 1983. Water 2000. A perspective in Australia's water resources to the year 2000, Commonwealth Department of Resources and Energy, Australian Government Publishing Service, Canberra.

Dury, G. H. 1983. Step-functional incidence and impact of drought in pastoral Australia, Australian Geographical Studies 21(1): 69–96.

230

Gibbs, W. J. & Maher, J. V. 1967 Rainfall deciles as drought indicators, Commonwealth of Australia Bureau of Meteorology Bull. No. 48, Melbourne, 1967.

Glantz, M., Katz, R. & Kranz, M. 1987. The Societal Impacts associated with the 1982–83 Worldwide Climate Anomalies, National Center for Atmospheric Research, Boulder, USA, 1987.

Gregory, G. 1984. Country towns and the drought, Australian Rural Adjustment Unit, University of New England, Armidale.

Heathcote, R. L. 1969. Drought in Australia: a problem of perception, Geographical Review 59: 175–194.

Heathcote, R. L. 1981. Goyder's Line – a Line for all Seasons?, in D. J. & S. G. M. Carr (eds.), People and Plants in Australia, Academic Press, Sydney, 295–321.

Heathcote, R. L. 1986. Drought mitigation in Australia: reducing the losses but not removing the hazard, Great Plains Quarterly 6: 225– 237.

Heathcote, R. L. 1987. Images of a desert? Perceptions of arid Australia, Australian Geographical Studies 25: 3–25.

Heathcote, R. L. 1988. Drought in Australia: still a problem of perception?, GeoJournal 16(4): 387–397.

Heathcote, R. L. 1990. Drought follows the plough? The Australian experience, in Glant M. H. (ed.), Drought follows the Plough. Food First Group, San Francisco (in press).

Heathcote, R. L. & Thom, B. G. (eds.) 1979. Natural Hazards in Australia, Australian Academy of Science, Canberra.

Hunt, B. G. (ed.) 1985. Report on Drought Research in Australia, Meeting held in Melbourne, 1985, CSIRO, Division of Atmospheric Research, Aspendale.

Kraft, D. & Piggott R. 1989. Why single out drought?, Search 20(6): 189–192.

Macdonald Holmes J. 1960. The Geographical Basis of Keyline. Angus & Robertson, Sydney.

Martin, B. R. 1983. The 1982 drought in Australia, Desertification Control Bull., No. 9, United Nations Environment Program, Nairobi.

Minor, J. & Pickup, G. 1980. Assessment of research and practice in Australian natural hazards management. North Australian Research Unit Bulletin, No. 6, Darwin, 1980.

NDCC. 1985. Report by the National Drought Consultative Committee on Drought Policy, Commonwealth Department of Primary Industry, Canberra.

NDCC. 1986. Note on the nature of drought and issues in drought declaration and revocation procedures: drought relief and policy committee, Mimeo, National Drought Consultative Committee, Canberra, 1986.

Nicholls, N. 1983, Predictability of the 1982 Australian Drought, Search 14: 154–155.

Nicholls, N. 1985a. Towards the prediction of major Australian droughts, Australian Met. Mag. 33: 161–166.

Nicholls, N. 1985b. Impact of the Southern Oscillation on Australian crops. J. Climatology 5: 553–560.

Nicholls, N. 1986. Use of the Southern Oscillation to predict Australian sorghum yield. Agric. For. Met. 38: 9–15.

Oliver, J. (ed.) 1980. Response to disaster. James Cook University, Townsville.

Oliver, J. 1986. Natural hazards, in Jeans D. N. (ed.), The Natural Environment, Australia- A Geography, Vol. 1, pp. 283–314, Sydney University Press, Sydney.

Pickup, G. (ed.) 1978. Natural Hazards Management in North Australia. North Australian Research Unit, Darwin.

Pittock, A. B. 1984. On the reality, stability, and usefulness of southern hemisphere teleconnections, Australian Met. Mag. 32: 75–82.

Pittock, A. B. 1986. Climate predictions and social responsibility – guest editorial, Climatic Change 8: 203–207.

Powell, A. A. 1963. A national fodder reserve for the wool industry: an economic and statistical analysis. University of Sydney, Dept. Agri. Econ. Paper, No. 3.

Reid, J. A. (ed.) 1979. Planning for People in Natural Disaster. James Cook University, Townsville.

RMSAB. 1986. Report and Recommendations of the Drought Workshop, May 1986, Royal Meteorological Society (Australian Branch), Melbourne.

SAGRIC (n.d.). Changing Horizons, SAGRIC International, Adelaide.

Smith, D. I. 1989. Should there be drought subsidies? Droughts and drought policy in Australia. Search 20(6): 188–9.

Smith, D. I. & Callahan, S. D. 1988. Climatic and Agricultural Drought, Payments and Policy, A Study of New South Wales, CRES Working Paper, 1988/16, Centre for Resource and Environmental Studies, ANU, Canberra.

Smith, D. I. et al. 1979. Flood damage in the Richmond River Valley New South Wales: an assessment of tangible and intangible damages, Northern Rivers College of Advanced Education, Lismore and Centre for Research and Environmental Studies, Australian National University, Canberra.

Verhagen, A. M. W. & Hirst, F. 1961. Waiting times for drought relief in Queensland, CSIRO Division of Mathematical Statistics, Technical Paper No. 9.

Voice, M. E. & Hunt B. G. 1984. A study of the dynamics of drought initiation using a global general circulation model. J. Geophys. Res. 89(D6): 9504–9520.

Whetton, P. 1989. SST Anomalies, Drought Network News 1(2): 4–5.

Williams, D. B. 1986. Agriculture in the Australian Economy, in Martinelli, L. W. (ed.), Science for Agriculture: the Way Ahead, pp. 1–19, Australian Institute of Agricultural Science, Melbourne.

Williams, M. 1974. The Making of the South Australian Landscape. Academic Press, London.

Vegetatio **91**: 231–238, 1991.

Index

Author index

Subject index